Climate Change Mitigation and Adaptation to Improve Food Security in South Asia

Editors

Rafiq Islam
Soil, Water, and Bioenergy Resources,
The Ohio State University, Columbus, Ohio, USA

AHM Mustafizur Rahman
Department of Soil, Water and Environment
University of Dhaka, Bangladesh

Heulin Thierry
LEMiRE/BIAM, CEA Cadarache
Saint-Paul-lez-Durance, France

Mannava Sivakumar
Chemin des Corbillettes, Geneva, Switzerland

CRC Press
Taylor & Francis Group
Boca Raton London New York

CRC Press is an imprint of the
Taylor & Francis Group, an **informa** business
A SCIENCE PUBLISHERS BOOK

First edition published 2026
by CRC Press
2385 NW Executive Center Drive, Suite 320, Boca Raton FL 33431

and by CRC Press
4 Park Square, Milton Park, Abingdon, Oxon, OX14 4RN

Library of Congress Cataloging-in-Publication Data (applied for)

ISBN: 978-1-032-62849-3 (hbk)
ISBN: 978-1-032-62859-2 (pbk)
ISBN: 978-1-032-62869-1 (ebk)

DOI: 10.1201/9781032628691

Typeset in Times New Roman
by Prime Publishing Services

Preface

Climate change presents a monumental challenge to global food security, especially in South Asia, a region with dense populations and a profound reliance on agriculture. The Intergovernmental Panel on Climate Change highlights that food production systems are both significant contributors to and victims of climate change, accounting for up to 37% of global greenhouse gas emissions. At the same time, these systems face the adverse effects of rising temperatures, erratic precipitation, flooding, droughts, and secondary salinity. These conditions exacerbate land degradation and fuel the proliferation of weeds, pests, and diseases, severely threatening agricultural productivity and food availability.

In South Asia, these vulnerabilities are magnified by rapid population growth and the increasing strain on agricultural resources. By 2050, food production must double to meet global demand—a daunting challenge under existing environmental pressures. Alarmingly, rising sea levels and coastal erosion threaten to reduce the region's landmass by 17% and food production by 30%, compounding the crisis further. Addressing these issues requires innovative and sustainable solutions that balance agricultural intensification with environmental stewardship.

This book, **Climate Change Mitigation and Adaptation to Improve Food Security in South Asia,** serves as a vital resource in this context. Drawing from the proceedings of the 2022 International Conference on Climate Change and Food Security held in Dhaka, Bangladesh, it provides a comprehensive exploration of the challenges and transformative strategies needed to secure food systems amidst climate adversity. Through 23 rigorously curated chapters, this volume brings together leading experts to discuss topics ranging from climate variability, abiotic and biotic stresses, and sustainable farming practices to soil and nutrient management, climate-resilient crops, and adaptive policy frameworks. Each chapter presents actionable strategies to mitigate climate impacts, enhance agricultural resilience, and improve food security for vulnerable populations.

This book is a testament to the region's resilience and ingenuity. It blends cutting-edge research, practical case studies, and forward-looking recommendations. Key highlights include discussions on soil carbon sequestration, integrated nutrient management, vermicomposting innovations, and the utilization of coastal landraces for salt-tolerant breeding. Contributors also share insights into biopesticides, climate-smart agricultural practices, and strategic research agendas designed to address South Asia's unique challenges.

With its broad scope and practical orientation, this volume is an indispensable resource for policymakers, researchers, agricultural practitioners, and development professionals. It provides a roadmap for implementing climate-smart agricultural systems, fostering adaptive resilience, and achieving sustainable food security in South Asia and beyond.

The publication of this book and the success of the International Conference on Climate Change and Food Security in South Asia were made possible through the collaboration and support of the organizing and sponsoring institutions. The Dhaka University, Bangladesh, and the Centre national de la recherche scientifique (CNRS), France, organized the conference in partnership with the World Meteorological Organization, the Food and Agriculture Organization, and the Embassy of France at Dhaka, Bangladesh.

Special acknowledgment is extended to Prof. Dr. Md. Akhtaruzzaman, Vice-Chancellor of Dhaka University, as President of the Organizing Committee. Prof. Dr. ASM Maksud Kamal, Pro-Vice-Chancellor, served as Vice President, while Prof. Mamtaz Uddin Ahmed, Treasurer, acted as Treasurer of the Organizing Committee. Prof. Dr. Md. Zillur Rahman, Dean of Earth & Environmental Sciences, and Prof. Dr. A.K.M. Mahbub Hasan, Dean of Biological Sciences, served as members. Prof. SMA Faiz, former Vice-Chancellor, contributed as Advisor. Their instrumental roles were vital in facilitating the event.

The editors express their heartfelt gratitude to the contributors for their invaluable expertise and dedication. This work stands as both a reflection of collective knowledge and a call to action—inviting collaboration across disciplines to tackle climate challenges and secure a sustainable future.

Special thanks go to all others for their hard work, commitment, and excellence in finalizing this book. Finally, we extend our profound thanks to the team at CRC Press for their support and guidance in bringing this book to fruition. Our deepest gratitude goes to Vijay Primlani, whose encouragement and meticulous guidance have been integral to the successful completion and publication of this book.

Foreword

The international symposium on 'Climate Change and Food Security in South Asia' is jointly organized by the University of Dhaka, Bangladesh and CNRS (Centre National de la Recherche Scientifique), France, in collaboration with WMO (World Meteorological Organization). This event serves as a continuation of the symposium held in 2008. This year, we are honored to welcome leading experts from various disciplines, including meteorology, agriculture, livestock, and forestry, representing South Asia, Europe, and other parts of the globe. We are delighted to announce that selected deliberations from the conference will be published by CRC Press, USA, in the form of a book entitled *Climate Change Mitigation and Adaptation to Improve Food Security in South Asia.*

The effects of climate change will impact all four pillars of food security: availability, accessibility, utilization, and system stability. It will have profound implications for human health, livelihood assets, food production and distribution channels, purchasing power, and market flows.

While scientific research and teaching are of paramount importance, creative thinking in the humanities and social sciences is equally essential to address the economic, social, political, legal, philosophical, and cultural dimensions of environmental problems. We hope that a multi-sectoral and multidimensional approach to climate change and food security will enable global communities to achieve shared targets and objectives outlined in the Sustainable Development Goals by 2030.

The selected papers presented at this conference have been compiled in this book. We trust it will serve as a significant source of information and a reference for all agencies and organizations interested in climate change and food security.

We extend our sincere gratitude to CRC Press for publishing this book for global users. Our heartfelt thanks also go to the Editorial Board, the National Organizing Committee, and the University of Dhaka authorities, particularly the Chief Coordinator, Professor Mustafizur Rahman, of Dhaka University, and Dr. Thierry Heulin, Director of Research at LEMiRE, CNRS, France, for their efforts in ensuring the success of this event.

Secretary General, WMO | **Vice-Chancellor, University of Dhaka**

Contents

List of Contributors

Ahmed, Niaz
Department of Soil Science, Bahauddin Zakariya University, Multan, Pakistan.

Ahsan, S.
Metropilitan State University of Denver, Dever, Colorado, USA.

Akhter, S.
Tuber Research Center, Bangladesh Agricultural Research Institute, Gazipur, Bangladesh.

Aksakal, E.L.
Dept of Soil Science and Plant Nutrition, Ataturk University, Erzurum, Turkey.

Alam, Khandakar Mohiul
Bangladesh Sugarcrop Research Institute, Jamalpur.

Alam, M.K.
Commonwealth Scientific and Industrial Research Organization, Canberra, Australia.

Alam, Md. Khairul
Commonwealth Scientific and Industrial Research Organization, Wembley, WA, Australia.

Ali, Md. Mohsin
Soil Science Division, Bangladesh Institute of Nuclear Agriculture (BINA), Mymensingh.

Bandaogo, A.
Institute of the Environment and Agricultural Research, Bobo-Dioulasso, Burkina Faso.

Banna, Mostafa Faruq Al
FPMU, Ministry of Food, Dhaka, Bangladesh.

Barik, K.
Dept of Soil Science and Plant Nutrition, Ataturk University, Erzurum, Turkey.

Basnyat, Divas B.
Nepal Development Research Institute, Manbhawan, Lalitpur, Nepal.

Bayhan, Y.
Biosystems Engineering Dept., University of Namik Kemal, Tekirdag, Turkey.

Bell, Richard W.
Agriculture Discipline, Murdoch University, Perth, WA, Australia.

Bhan, S.C.
India Meteorological Department.

Biswas, J.C.
Krishi Gobeshona Foundation.

Biswas, Wahidul K.
Sustainable Engineering Group, Curtin University, WA, Australia.

Chatterjee, R.
International Rice Research Institute South Asia Regional Center, Uttar Pradesh, India.

Das, Shamal Chandra
Bangladesh Water Development Board, Bangladesh.

Didenko, N.O.
Institute of Water Problems and Land Reclamation, Kyiv, Ukraine.

Farhan, M.J.
Dept. of Soil Science and Water Rescores, University of Tikrit, Tikrit, Iraq.

Goroshi, Sheshakumar
India Meteorological Department.

Gupta, Dipali Rani
Institute of Biotechnology and Genetic Engineering, Bangabandhu Sheikh Mujibur Rahman Agricultural University, Bangladesh

Gurung, Jaya K.
Nepal Development Research Institute, Manbhawan, Lalitpur, Nepal.

Haiju, Rojina
Nepal Development Research Institute, Manbhawan, Lalitpur, Nepal.

Haque, M.A.
Missing

Haque, Md. Aminul
Bangladesh; Bangladesh Sugarcrop Research Institute, Pabna, Bangladesh.

Haque, Md. Azizul
Soil Science Division, Bangladesh Institute of Nuclear Agriculture (BINA), Mymensingh.

Heulin, Thierry
Lab Microbial Ecology of the Rhizosphere (LEMiRE), Aix Marseille Univ, CEA, CNRS, UMR 7265 BIAM, F-13115 Saint Paul-Lez-Durance, France.

Hossain, Gazi Md. Akram
Bangladesh; Bangladesh Sugarcrop Research Institute, Pabna, Bangladesh.

Hossain, Md. Amzad
Bangladesh; Bangladesh Sugarcrop Research Institute, Pabna, Bangladesh.

Hossain, Md. Billal
Bangladesh Hi-Tech Park Authority, Information and Communication Technology Division, ICT Tower, Dhaka-1207, Bangladesh.

Hussain, Muhammad Baqir
Department of Soil and Environmental Sciences, MNS University of Agriculture, Multan, Pakistan.

Irkitbey, A.
Kazakh National Agrarian Research University, Almaty, Kazakhstan.

Islam, K.R.
The Ohio State University, Columbus, Ohio, USA.

Islam, Md. Shafiqul
Bangabandhu Sheikh Mujibur Rahman Agricultural University, Gazipur, Bangladesh.

Islam, Rafiq
The Ohio State University, Columbus, USA.

Islam, Tofazzal
Institute of Biotechnology and Genetic Engineering, Bangabandhu Sheikh Mujibur Rahman Agricultural University, Bangladesh

Jahangir, M.M.R.
Bangladesh Agricultural University, Mymensingh, Bangladesh.

Kamal, F.H.
Institute of Water Modelling, Bangladesh.

Kamal, S.M. Rezaul Mostofa
Ministry of Home Affairs, Government of Bangladesh.

Karim, A.J.M. Sirajul
Bangabandhu Sheikh Mujibur Rahman Agricultural University, Gazipur, Bangladesh.

Karim, Md. Abdul
Bangabandhu Sheikh Mujibur Rahman Agricultural University, Bangladesh.

Khairo, A.M.
Dept. of Soil Science and Water Rescores, University of Tikrit, Tikrit, Iraq.

Khan, Z.H.
Institute of Water Modelling, Bangladesh.

Khatun, M. Rokeya
Bangladesh Agricultural Research Institute, Gazipur, Bangladesh.

Lavrenko, S.
Kherson State Agrarian and Economic University, Kherson, Ukraine.

Masood, Muhammad
Bangladesh Water Development Board, Bangladesh.

Masood, Sajid
Department of Soil Science, Bahauddin Zakariya University, Multan, Pakistan.

Masud, M.M.
Soil Science Division, Bangladesh Agricultural Research Institute, Bangladesh.

Mondal, M.K.
International Rice Research Institute, Bangladesh and Philippines.

Nahar, Kamrun
Soil Science Division, Bangladesh Institute of Nuclear Agriculture (BINA), Mymensingh.

Palanisamy, Ramaraj
India Meteorological Department.

Parvin, S.
Bangladesh Rice Research Institute, Bangladesh.

Rahman, Mahbub
School of Agriculture, The University of Adelaide, Australia.

Rahman, Mustafizur
Department of Soil, Water and Environment, University of Dhaka, Bangladesh.

Ray, Kamaljit
Ministry of Earth Sciences, New Delhi, India.

Razzaghi, S.
Dept. of Soil Science and Plant Nutrition, Erciyes University, Kayseri, Turkey.

Ritu, S.P.
Bangladesh Rice Research Institute, Bangladesh.
Sylhet Agricultural University, Bangladesh.

Sarker, Rakhi R.
Bangladesh Institute of Nuclear Agriculture, Mymensingh, Bangladesh.

Sarwar, A.K.M. Golam
Department of Crop Botany, Bangladesh Agricultural University, Mymensingh, Bangladesh.

Sen, R.
Soil Science Division, Bangladesh Agricultural Research Institute, Bangladesh.

Seraj, Zeba I.
Department of Biochemistry and Molecular Biology, University of Dhaka, Bangladesh.

Singh, K.K.
India Meteorological Department.

Sivakumar, M.V.K.
Weather and Climate Extrêmes (Elsevier), Switzerland

Solaiman, A.R.M.
Bangabandhu Sheikh Mujibur Rahman Agricultural University, Gazipur, Bangladesh.
Present address: Mawlana Bhashani Science and Technology University, Bangladesh.

Sutradhar, A.
Institute of Water Modelling, Bangladesh.

Thapa, Srijana
Nepal Development Research Institute, Manbhawan, Lalitpur, Nepal.

Yadav, S.
International Rice Research Institute, Bangladesh and Philippines.
The University of Queensland, Australia.

Yelikbayev, B.K.
Satbayev University, Almaty, Kazakhstan.

CHAPTER 1

Climate Change, Agriculture, Carbon, and Food Security

Islam, K.R.[1*], *Khairo, A.M.*[2], *Farhan, M.J.*[2], *Irkitbey, A.*[3], *Yelikbayev, B.K.*[4], *Didenko, N.O.*[5], *Lavrenko, S.*[6], *Bayhan, Y.*[7], *Barik, K.*[8], *Aksakal, E.L.*[8], *Razzaghi, S.*[9], *Ahsan, S.*[10], *Bandaogo, A.*[11] and *Chatterjee, R.*[12]

Introduction

Throughout the course of Earth's existence, it has undergone alternating phases of warm periods and ice ages, resulting in a natural greenhouse effect. This effect is crucial for maintaining liquid water and supporting life, as the mean global annual temperature would have been –18°C in its absence (Voet and Voet, 2010). However, since the 1800s, human activities have emerged as the primary catalyst for an increase in the concentration of greenhouse gases (GHGs) in the troposphere, leading to significant impacts on global warming and climate change. These activities, such as the combustion of fossil fuels, household, commercial, and industrial practices, as well as deforestation and conventional agricultural activities, contribute to the release of heat-trapping gases (USEPA, 2023). The key GHGs include carbon dioxide (CO_2), carbon monoxide (CO), methane (CH_4), nitrous oxide (N_2O), water vapor, ozone (O_3), and fluorinated gases (Table 1).

[1] The Ohio State University, Columbus, Ohio, USA.
[2] Dept. of Soil Science and Water Resources, University of Tikrit, Tikrit, Iraq.
[3] Kazakh National Agrarian Research University, Almaty, Kazakhstan.
[4] Satbayev University, Almaty, Kazakhstan.
[5] Institute of Water Problems and Land Reclamation, Kyiv, Ukraine.
[6] Kherson State Agrarian and Economic University, Kherson, Ukraine.
[7] Biosystems Engineering Dept., University of Namik Kemal, Tekirdag, Turkey.
[8] Dept. of Soil Science and Plant Nutrition, Ataturk University, Erzurum, Turkey.
[9] Dept. of Soil Science and Plant Nutrition, Erciyes University, Kayseri, Turkey.
[10] Metropilitan State University of Denver, Colorado, USA.
[11] Institute of the Environment and Agricultural Research, Bobo-Dioulasso, Burkina Faso.
[12] International Rice Research Institute South Asia Regional Center, Uttar Pradesh, India.
* Corresponding author: islam.27@osu.edu.

Table 1 Greenhouse gas emission, their global warming potential and tropospheric lifetime. (*Source:* IPCC, 2014)

Greenhouse Gas	*Chemical Formula*	*Emission Contrib. (%)*	*GWP in a 100 Yr. Time Span*	*Tropospheric Lifetime (Yr.)*
Carbon Dioxide	CO_2	76	1	100
Methane	CH_4	16	25	12
Nitrous Oxide	N_2O	6	265	121
Fluorinated gases (F-gases)		2		
Chlorofluorocarbon-12 (CFC-12)	CCl_2F_2		10,200	100
Hydrofluorocarbon-23 (HFC-23)	CHF_3		12,400	222
Sulfur Hexafluoride	SF_6		23,500	3,200
Nitrogen Trifluoride	NF_3		16,100	500

With global warming potentials (GWPs) of emitted GHGs typically ranging from thousands to tens of thousands, they are termed high-GWP gases because, for a given mass, they trap more heat than CO_2. The GWP of CH_4 is 25 times greater, and N_2O is 296 times greater than CO_2 (GWP=1). However, the Earth's troposphere contains significantly more CO_2 than CH_4 or N_2O (Table 1). The concentration of CO_2 has risen from the pre-industrial level of 280 μmol mol^{-1} to the current level of 419 μmol mol^{-1} and is projected to reach 700–800 μmol mol^{-1} by 2100 (IPCC, 2021).

Human-induced global warming and climate change associated with the increasing accumulation of GHGs in the troposphere have broader impacts (Fig. 1). These include heightened frequency of extreme weather events, especially erratic precipitation, temperature rise, heatwaves, drought, and wet desertification, posing significant threats to ecosystem services and ecological functions (USEPA, 2023; Zhao et al., 2023). Notably, global temperatures have increased by 40% in the past 150 years, with half of this increase occurring in the last 30 years (IPCC, 2014). This could result in an average global temperature increase of 1–1.5°C sometime in the next century. As a response, climate zones may shift, polar ice may melt, and crops, forests, and animals may react differently by species. Some effects could be catastrophic.

Agriculture significantly contributes to the global carbon footprint, responsible for over 25% of anthropogenic GHG emissions linked to climate change effects on terrestrial ecosystems (IPCC, 2018; Zhao et al., 2023). While conventional agricultural practices generate substantial quantities of food, feed, and fiber (Fig. 2), they adversely impact soil health, water quality, and agroecosystem services (Islam and Didenko, 2019). Persistent annual tillage operations have demonstrated negative effects on biodiversity and efficiency, accelerating erosion, edge-of-field nutrient loss, compromising structural stability, increasing GHG emissions, diminishing soil organic carbon (SOC) content, and subsequently leading to soil health and land quality degradation (Aziz et al., 2013; Islam et al., 2021; Rahman et al., 2021). Specifically, accelerated soil organic matter (SOM) decomposition and crop residue burning, common in soil-crop management practices, are primary sources of CO_2 emissions in agriculture (Zhao et al., 2023). Methane emissions result from puddled and flooded rice production, intestinal fermentation in livestock digestive systems,

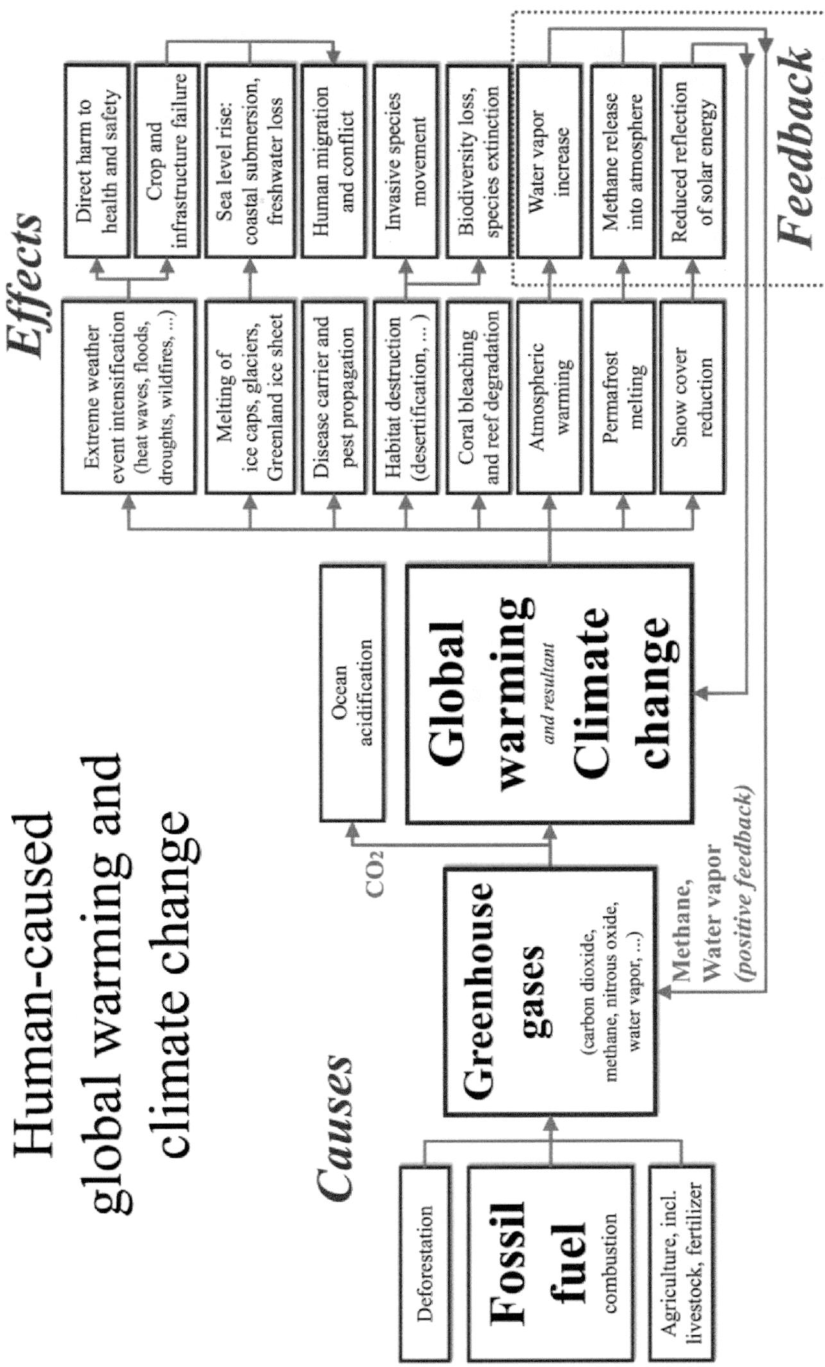

Fig. 1 Human-induced climate change effects on ecosystems (https://en.m.wikipedia.org/wiki/ File:20200101_Global_warming_-_climate_change_-_causes_effects_feedback.png).

and the decomposition of feces and crop residues in moist conditions. Similarly, N_2O emissions arise from intensively chemically fertilized soils, especially through nitrogen (N) fertilization, as well as from manure and compost amendments (Fig. 2). Globally, economic and population growth are pivotal factors driving the increase in GHG emissions, with projections indicating their continued rise, influencing future climate change (Jat et al., 2022).

Land degradation linked to agriculture and other activities poses a threat to the distribution of productive soils worldwide (https://www.globalagriculture.org). On a global scale, 10–20% of drylands and 24% of productive lands are degraded. Annually, an estimated 24 billion tons of fertile soil are lost due to erosion, equating to 3.4 tons per person. Water (sheet and rill) erosion rates averaged 6.6 tons per hectare in 2017, while wind erosion rates averaged 4.8 tons per hectare in the U.S. (USDA, 2020). Furthermore, approximately 810 hectares of agricultural land in the United States are paved over, fragmented, and converted for uses that jeopardize farming (www.farmland.org).

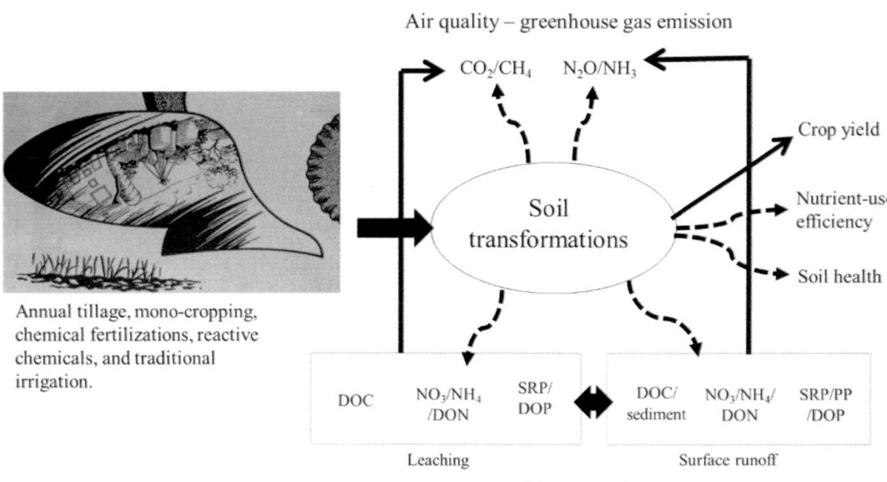

Fig. 2 Conventional agriculture's impact on crop yield, soil health, greenhouse gas emission, and water quality associated with ecosystem services (DOC=dissolved organic C, DON= dissolved organic N, DOP=dissolved organic P, SRP=Soluble reactive P; and PP=particulate P). *Source*: Rafiq Islam.

The European Union experiences an annual soil loss of 970 million tons due to water erosion, with the current cost of land degradation estimated at approximately US $490 billion each year. Globally, 40% of degraded land is concentrated in regions with high poverty rates, directly affecting the health and livelihoods of around 1.5 billion people (https://www.globalagriculture.org/). A study supported by FAO, UNDP, and UNEP revealed that South Asian countries (India, Pakistan, Bangladesh, Afghanistan, Nepal, Sri Lanka, Bhutan) incur an annual loss of at least US $10 billion due to land degradation (Khor, 2008; Spijkers, 2011). This accounts for 2% of the region's GDP or 7% of its agricultural output. The study indicated that 43% of the region's total agricultural land, approximately 140 million hectares, suffered from one or more forms of degradation. Among these, 31 million hectares were highly degraded, and 63 million hectares were moderately degraded. Bangladesh ranked as the most severely affected country, with 75% of its agricultural land degraded, followed by Pakistan (61%), Sri Lanka (44%), Afghanistan (33%), Nepal (26%), India (25%), and Bhutan (10%) (Khor, 2008; Spijkers, 2011).

Climate Change and Agricultural Production

Agriculture stands as one of the most vulnerable sectors to the impacts of global warming and climate change (IPCC, 2021). The dependence of food production on climatic variables such as temperature, radiation, rainfall, moisture, and CO_2 concentration makes crop productivity susceptible to climate change (Nelson et al., 2009). Changes in climatic factors associated with crop production pose diverse challenges to addressing food security (Smith and Gregory, 2013; Lipper et al., 2014; Ziska, 2014).

Studies indicate an increase in the frequency and intensity of extreme events like droughts, heavy waves, intense rainfall, flooding, and high maximum temperatures, which are already occurring and expected to accelerate globally (IPCC, 2014; Lipper et al., 2014). Average and seasonal maximum temperatures are projected to rise with greater variability, while rainfall will become intensively variable, albeit unevenly distributed. Water scarcity and drought in dry regions are expected to increase by the end of the 21st century (IPCC, 2014). The impact of increased temperature on crops will vary based on each crop's optimal temperature for growth and reproduction (Hatfield et al., 2014). While some regions may benefit from global warming, others may experience adverse effects, impacting grain yield and food quality.

Studies suggest that despite short-term yield gains in some crops, increased temperatures will have a negative long-term impact, eventually reducing crop yields under future climate change (Bruinsma, 2003; IPCC, 2007). The effects of climate change on global agricultural production are expected to be unevenly distributed, with lower latitude and developing countries suffering more due to their geographical location, higher agricultural share in economies, and limited capacity to adapt. In contrast, high latitude regions may benefit from climate change (ADB, 2006). Global estimates predict a 15.9% reduction in agricultural productivity by the 2080s if global warming continues unabated, with developing countries experiencing a larger decline of 19.7% (Cline, 2007). According to Lipper et al. (2014), climate change has already reduced global maize and wheat yields by 3.8% and 5.5%, respectively. Paddy loss due to floods in Bangladesh and India is estimated at 4 million tons per year, enough to feed 30 million people (Asada and Matsumoto, 2009).

Increasing CO_2 concentration is expected to dilute N concentration, reducing protein quality via amino acids composition. Moreover, the concentration of essential nutrients in crops like wheat, soybeans, and rice may be affected, impacting food quality and posing a potential threat to human health and animal nutrition. Climate change is anticipated to encourage weed and pest proliferation, leading to increased likelihood of crop failures in the short-term and declines in food production in the long-term (Nelson et al., 2009; Ziska, 2014). Weeds, pests, and fungi are expected to thrive under warmer temperatures and wetter climates, posing new challenges to crops previously unexposed to these conditions. Livestock faces risks from direct heat stress and reduced feed quality, affecting pasture and rangeland sustainability. Human health may be affected by reduced food quality and increased use of reactive chemicals (e.g., pesticides) due to heightened weed and pest pressures, alongside reductions in the efficacy of pesticides (Ziska, 2014). The IPCC (2021) reported that climate change negatively affects crop production more commonly than positively, with developing countries highly vulnerable to further negative impacts. Climate change is likely to affect agricultural production and food security globally, regionally, and locally by influencing soil-plant-water ecosystems (Brown et al., 2014; https://www.globalagriculture.org).

While climate change poses a significant risk to global agriculture, crop production in South Asia is particularly vulnerable in addressing food security (Aryal et al., 2020). Climatic variability, especially temperature and rainfall, explains 60% of crop yield variability and is a crucial factor influencing food production (Matiu et al., 2017; Aryal et al., 2020). The South Asian region is projected to experience a temperature rise of 0.5–1.2°C by 2020, 0.88–3.16°C by 2050, and 1.56–5.44°C by 2080. Heat-stressed areas in the region could increase by 21% in 2050, depending on future scenarios (IPCC, 2007; Tesfaye et al., 2017). These projected long-term changes in temperature and rainfall patterns are expected to impact land preparation, cropping seasons, crop types, planting suitability, fertilization, irrigation, disease and pest incidence, and weed infestations. For example, the growing periods for spring maize in Pakistan have shifted an average of 4.6 days per decade earlier between 1980-2014, while the planting of autumn maize has been delayed by 3 days per decade, significantly affecting crop yields (Abbas et al., 2017). Additionally, water availability, through rainfall and groundwater recharge, is anticipated to decline due to climate change, while agricultural water consumption is predicted to increase by 19% in 2050 (UN-Water, 2013).

Several studies, as summarized by Aryal et al. (2020), have reported the impacts of climate change on major food crops, including rice, wheat, and maize in South Asia (Tables 2, 3). Results show that crop yields in the region have been significantly influenced by changes in temperature and rainfall variability (Tables 2, 3). Comparing crop yields in the 1990s without climate change effects to those in the 2020s and 2050s, Parry et al. (2004) reported a reduction of 2.5–10% in several regions of Asia in the 2020s and 5–30% in the 2050s. The impacts of temperature and water stresses on agriculture, as reported by Lal (2011), indicate that summer crop productivity in South Asian countries is expected to decrease due to heightened climate variability and the widespread incidence and virulence of pests after 2050. Winter crops are expected to be more adversely affected by a temperature rise of 2°C. By the end of the 21st century, the net cereal production in South Asia is projected to decrease by at least 4–10% if the temperature increases by 3°C. Knox et al. (2012) suggested that the average yield of eight major crops would be reduced by 8% by the 2050s in South Asia. Specifically, maize and sorghum are anticipated to experience yield reductions of 16% and 11%, respectively, while rice is expected to be unaffected.

Aryal et al. (2020) documented a broad range of yield losses resulting from the impact of temperature changes on wheat, rice, and maize productivity in South Asia (Table 2). In India, crop yields decreased by 5%, 6–8%, and 10–30% in wheat, rice,

Table 2 Impact of change (increase) in temperature (Δ T) on crops in South Asian countries.

Country	*ΔT (+)*	*Wheat*	*Rice*	*Maize*	*References**
India	<2°C	Yield decrease by 0.45 Mg/ha (1); yield decrease by 0.4 Mg/ha (2); yield decrease by 5.2% (6)	Yield decrease by 6% (3); yield decrease by up to 8.2% (4)	Yield decrease by 10–30% under 350 ppm CO_2 (5)	[1]Sinha & Swaminathan (1991); [2]Morey & Sadaphal (1981); [3]Saseendran et al. (2000); [4]Mathauda et al. (2000); [5]Kalra et al. (2007); [6]Gupta et al. (2017)
	2–3°C	Decrease in all regions (1)	Yield decrease by 0.75 Mg/ha (2); in % yield decrease by 8.4% (3)	Yield decrease by 10–30% under 350 ppm CO_2 (4)	[1]Aggarwal & Sinha (1993); [2]Sinha & Swaminathan (1991); [3]Mathauda et al. (2000); [4]Kalra et al. (2007)
	>3°C			Yield decrease by 10–30% under 350 ppm CO_2 (3)	[3]Kalra et al. (2007)
Bangladesh	<3°C	Yield loss exceed 60% (1)	Yield decrease by 2.6–13.5% (2)		[1]Karim et al. (1996); [2]Basak et al. (2009)
	>3°C	Yield loss exceed 60% (1)	Yield decrease by 0.11–28.7% (2)		[1]Karim et al. (1996); [2]Basak et al. (2009);
Nepal	4°C	Yield increases by 18.4% due to CO_2 fertilization and by 8.6% with an increase in temperature (1)	Yield decrease by 1.8% in Terai and increase in hills by 5.3% and mountainous by 33.3% (1)	Yield decrease by 26.4% in Terai and by 9.3% in hills (2)	[1]Malla (2008); [2]Prasai (2010)
Pakistan	<3°C	Yield decrease by 5–7% (1); Yield decrease by 7% in Swat district and 14% in Chitral district (2); yield decrease by 6–9% in sub-humid, semiarid, and arid zones (3)			[1]Aggarwal & Sivakumar (2010); [2]Hussain & Mudasser (2007); [3]Sultana & Ali (2006)

Contd.

Table 2 *Contd.*

Country	ΔT (+)	*Wheat*	*Rice*	*Maize*	*References**
	>3°C	Yield decrease in arid, semiarid, and sub-humid zones, but increases in humid zone (1); yield decrease by 21% in Swat district and 23% in Chitral district (2)			[1]Sultana et al. (2009); [2]Hussain & Mudasser (2007)
Sri Lanka	≤1°C		A loss of 6% rice output (1); yield decrease by 1–5% (2)		[1]GoSL (2000); [2]Vidanage & Abeygunawardane (1994)

Note: [a]Number in the parentheses refers to the corresponding reference mentioned (*Source:* Aryal et al., 2020).

and maize, respectively. Despite variations in input use and crop management in India, both season-long and terminal-heat stress negatively affected rice and wheat, with wheat proving more sensitive than rice (Arshad et al., 2017). In Nepal, rising temperatures positively impacted crop yields, especially for wheat. However, rice and maize yields in the mid-hills and Terai regions significantly decreased with increasing temperatures. In Bhutan, the impacts of climate change amplified crop losses due to unusual outbreaks of diseases and pests, erratic rainfall, windstorms, hailstorms, droughts, flash floods, and landslides (Chhogyel and Kumar, 2018). Similar responses to climate change-induced crop yield losses were observed in Bangladesh, Pakistan, and Sri Lanka. It is projected that almost half of the Indo-Gangetic Plains (Bangladesh and India), the major food basket of the South Asian region, may face severe wheat production challenges by 2050 due to heat stress (Ortiz et al., 2008; Dey et al., 2017).

Climate change, closely tied to rising temperatures, has a direct impact on water availability, significantly influencing crop productivity. For instance, if India continues to deplete its groundwater for irrigation, the adverse effects of increased warming and other climatic variations on crop production are expected to intensify by half (Fishman, 2018). Kumar et al. (2004) demonstrated that a 19% reduction in summer monsoon rainfall results in an approximately 18% decrease in food grain production (Table 3). Even a modest temperature rise (e.g., 1.5–2°C) in South Asia can critically affect water resources due to heightened monsoon variability and glacial meltwater, posing a substantial threat to future agricultural productions (Vinke et al., 2017).

Looking ahead to 2050, the global population is projected to surpass 10 billion, with approximately 10% of the current 8 billion people on Earth already facing undernourishment, and 3 billion lacking access to healthy diets (Nuccitelli, 2022). A staggering 75% of the world's poor reside in rural areas, where agriculture stands as their primary source of income. Schmidhuber and Tubiello (2007) reported that by 2080, anywhere between 5–170 million people worldwide may confront severe food scarcity, with over 95% of these individuals residing in developing countries that are expected to be most affected by climate change. Concurrently, per capita food consumption is on the rise, anticipated to lead to a doubling of global meat consumption and a 60% increase in cereal consumption by 2050 (Zhai and Zhuang, 2009). Agriculture plays a pivotal role in Southeast Asia, contributing to more than 10% of the gross domestic product in most regional economies and offering employment to over one-third of the working population in the region (Zhai and Zhuang, 2009).

Table 3 Impacts of change rainfall variability on crops in South Asian countries.

Country	*Impacts*	*Crops*	*References**
Bangladesh	Positive and statistically significant effect	Rice	Sarker et al. (2012)
Bangladesh	Negatively affects rice yields	Rice	Alauddin and Hossain (2001)
Bangladesh	Variability in rice yields leading to Yield decrease by 8–17% by 2050	Rice	Rahman et al. (2009)
India	More rainfall leads to higher mNDVI in the drier western parts of the basin and lower mNDVI in the eastern parts of the basin (*)	Crops (vegetation)	Siderius et al. (2014)
India	Annual total yield were significantly correlated with all-India summer monsoon rainfall	Wheat and Rice	Kumar et al. (2007)
India	Statistical analysis of state-level Indian data confirms that drought and extreme rainfall negatively affected rice yield (harvest per hectare)	Rice	Aufhammer et al. (2012)
Nepal	The yield of individual cereals is correlated with the seasonal rainfall data	Paddy, maize, millet, wheat, and barley	Bhandari (2013)
India	All-India crop yield index shows a strong relationship with all-India summer monsoon rainfall	Food grains	Prasanna (2014)
Pakistan	Effect of rainfall on the yield of rice, wheat, and maize is negative and nonsignificant except for wheat, which is significant	Rice, wheat, and maize	Ali et al. (2017)
Pakistan	Normal precipitation during vegetative and maturity stages and their deviations from the historical mean (positive) exert a positive impact on the wheat yield	Wheat	Ahmad et al. (2014)

Note: NDVI is Normalized Difference Vegetation Index (*); for the Rabi season, the relationship for those areas with a significant impact is mostly positive (*Source:* Aryal et al., 2020).

Addressing the food security challenge in developing countries necessitates a substantial increase in cereal yields by 40%, along with a corresponding rise in irrigation water requirements by 40–50%. Additionally, there may be a need for an additional 100–200 million hectares of land, particularly in Asia, sub-Saharan Africa, and Latin America (Lipper et al., 2014). The demographic landscape in developing countries is expected to witness an influx of an extra 2.4 billion people, concentrated in South Asia and sub-Saharan Africa, by 2050. While agriculture remains a pivotal economic sector and primary source of employment in these regions, the current reality is that over 20% of the population, on average, grapples with food insecurity (Wheeler and Braun, 2013; Lipper et al., 2014).

Projections underscore the global imperative to increase agricultural production by 60% by 2050 to meet escalating demand, with a major share of this augmentation required from enhanced crop productivity. However, an in-depth analysis based on South Asian countries' agricultural land area, population growth, and agricultural land per capita does not reveal economically viable options for the horizontal expansion of agriculture. Instead, the emphasis should be on achieving a vertical increase in crop productivity per unit of land (Table 4).

Table 4 Total and agricultural land distribution and population density in South Asian countries.

Country	*Tot. Land Area (km^2)*	*Agric. Land (km^2)*	*Agric. Land/Total Land (%)*	*Total Population*	*Pop. Density (Person/km^2)*	*Agric. Land/ Person*
Afghanistan	652230	383130	58.7	42,239,854	65	0.91
Bangladesh	130170	100680	77.3	172,954,319	1329	0.06
Bhutan	38140	5130	13.5	787,424	21	0.65
India	2973190	1785279	60.0	1,428,627,663	481	0.12
Maldives	300	59	19.7	521,021	1737	0.01
Nepal	143350	41210	28.7	30,896,590	216	0.13
Pakistan	770880	363030	47.1	240,485,658	312	0.15
Sri Lanka	61860	28120	45.5	21,893,579	354	0.13
Total	4,708,260	2,706,638	–	1,938,406,108	412	–
Average	–	–	57.2	–	–	0.14

Source: https://www.worldometers.info/world-population/southern-asia-population. https://data.worldbank.org/indicator/SP.POP.TOTL?locations=8S.

By 2050, the global demand for food security is set to double, placing existing farming practices in a heightened dependency on reactive chemicals, freshwater, and energy inputs (Islam and Didenko, 2019). This intensification of farming under climate change is anticipated to have adverse consequences for agroecosystem services. The imperative to sustainably manage the world's climate-stressed and pollution-degraded agricultural system is paramount to meet the needs of the expected 9 billion people by 2050 (Nuccitelli, 2022). Addressing this challenge involves reducing GHG emissions to mitigate climate change effects and ensure food security for the growing population, making it one of the foremost challenges of the 21st century (Ziska et al., 2012). While increasing agricultural production is achievable, it is likely to come at a higher environmental cost and pose threats to biodiversity unless proactive measures are taken for climate change mitigation and adaptation.

Climate Change Mitigation and Adaptation

Climate-Smart Agriculture

Climate-smart agriculture (CSA) represents an integrated approach aimed at transforming and reorienting agricultural production systems to sustain food security in the face of climate change (Lipper et al., 2014). This approach harmonizes economic viability, social acceptability, and environmental compatibility to collectively address climate change-related threats to food security within the framework of sustainable agricultural development (FAO, 2013). The key objectives of CSA encompass:

1. **Increased Productivity:** Producing more high-quality food to enhance nutrition security and boost incomes, especially for the 75% of the world's low-income population residing in rural areas reliant on agriculture for their livelihoods.
2. **Enhanced Resilience:** Reducing vulnerability to drought, pests, diseases, and other climate-related risks and shocks, while improving the capacity to adapt and thrive in the face of longer-term stresses such as shortened seasons and erratic weather patterns.
3. **Reduced Emissions:** Pursuing lower emissions for each calorie or kilogram of food produced, avoiding deforestation from agriculture, and identifying ways to sequester tropospheric CO_2 via plants as SOM to enhance soil health and sustain crop production for food security.

Carbon within SOM plays a crucial role in the earth's carbon cycle and is vital for mitigating the effects of climate change (Sundermeier et al., 2011; Raphael, 2024). Mitigating emitted CO_2 as SOM is a pivotal strategy to enhance soil health and increase crop production within the context of climate-smart agriculture.

Conservation Tillage

While conventional tillage (CT) offers benefits such as field workability, seed bed preparation, weed and soil-borne disease suppression, and higher crop yields, it adversely affects soil biology by fragmenting fungal hyphae and mycorrhizal associations (Islam and Weil, 2000; Sundermeier et al., 2011). CT degrades soil health and accelerates GHG emissions, especially CO_2 (Sundermeier et al., 2011). Transitioning from CT to no-till (NT) or minimum tillage emerges as a proactive strategy to enhance soil functional capacity in the CO_2 source-to-sink pathway (Sundermeier et al., 2011; Islam et al., 2022). Under NT, crop residues accumulate on the surface, acting as mulch to mitigate rising soil temperature, reduce evaporative demand, control soil erosion, retain moisture longer, and limit O_2 diffusion. Additional benefits encompass increased biodiversity with fungal dominance (higher C-use efficiency), slower decomposition of residues and native SOM, augmented aggregate formation and stability, greater SOC sequestration, and improved soil quality (Islam and Weil, 2000; Islam et al., 2022).

Conservation tillage, a practice that mandates leaving a minimum of 30% of crop residues, is pivotal for controlling edge-of-field soil loss (Kanicki, 2014; Bergtold and Sailus, 2020). With approximately 40% of crop residues remaining in the soil under conservation tillage (30% recommended residues and 10% as root biomass),

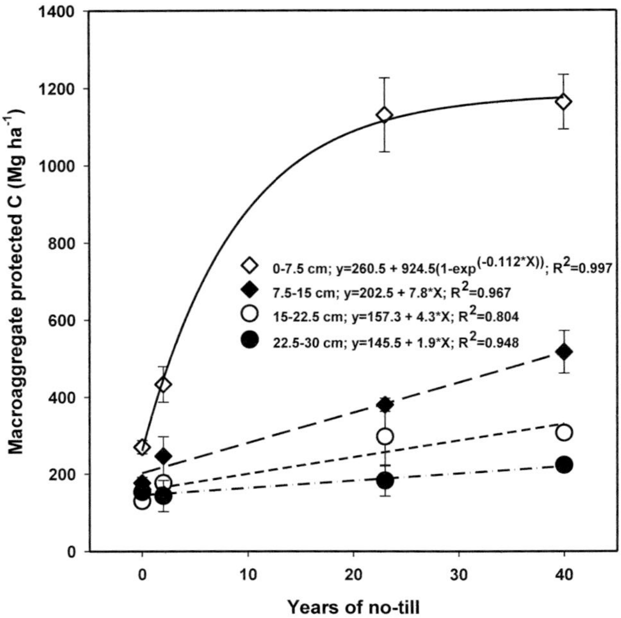

Fig. 3 Impact of long-term no-till on aggregate protected carbon sequestration at different depths of soil. (*Source:* Sundermeier et al. (2011))

equivalent to 1.6 tons of residue per hectare annually. Calculating the conversion rate of crop residues to SOM at 30% (Hann, 1977), SOM amounts to 0.48 tons per hectare per year. Given that SOM comprises about 58% total carbon (C), SOC equates to 0.28 tons per hectare per year. Using the coefficient "*r*" as the percentage of native SOC left in the second year after the decomposition of SOC formed in the first year, assuming *r* as 99% (Hann, 1977), the SOC stock in 10 years will be:

$$Y = [SOC * (1 - r^n)]/(1 - r)$$

$$Y = [0.28 * (1 - 0.99^{10})]/(1 - 0.99)$$

$$Y = 2.68 \text{ tons C/ha.}$$

Results indicate that aggregate-protected C was 24%, 80%, and 92% higher in 2, 23, and 44-year no-till (NT), respectively, compared to conventional tillage (CT) (Sundermeier et al., 2011). Over a 50+ year period, NT at 10-year intervals resulted in aggregate-protected C sequestration rates of 92, 63, 47, and 37 kg/ha/yr. at depths of 0–7.5, 7.5–15, 15–22.5, and 22.5–30 cm, respectively, when compared to CT (Fig. 3). It has been reported that C sequestration associated with soil health improvement sustains agricultural productivity (Islam et al., 2022). Reduced tillage with residue management has proven beneficial in increasing labile C and SOC stocks in intensive rice-based cropping systems in Bangladesh (Sarker et al., 2023).

In the South Asian region, especially Bangladesh, India, Nepal, and Pakistan, rice-wheat rotation dominates, covering more than 13.5 million hectares of CT lands (Zahid et al., 2020). In recent years, the productivity of rice-wheat has declined due to repeated plowing, unbalanced chemical fertilization, and increased pest and soil-borne disease pressures (Bhatt et al., 2016). It has been reported that average wheat yield and irrigation efficiency in zero-till were higher than in CT for rice-wheat rotation and rice-wheat systems with the inclusion of legumes, owing to enhanced wheat rooting after unpuddled rice (Chaki et al., 2021).

Cropping Diversity and Cover Crops

Cropping diversity emerges as a solution to mitigate climate change while addressing the demand for food security (Moss et al., 2020). Diversifying crops through mixed cropping, intercropping, relay crops, or crop rotation with cover crops reduces the environmental impact of agriculture by enhancing microbial efficiency, labile carbon, and nutrients, thereby improving soil health to sustain crop productivity (Moss et al., 2020; Islam et al., 2021; Sarker et al., 2023).

Plants primarily fix tropospheric CO_2 through two photosynthetic processes: The C_3 and C_4 pathways, C_3 plants, such as rice, wheat, soybeans, small grains, and legumes, respond more effectively to increasing CO_2 concentration than C_4 plants, such as maize, millet, sorghum, sugarcane, and many grasses, as C_4 plants are already efficient users of CO_2 at ambient levels. An increase in CO_2 concentration positively impacts C_3 crop yields by stimulating photosynthesis and reducing water loss via transpiration. It is reported that crop yield may increase by 10–25% for C_3 crops and by 0–10% for C_4 crops when CO_2 levels reach 550 ppm (IPCC, 2007). Therefore, managed agroecosystems containing both C_3 and C_4 plants may provide a competitive advantage for C_3 plants under increased CO_2 in the troposphere. A meta-analysis revealed that cropping diversity with cover crops reduced GHG emissions by 8% while increasing SOC sequestration (Poeplau and Don, 2015). Cover crops also suppress weeds, capture residual N, and improve crop yield (Islam et al., 2021). The benefits and services provided by cropping diversity with cover crops are outlined in Fig. 4 and Table 5.

In South Asia, agriculture is characterized by rice monocropping or rice-wheat rotation, with or without cover crops. Numerous studies have highlighted the substantial benefits of diversified legume-based cropping systems, including significantly higher crop yields and SOC content. These positive outcomes are attributed to the N recycling effects facilitated by leguminous crop residues. The existing lack of crop diversification emerges as a key factor contributing to the gradual decline in agricultural production over time. This decline serves as a crucial warning sign for stakeholders in South Asia, emphasizing the need to formulate appropriate national policies. These policies should aim at fostering agricultural diversification as a strategic approach to ensure sustainable food and nutritional security in the region.

Chemical Fertilization

The Green Revolution, facilitated by chemical fertilization, has significantly contributed to a global increase of approximately 50% in crop yields, with a more pronounced impact in developing countries (Yousaf et al., 2017; Anonymous, 2019). Optimizing NPK fertilizations has been reported to result in a notable 19–41% increase in rice yields and a substantial 61–76% boost in rapeseed yields (Yousaf et al., 2017). Without N fertilizations, maize yields could have experienced a 40%

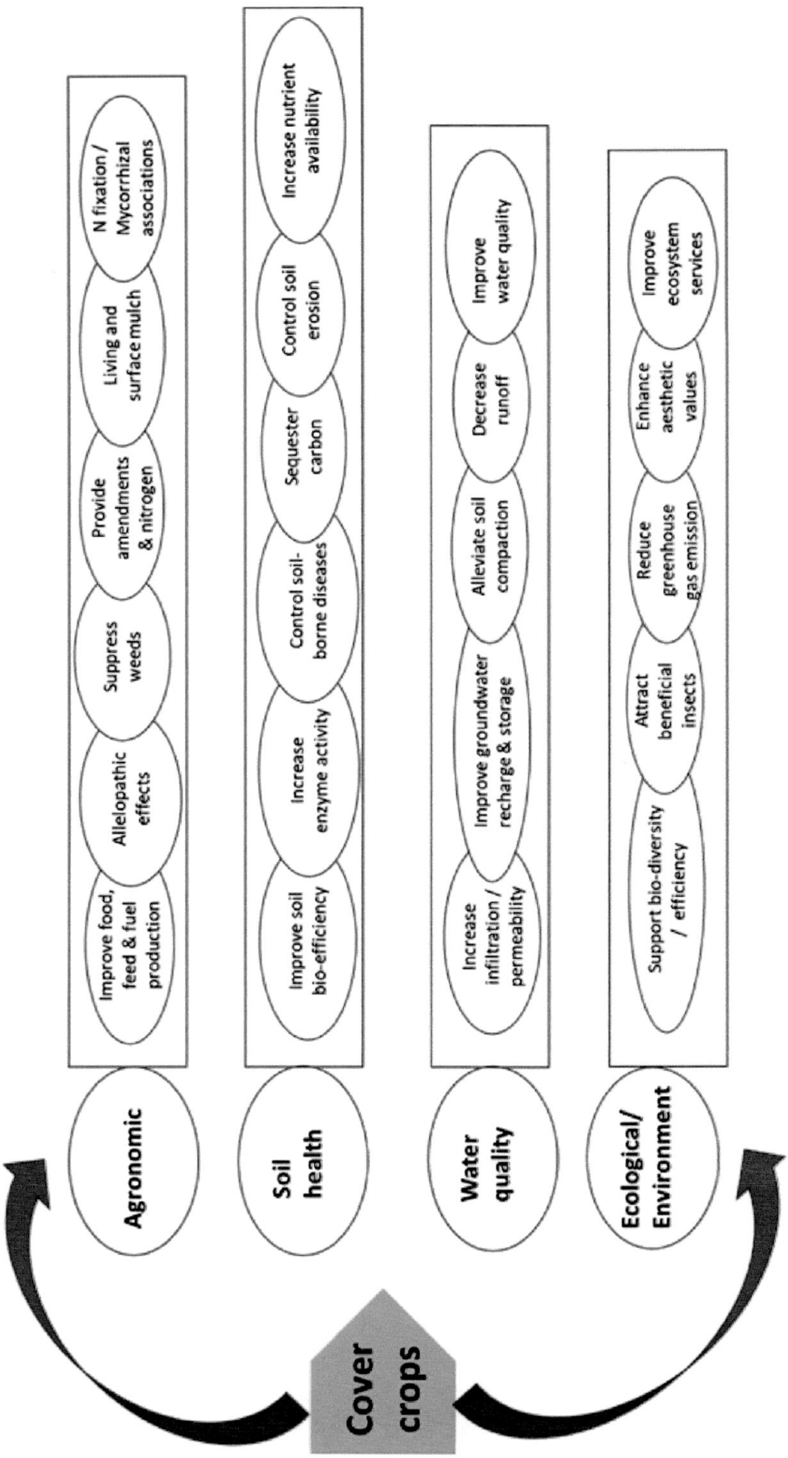

Fig. 4 Cover crops functions and benefits in agroecosystems (***Source:*** Islam and Sherman, 2021).

Table 5 Biomass production, nitrogen and carbon contents, and biomass nitrogen contribution and/or recycling by legume and non-legume cover crops.

Cover Crop	*Dry Biomass (Mg/ha)*	*Nitrogen (%)*	*Carbon (%)*	*C: N of Biomass*	*Biomass N (kg/ha)*
Legumes					
Alfalfa	4.8	3.72	41.9	11.3	400
Cowpea	3.2	2.97	42.6	14.3	101
Crimson clover	1.8	1.97	41.7	21.2	36
Hairy vetch	2.1	2.33	44.3	19.0	53
Jumbo Ladino clover	0.7	3.26	42.8	13.1	55
Medium red clover	0.9	2.77	43.7	15.8	53
Mammoth red clover	0.8	2.00	43.5	21.7	35
Mung bean	2.9	3.92	40.4	10.7	124
White clover	1.3	2.92	43.3	14.8	38
Austrian winter pea	3.1	3.60	44.1	12.3	111
Non-legumes					
Annual ryegrass	1.3	2.11	43.2	20.5	62*
Cereal rye	1.4	0.89	42.6	48.1	26*
Coriander	0.5	1.52	41.4	27.3	18*
Oilseed radish	3.9	2.11	41.1	19.5	82*
Sorghum Sudan grass	12.2	0.68	43.3	63.3	85*
Spelt	2.1	0.99	42.7	43.3	45*

(*Source:* Rafiq Islam).

decline, and long-term studies suggest a potential 40–57% decrease in wheat yield without chemical fertilizations (Anonymous, 2019).

Despite these positive outcomes, it is crucial to recognize that the excessive or unbalanced use of chemical fertilization can contribute to significant carbon and N footprints, accelerating GHG emissions and exacerbating the effects of climate change. Therefore, optimizing chemical fertilization practices for crop growth is essential to minimize carbon and N footprints and mitigate associated GHG emissions.

In a study by Ortas and Islam (2018), the optimization of phosphorus (P) fertilization emerged as a crucial factor in supporting economically viable crop production. The research revealed a significant non-linear increase in corn yields with higher rates of P fertilization. Corn yields showed a notable improvement, increasing from 4.3 Mg/ha in the control to 5.6, 5.7, and 6.1 Mg/ha with P fertilization rates of 50, 100, and 200 kg of P_2O_5/ha, respectively. However, the extrapolation of the relationship suggested that 100 kg P_2O_5/ha was optimal for achieving near-maximum economic corn production, approximately 95%. Similarly, the relationship indicated that applying 50 kg P_2O_5/ha resulted in achieving 90% of the corn yield.

A 2-year field study evaluating variable combinations of NP fertilization on irrigated beans demonstrated consistent grain yield increases of 25–30% under $N_{45}P_{45}$

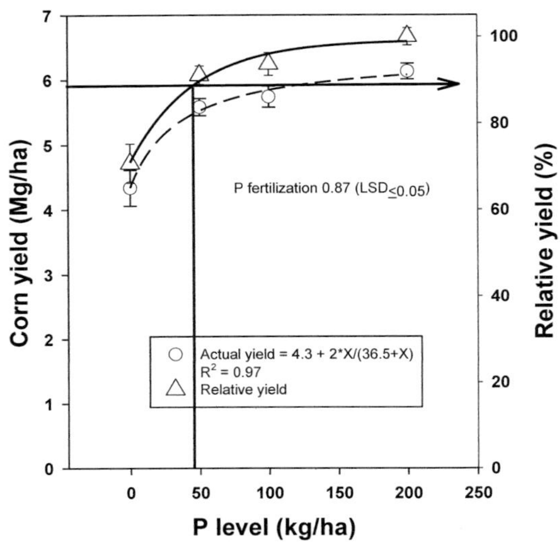

Fig. 5 Impact of phosphorus fertilization on corn yields (*Source:* Ortas and Islam, 2018).

(kg/ha) and $N_{90}P_{90}$ (kg/ha) compared to the control (Lavrenko et al., 2021). The $N_{45}P_{45}$ combination produced near-maximum economic bean production (95%) compared to $N_{90}P_{90}$. Additionally, under $N_{45}P_{45}$, there was a reduction of 45 kg/ha in both N and P fertilizers under both deep and shallow tillage systems. Furthermore, the water-use efficiency of beans increased by 22–27% due to the optimization of NP fertilization.

Soil Amendments – Gypsum, Lime, Biochar, and Compost

To minimize the impacts of climate change and conventional agriculture, proactive recycling of both inorganic and organic soil amendments, including gypsum, lime, biochar, and compost, has been identified as a sustainable alternative (Amoakwah et al., 2020; Wright et al., 2022; Islam et al., 2022; Khangura et al., 2023).

Gypsum application, when combined with cover crops, was found to enhance bio-efficiency, chemical balancing, and physical stability, leading to improved soil health under corn-soybean rotation (Islam et al., 2021). The application of gypsum at 2.2 Mg/ha increased microbial biomass, active carbon, carbon pool index (carbon accumulation), and influenced the carbon management index (carbon accumulation and lability) across degraded fields in Alabama, Indiana, and Ohio (Islam et al., 2022). The results indicated that SOC accumulation increased, contributing to enhanced soil health on degraded or marginal lands.

Biochar and compost, as carbon-enriched recyclable soil amendments, have shown promise in improving soil health and sustaining crop productivity (Amoakwah et al., 2020; Khairo and Farhan, 2024). Biochar, when applied at 30 tons/ha, increased the fungal-to-bacterial ratio, promoting greater SOC sequestration due to the higher C-use efficiency of fungi compared to bacteria (Amoakwah et al., 2020). In a study in Iraq, biochar applied at 5% of the soil significantly increased maize yield, harvest index, N-use efficiency, and water-use efficiency compared to the control. Similarly, compost, when applied at 5%, led to a significant increase in maize yield, harvest

Table 6 Compost and biochar amendments effect on grain yield, grain nutrient content, and nitrogen- and water-use efficiency of irrigated maize.

Compost (% Soil)	*Biochar (% Soil)*	*Irrigation (% FMC)*	*Grain Yield (g/plant)*	*Harvest Index*	*Total N (%)*	*Total P (%)*	*Total K (%)*	*NUE (kg/kg)*	*WUE (kg/m³)*
0			26.3c	0.35c	1.97b	0.31c	2.87c	0.38b	1.27c
2			33.9b	0.40b	2.73a	0.39b	3.46b	0.86a	1.48b
5			41.1a	0.43a	2.91a	0.58a	4.10a	0.93a	1.69a
	0		31.6y	0.38x	2.31z	0.38x	3.21y	0.59x	1.39y
	2		32.2xy	0.39x	2.40y	0.37x	3.27xy	0.61x	1.42xy
	5		33.0x	0.39x	2.44x	0.39x	3.33x	0.60x	1.45x
		50	30.3C	0.38B	2.09C	0.33B	3.25B	0.69B	2.13A
		75	33.0B	0.39B	3.11A	0.59A	3.95A	0.62B	1.32B
		100	37.9A	0.41A	2.41B	0.35B	3.23B	0.86A	0.98C

index, N-use efficiency, and water-use efficiency. The combined application of compost and biochar demonstrated positive effects on growth, yield, N-use efficiency, and water-use efficiency of irrigated corn, particularly under the expected climate change effects (Khairo and Farhan, 2024).

Yelikbayev et al. (2023) found that after 3, 11, 18, and 29 years, SOC storage increased by 5.3–6.2 times under alfalfa-organic amendments compared to the control. SOC sequestration rates were observed at 614 ± 129, 710 ± 69, and 744 ± 161 kg/ha/yr. at 0–10 cm depth under alfalfa, alfalfa-manure, and alfalfa-vermicompost treatments, respectively, in contrast to the control. However, the SOC rates were lower at 10–20 cm depth. Particularly, the alfalfa-vermicompost treatment significantly accumulated SOC, indicating the potential of rejuvenating degraded soils derived from loess materials in Kazakhstan through SOC sequestration to mitigate the effects of climate change.

Chemical Inducing/Acquired Resistance

Climate change poses threats to agricultural sustainability through abiotic and biotic stresses on plants. Chemical inducers have been proposed as environmental stress alleviators in response to climate change effects and soil-crop environmental degradations (Khan et al., 2015; Didenko et al., 2019; Irkitbay et al., 2023). Studies have reported that dilute solutions of salicylic acid improve abiotic and biotic stress tolerance in plants, sustaining crop growth and yield (Khan et al., 2015; Yakhin et al., 2017; Islam and Didenko, 2019). In Ukraine, the application of 0.02 M salicylic acid as a chemical inducer exerted positive effects on crop productivity under both CT and NT systems. Soybean yield increased by 14% with salicylic acid treatment compared to the control, with a more pronounced effect under CT than NT (Didenko et al., 2019).

Irkitbay et al. (2023) reported that both salicylic acid (at concentrations of 0.25 and 0.50 mM) and oxalic acid (at concentrations of 0.10 and 0.20 mM) treatments influenced the abiotic stress tolerance of wheat. The effect of oxalic acid was more pronounced than that of salicylic acid in improving stress tolerance of wheat (Fig. 5). Wheat grain yield increased by 5–8% with 0.25- and 0.50-mM salicylic acid, respectively, compared to the control. Similarly, grain yield significantly increased by 9–19% with 0.10- and 0.20-mM oxalic acid, respectively. Hence, chemical inducing of crops could offer a novel and integrated approach to improve crop growth and yield in dry and seasonally drought-prone areas in South Asia.

Water Management

Effective water management, involving soil moisture replenishment and irrigation, plays a crucial role in mitigating climate change effects and enhancing crop productivity. According to Khairo et al. (2024), irrigating corn at 100% field-moisture capacity (FMC) significantly increased corn yield (10–25%) and harvest index compared to corn irrigated at 50% FMC. However, deficit irrigation at 50% FMC notably enhanced water-use efficiency (35–54%) of corn compared to higher irrigation rates.

In South Asia, improved water management practices, such as managing soil under alternate wetting-drying cycles and avoiding continuous water-logging, can reduce

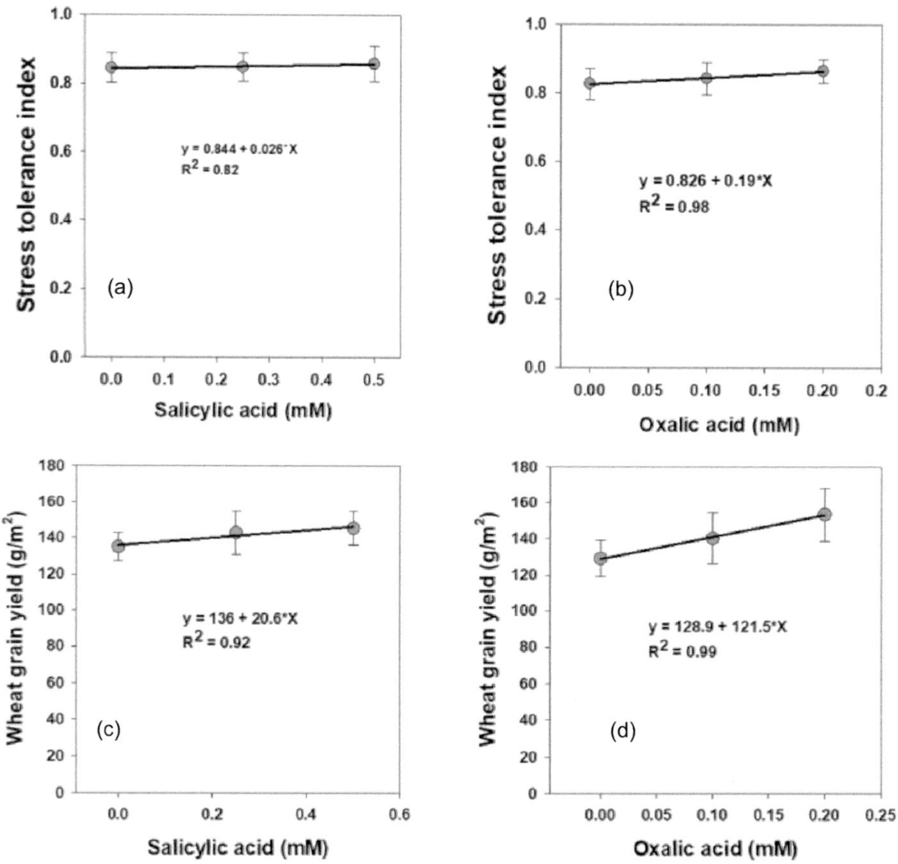

Fig. 6 Relationship of salicylic acid with (a) stress tolerance and (c) yield of wheat. Oxalic acid treatments with (b) stress tolerance and (d) yield of wheat.

off-rice season greenhouse gas emissions (Zhao et al., 2023). Large areas in South Asia, especially in India, Bangladesh, and Pakistan, depend heavily on groundwater for irrigated rice agriculture. However, increasing climatic aridity poses challenges to soil health and crop yields in these regions. In areas like western India, northern Bangladesh, and Pakistan, the scarcity of freshwater for dry-season irrigation hinders agricultural production (Rushton et al., 2020). In Bangladesh, approximately 79% of irrigated lands rely on groundwater, with the northwest region utilizing groundwater for about 95% of Boro rice irrigation (Peña-Arancibia et al., 2020). Boro rice, crucial for food security, accounts for over 55% of total rice production in Bangladesh. Excessive water use for Boro rice irrigation, up to 45%, contributes to water loss. Proactive strategies, including cultivating drought-resistant Boro rice varieties, precision irrigation, and chemical inducing (using salicylic acid), are essential for ensuring groundwater sustainability and sustaining dry-season agriculture in South Asia, particularly in semi-arid regions of Bangladesh.

Grasslands and Grazing Management

Grassland ecosystems cover one-third of the global terrestrial carbon stocks, and maintaining or restoring their health is crucial for mitigating climate change effects. Human disturbances, such as overgrazing and agriculture, and climate change threaten grasslands, leading to CO_2 emissions (Ontl and Janowiak, 2017). To enhance natural climate solutions in global grasslands, best management practices are recommended (Bai and Cotrufo, 2022). Restoring degraded grasslands through limiting soil disturbance, improving grazing management, using prescribed fire, increasing biodiversity, and planting native species can boost SOC stocks (Ontl and Janowiak, 2017; Chatterjee, 2021).

Grazing livestock, especially in rotational systems, can contribute to soil carbon sequestration by encouraging deeper roots and stabilizing biomass carbon in the ground. Strategies like planting grass with legumes as cover crops improve root production and soil fertility. Fertilization and irrigation practices positively impact plant diversity and SOC stocks, particularly in grasslands managed as pasture. Prescribed fire helps prevent invasive species and increases plant productivity. The global potential for carbon sequestration in grasslands is significant, ranging from 2.3–7.3 billion tons of CO_2e/yr. for biodiversity restoration, 148–699 megatons of CO_2e/yr. for improved grazing management, and 147 megatons of CO_2e/yr. for planting legumes in pasturelands (Bai and Cotrufo, 2022). Livestock raised on grasslands contribute to food security, supporting livelihoods and dietary diversity globally, including in South Asia.

Forest Management – Reforestation and Afforestation

Photosynthesis by terrestrial vegetation absorbs approximately 120 billion tons of carbon from the troposphere annually, surpassing the carbon released by fossil fuel burning by over 20 times and constituting about a sixth of the total carbon in today's troposphere. Forest ecosystems have been proposed as significant repositories of the "missing carbon" in the global carbon budget imbalance (Dixon et al., 1994).

The carbon fertilization effect resulting from increased tropospheric CO_2 concentration strongly influences C_3 plants, including most forest species, which exhibit a lower rate of photosynthetic efficiency compared to C_4 plants. The competition between CO_2 and O_2 for the active site of the enzyme Ribisco/bisphosphate-carboxylase oxygenase (Rubisco) plays a pivotal role. An increase in CO_2 shifts the balance towards carboxylation rather than oxygenation, reducing photorespiration in C_3 plants. This process conserves carbon that would otherwise be oxidized by the glycolate pathways, leading to an increased quantum yield of CO_2 assimilation in C_3 plants, where Rubisco is the sole CO_2-fixing enzyme. Most forests utilize the C_3 pathway, acting as net carbon sinks to mitigate greenhouse effects (Musselman and Fox, 1991).

For instance, approximately 1.25 tons of residues per hectare are deposited annually on the surface of temperate forests. Assuming a conversion rate of residues to SOM of about 50% (Gabriel and Michelis, 1991), the SOM formation is 1.25 * 0.5 => 0.625 tons. As SOM contains about 58% carbon, the SOC content will be 0.625 * 0.58 => 0.3625 tons per hectare per year. Considering the SOC decomposition coefficient (*r*) as 99.3% in forest ecosystems (Gabriel and Michelis, 1991), the amount of SOC stored in a 10-year cycle will be:

$$Y = [0.3625 * (1 - 0.993^{10}) / (1 - 0.993)] \Rightarrow 3.51 \text{ tons C/ha}.$$

Therefore, forest soils can accumulate substantial amounts of carbon over time to mitigate climate change effects. Additionally, the current level of CO_2 in the troposphere is not sufficient to increase SOC accumulation beyond 40% of its capacity (Wallace, 1994).

It has been reported that forests, especially mangrove forests in the coastal areas of tropical and subtropical regions, play a crucial role in protecting coastlines from erosion and storm surges. They provide a vital habitat for a wide variety of plant and animal species and serve as a sink of CO_2 (https://vnvadvisory.com/mangrove-reforestation). In the South Asian region, particularly in Southern Bangladesh and India, a significant portion of the coastal region is covered under mangrove forests. In Bangladesh, afforestation and reforestation programs to green up newly raised islands (char lands) within the river channel/estuary or attached land to the riverbanks and the eroded mangrove forests primarily using Sunderi (*Heritiera fomes*) and the mangrove apple (*Sonneratia caseolaris*), all native tropical and mangrove species, have been successfully implemented over the years (https://www.earthday.org/campaign/the-canopy-project-bangladesh). The average annual C sequestration rate for mangrove forests is estimated to range between 6–8 Mg CO_2e/ha. These rates are about two to four times greater than the global rates observed in mature tropical forests.

Carbon finance can play a crucial role in mangrove afforestation and reforestation programs by providing a financial incentive for the conservation and restoration of these fragile ecosystems. Carbon finance mechanisms, such as carbon offsetting and carbon credits, allow for the generation of income from the carbon sequestered by the trees and in soils of the afforested and reforested areas. This income can be utilized to cover the costs of afforestation and reforestation activities. Despite the immense pressure from global population growth and urbanization to convert forests to agriculture and from agriculture to urban uses, resulting in increased ecosystem disservices, sustainable forest management holds the potential to positively address these challenges with greater economic viability and environmental compatibility.

Conclusions

The global carbon cycling involves a mix of natural steady-state rates and reservoirs, along with changing rates and reservoirs affected by human activities. Human activities release billions of tons of GHGs, especially CO_2, annually into the troposphere through fossil fuel combustion, deforestation, agriculture, and urbanization, contributing to global warming. While agriculture accounts for approximately 24% of anthropogenic GHG emissions, climate change adversely affects agriculture and food security. Mitigating and adapting to climate change are crucial for sustainable agriculture to ensure global food security amidst a growing population. Agriculture must be economically viable, environmentally compatible, and socially acceptable, with a focus on proactive climate-smart practices, drought-tolerant crops, chemical inducing, precision fertilization, optimized irrigation, carbon sequestration, and soil health improvement. Although South Asian countries have formulated national and local policies to address climate change challenges, there is a need for enhanced coordination across different institutions to implement effective adaptation policies in the future.

References

Abbas, G., Ahmad, Sh., Ahmad, A., Nasim, W., Fatima, Z., Hussain, S., Rehman, M.H., Khan, M.A., Hasanuzzaman, M., Fahad, S., Boote, K.J., Hoogenboom, G. (2017). Quantification of the impacts of climate change and crop management on phenology of maize-based cropping system in Punjab, Pakistan. *Agricultural and Forest Methodology*, 247: 42–55.

Aggarwal, P.K., Sinha, S.K. (1993). Effect of probable increase in carbon dioxide and temperature on wheat yields in India. *Journal of Agricultural Meteorology*, *48*: 811–14.

Aggarwal, P.K., Sivakumar, M.V. (2010). Global climate change and food security in South Asia: An adaptation and mitigation framework. *In*: Lal, R., Sivakumar, M.V.K., Faiz, S.M.A., Rahman, A.H.M.M., Islam, K.R. (Eds.), *Climate Change and Food Security in South Asia*, 253–75. Berlin: Springer.

Ahmad, M., Siftain, H., Iqbal, M. (2014). *Impact of Climate Change on Wheat Productivity in Pakistan: A District Level Analysis, MRPA 72859*. Climate Change Working Paper No. 1, Pakistan Institute of Development Economics, Islamabad, Pakistan.

Alauddin, M., Hossain, M. (2001). *Environment and Agriculture in a Developing Economy:Problems and Prospects for Bangladesh*. Cheltenham: Edward Elgar Publishing Ltd. UK.

Ali, A., Rahut, D.B., Mottaleb, K.A., Erenstein, O. (2017). Impacts of changing weather patterns on smallholder well-being: Evidence from the Himalayan region of northern Pakistan. *International J. Climate Change Strategies and Management*, *9*: 225–40.

Amoakwah, E., Arthur, E., Frimpong, K.A., Parikh, S.J., Islam, R. (2020). Soil organic carbon storage and quality are impacted by corn cob biochar application on a tropical sandy loam. *Journal of Soils and Sediments*, *20*: 1960–69.

Anonymous (2019). Sustainable Forestry in South Asia – Current Status, Science, and Conservation Priorities. 7–9th Nov., New Delhi, India (https://sari.umd.edu/meetings/sustainable-forestry-south-asia).

Arshad, M., Amjath-Babu, T., Krupnik, T.J., Aravindakshan, S., Abbas, A., Kächele, H., Müller, K. (2017). Climate variability and yield risk in South Asia's rice–wheat systems: Emerging evidence from Pakistan. *Paddy and Water Environment*, *15*: 249–61.

Aryal, J.P., Sapkota, T.B., Khurana, R., Khatri-Chhetri, A., Rahut, D.B., Jat, M.J. (2020). Climate change and agriculture in South Asia: Adaptation options in smallholder production systems. *Environ. Dev. Sustain*. *22*: 5045–75.

Asada H., Matsumoto J. (2009). Effects of rainfall variation on rice production in the Ganges-Brahmaputra Basin. *Climate Research*, *38*: 249–60.

Asian Development Bank (ADB). 2006. *Urbanization and Sustainability in Asia: Good Practice Approaches in Urban Region Development*. Manila: ADB.

Aufhammer, M., Ramanathan, V., Vincent, J.R. (2012). Climate change, the monsoon, and rice yield in India. *Climatic Change*, *111*: 411–24.

Aziz, I., Mahmood, T., Islam, K.R. (2013). Effect of long term no-till and conventional tillage practices on soil quality. *Soil and Tillage Research*, *131*: 28–35.

Bai, Y., Cotrufo, M.F. (2022). Grassland soil carbon sequestration: Current understanding, challenges, and solutions. *Science*, *377* (6606): 603–608.

Basak, J.K., Ali, M.A., Islam, M.N., Alam, M.J.B. (2009). Assessment of the effect of climate change on boro rice production in Bangladesh using CERES-Rice model. *In*: *Proc. Intern. Conf. Climate Change Impacts and Adaptation Strategies for Bangladesh*, 103–13.

Bergtold, J., Sailus, M., Jackson, T. (2020). *Conservation Tillage Systems in the Southeast: Production, Profitability, and Stewardship*. USDA: Sustainable Agriculture Research & Education.

Bhandari, G. (2013). Effect of rainfall on the yield of major cereals in Darchula district of Nepal. *International Journal of Environment*, *3*: 205–13.

Bhatt, R., Kukal, S.S., Busari, M.A., Arora, S., Yadav, M. (2016). Sustainability issues on rice–wheat cropping system. *Int. Soil Water Conserve*. Res., *4*: 64–74.

Brown, E., Cloke, J., Gent, D., Johnson, P.H., Hill, C. (2014). Green growth or ecological commodification: Debating the green economy in the global south. *Geografiska Annaler, Series B: Human Geography*, *96*: 245–59.

Bruinsma, J. (2003). *World Agriculture: Towards 2015/2030, an FAO Perspective*. London: Earthscan Publications.

Chaki, A.K., Gaydon, D.S., Dalal, R.C., Bellotti, W.D., Gathala, M.K., Hossain, A., Rahman, M.A., Menzies, N.A. (2021). Conservation agriculture enhances the rice-wheat system of the Eastern Gangetic Plains in some environments, but not in others. *Field Crops Research*, *265*: 108109.

Chatterjee, R. (2021). *Cover Crops and Sustainable Agriculture*. (1st Edn.), 25 pp. Imprint CRC Press., eBook ISBN9781003187301.

Chhogyel, N., Kumar, L. (2018). Climate change and potential impacts on agriculture in Bhutan: A discussion of pertinent issues. *Agriculture and Food Security*, *7*: 79.

Cline, W. (2007). *Global Warming and Agriculture: Impact Estimates by Country*. Washington, DC: Center for Global Development and Peterson Institute for International Economics.

Dey, N.C., Saha, R., Parvez, M., Bala, S.K., Islam, A.K.M.S., Paul, J.K., Hossain, M. (2017). Sustainability of groundwater use for irrigation of dry-season crops in northwest Bangladesh. *Groundwater for Sustainable Development*, *4*: 66–77.

Didenko, N., Konovalova, V., Romashchenko, M., Islam, K.R. (2019). Using climate-smart agriculture to improve irrigated crop yield and soil quality in Ukraine. Paper presented at the *International Biological, Agricultural, & Life Science Congress*, Nov. 7–8, Lviv, Ukraine. www.bialic.org.

Dixon, R.K., Brown, S., Houghton, R.A., Solomon, A.M., Trexler, M.C., Wisniewski, J. (1994). Carbon Pools and Flux of Global Forest Ecosystems. *Science*, *263*: 185–90.

FAO (2013). *Climate-Smart Agriculture Sourcebook*. FAO: Rome, Italy, 2013.

Fishman, R. (2018). Groundwater depletion limits the scope for adaptation to increased rainfall variability in India. *Climatic Change*, *147*: 195–209.

Gabriel, S.D., Michelis, P. (1991). Soil Organic Matter and Water Erosion Process. *In*: Wilson, W.S. (Ed.), *Advances in Soil Organic Matter Research: The Impact on Agriculture and the Environment*. Royal Society of Chemistry, UK: 141–52.

GoSL (2000). *Initial National Communication under the United Nations Framework Convention on Climate Change*. Colombo: Government of Sri Lanka.

Gupta, R., Somanathan, E., Dey, S. (2017). Global warming and local air pollution have reduced wheat yields in India. *Climatic Change*, *140*: 593–604.

Hann, S.D. (1977). Humus, its formation, its relationship with the mineral part of the soil, and its significance for soil productivity. *In*: *Organic Matter Studies* (Vol. II), 21–30. IAEA.

Harrison, K.G., Broecker, W.S. and Bonani, G. (1993). The Effect of Changing Land Use on Soil Radiocarbon. *Science*, *262*: 725–76.

Hatfield, J.L., Prueger, J.H. (2014). Temperature extremes: Effect on plant growth and development. *Weather and Climate Extremes*, *10* (Part A): 4–10. http://dx.doi.org/10.1016/j. wace.2015.08.001i

Hussain, S.S., Mudasser, M. (2007). Prospects for wheat production under changing climate in mountain areas of Pakistan: An econometric analysis. *Agricultural Syst.*, *94*: 494–501.

IPCC (2007). *Climate Change 2007: The Physical Science Basis*. Contribution of Working Group I to the Fourth Assessment Report of the Intergovernmental Panel on Climate Change. Cambridge and New York: Cambridge University Press.

IPCC (2014). *Climate Change 2014*: *Impacts, Adaptation, and Vulnerability. Part A: Global and Sectoral Aspects*. Contribution of Working Group II to the Fifth Assessment Report of the Intergovernmental Panel on Climate Change. (Eds., Field, C.B., V.R. Barros, D.J. Dokken, K.J. Mach, M.D. Mastrandrea, T.E. Bilir, M. Chatterjee, K.L. Ebi, Y.O. Estrada, R.C. Genova, B. Girma, E.S. Kissel, A.N. Levy, S. MacCracken, P.R. Mastrandrea, L.L. White) Cambridge University Press, Cambridge, United Kingdom, and New York, USA, 1132 pp.

IPCC (2018). *Summary for Policymakers of IPCC Special Report on Global Warming of 1.5C Approved by Governments.* Cambridge University Press: Cambridge, UK.

IPCC (2021). *Climate Change 2021: The Physical Science Basis.* Contribution of Working Group I to the Sixth Assessment Report of the Intergovernmental Panel on Climate Change. Cambridge University Press, Cambridge, United Kingdom, and New York, NY, USA.

Irkitbay, A., Sapahkova, Z., Didenko, N., Madenova, A., Kanat, G., Islam, K.R. (2023). Evaluating salicylic and oxalic acid on the enzyme activities, yield, and stress tolerance of wheat (*Triticum aestivum*, L.). *Research on Crops*, *24*: 219–28.

Islam, K.R., Weil, R.R. (2000). Land Use Effects on Soil Quality in a Tropical Forest Ecosystem of Bangladesh. *Agriculture, Ecosystems and Environment*, *79*: 9–16.

Islam, K.R., Didenko, N. (2019). *Impact of Sustainable Agricultural Management Practices on Soil Quality and Crop Productivity.* Final report, Grant award # 63888/63886. Civilian Research and Development Foundation (CRDF)-Global, Washington DC, USA.

Islam, R., Didenko, N., Sherman, B. (2021). Cover crops and agroecosystem services. *In*: R. Islam, B. Sherman (Eds.). *Cover Crops and Sustainable Agriculture.* (1st Edn.), CRC Press, Baco Ratoon, FL, USA.

Islam, K.R., Roth, G., Rahman, M.A., Didenko N.O., Reeder, R.C. (2021). Cover crop complements flue gas desulfurized gypsum to improve no-till soil quality. *Communications in Soil Science and Plant Analysis*, *52*: 926–47.

Islam, K.R., Dick, W.A., Watts, D.B., Gonzalez, J.M., Fausey, N.R., Flanagan, D.C., Reeder, R.C., VanToai, T., Batte, M.T. (2022). Gypsum, crop rotation, and cover crop impacts on soil organic carbon and biological dynamics in rainfed transitional no-till corn-soybean systems. *PLOS One*, *17*: e0275198.

Jat, M.L., Chakraborty, D., Ladha, J.K., Parihar, C.M., Datta, A., Mandal, B., Nayak, H.S., Maity, P., Rana, D.S., Chaudhari, S.K., Gerard, B. (2022). Carbon sequestration potential, challenges, and strategies towards climate action in smallholder agricultural systems of South Asia. *Crop Environment*, 86–101.

Kalra, N., Chander, S., Pathak, H., Aggarwal, P., Gupta, N., Sehgal, M., Chakraborty, D. (2007). Impacts of climate change on agriculture. *Outlook on Agriculture*, *36*: 109–18.

Kanicki, D. (2014). Farm Equipment. https://www.farm-equipment.com/articles/10122-what-is-vertical-tillage-anyway.

Kant, P., Kleine, M., Acharya, L.P., Hossain, M.K., Lee, D.K. (2008). Rehabilitating forests and extending tree cover in South Asia: A synthesis. *In*: Lee, D.K. (Ed..), *Keep Asia Green* Vol. III, pp. 7–20. "South Asia" IUFRO World Series (Vol. 20-III), Vienna, Austria.

Karim, Z., Hussain, S.G., Ahmed, M. (1996). Assessing impacts of climatic variations on foodgrain production in Bangladesh. *In*: L. Erda, W.C. Bolhofer, S. Huq, S. Lenhart, and S.K. Mukherjee (Eds.), *Climate Change Vulnerability and Adaptation in Asia and the Pacific*, 53–62. Berlin: Springer.

Khairo, A.M., Farhan, M.J., Islam, K.R. (2024). Compost and biochar amendments improved growth, yield, nutrient uptake, nitrogen- and water-use efficiency of irrigated corn. *Journal of Soils and Sediments* (communicated).

Khan, M.I., Mehar, R., Fatma, T.S.P., Naser, A.A., Nafees, A.K. (2015). Salicylic acid induced abiotic stress tolerance & underlying mechanisms in plants. *J. Front. Plant Sci.*, *6*: 1–17.

Khangura, R., Ferris, D., Wagg, C., Bowyer, J. (2023). Practices and Mechanisms Used to Improve Soil Health. *Sustainability*, *15*: 2338.

Khor, M. (2008). *Food Crisis, Climate Change, and the Importance of Sustainable Agriculture.* Presentation at FAO Food Security Summit, Rome 4 June 2008. Third World Network. twnet@po.jaring.my

Knox, J., Hess, T., Daccache, A., Wheeler, T. (2012). Climate change impacts on crop productivity in Africa and South Asia. *Environmental Research Letters*, *7*: 034032.

Kumar, K.K., Kumar, R.K., Ashrit, R., Deshpande, N., Hansen, J. (2004). Climate impacts on Indian agriculture. *International Journal of Climatology: A Journal of the Royal Meteorological Society*, *24*: 1375–93.

Lal, R. (2011). Soil health and climate change: An overview. *In*: (Singh, B.P., Cowie, A.L. and Chan, K.Y. Eds.), *Soil Health and Climate Change. Soil Biology Series, 29*: 3–24. Springer-Verlag, Berlin, Germany.

Lavrenko, S.O., Lavrenko, N.M., Maksymov, D.O., Maksymov, M.V., Didenko, N.O., Islam, K.R. (2021). Variable tillage depth and chemical fertilization impact on irrigated common beans and soil physical properties. *Soil and Tillage Research, 212*: 105024.

Lipper, L., Thornton, P., Campbell, M.C., Baedeker, T., Braimoh, A., Bwalya, M., Caron, P., Cattaneo, A., Garrity, D., Henry, K., Hottle, R., Jackson, L., Jarvis, A., Kossam, F., Mann, W., McCarthy, N., Meybeck, A., Neufeldt, H., Remington, T., Sen, P.T., Sessa, R., Shula, R., Tibu, A., Torquebiau, E.F. (2014). Climate-smart agriculture for food security. *Nature Climate Change, 4*. www.nature.com/ natureclimatechange.

Malla, G. (2008). Climate change and its impact on Nepalese agriculture. *Journal of Agriculture and Environment, 9*: 62–71.

Mathauda, S., Mavi, H., Bhangoo, B., Dhaliwal, B. (2000). Impact of projected climate change on rice production in Punjab (India). *Tropical Ecology, 41*: 95–98.

Matiu M., Ankerst D.P., Menzel A. (2017). Interactions between temperature and drought in global and regional crop yield variability during 1961–2014. *PLOS One, 12*: e0178339.

Morey, D., Sadaphal, M. (1981). Efect of weather elements on yield of wheat at Delhi. PKV Research Journal (Punjabrao Krishi Vidyapeeth), *5*: 3.

Moss, C., Lukac, M., Harris, F., Outhwaite, C.L., Scheelbeek, P.F.D., Green, R., Berstein, F.M., Dangour, A.D. (2020). The effects of crop diversity and crop type on biological diversity in agricultural landscapes: A systematic review protocol. *Wellcome Open Res., 4*: 101.

Musselman, R.C., Fox, D.G. (1991). A Review of the Role of Temperate Forests in the Global CO_2 Balance. *J. Air and Waste Management Association, 41*: 798–807.

Nelson, G.C., Rosegrant, M.W., Koo, J., Robertson, R., Sulser, T., Zhu, T., Ringler, C., Msangi, C., Palazzo, A., Batka, M., Magalhaes, M., Valmonte-Santos, R., Ewing, M., Lee. D. (2009). *Climate Change. Impact on Agriculture and Costs of Adaptation* 19. Washington, DC: Intern. Food Policy Research Institute.

Nuccitelli, D. (2022). New IPCC report highlights urgency of climate change impacts. *Policy and Politics, Yale Climate Connections*. https://yaleclimateconnections.org/2022/02/new-ipcc-report-highlights-urgency-of-climate-change-impacts/.

Ontl, T., Janowiak, M. (2017). *Grassland and Carbon Management*. U.S. Department of Agriculture, Forest Service, Climate Change Resource Center. https://www.fs.usda.gov/ ccrc/topics/grassland-carbon-management.

Ortas, I., Islam, K.R. (2018). Phosphorus Fertilization Impacts on Corn Yield and Soil Fertility. *Communications in Soil Science and Plant Analysis, 49*: 1684–94.

Ortiz, R., Sayre, K. D., Govaerts, B., Gupta, R., Subbarao, G., Ban, T., Hodson, D., Dixon, J., Ortiz-Monasterio, I., Reynolds, M. (2008). Climate change: Can wheat beat the heat? *Agriculture, Ecosystems, and Environment, 126*: 46–58.

Parry M.L., Rosenzweig C., Iglesias A., Livermore, M., Fischer, G. (2004), Effects of climate change on global food production under SRES emission and socio-economic scenarios. *Global Environ. Change, 14*: 53–67.

Peña-Arancibia, J.L., Mainuddin, M., Ahmad, M.D., Hodgson, G., Ibn Murad, K.F., Ticehurst, C., Maniruzzaman, M., Mahboob, M., Kirby, J.M. (2020). Groundwater use and rapid irrigation expansion in a changing climate: Hydrological drivers in one of the world's food bowls. *Journal of Hydrology, 581*: 124300.

Poeplau, C., Don, A. (2015). Carbon Sequestration in Agricultural Soils via Cultivation of Cover Crops—A Meta-Analysis. *Agriculture, Ecosystems, and Environment, 200*: 33–41.

Prasai, B.K. (2010). National issue paper on agriculture sector (adaptation). United Nations Development Program. http://www.undpcc.org/docs/National%20issues%20papers/Agriculture%20 (adaptation)/17_Nepal%20NIP_%20agriculture%20adaptation.pdf.

Prasanna, V. (2014). Impact of monsoon rainfall on the total foodgrain yield over India. *Journal of Earth System Science, 123*: 1129–45.

Rahman, M.A., Jahiruddin, M., Kader, M.A., Islam, M.R., Solaiman Z.M. (2021). Sugarcane bagasse biochar increases soil carbon sequestration and yields of maize and groundnut in charland ecosystem. *Archives of Agronomy and Soil Science*, 1–15.

Rahman, A., Alam, M., Mainuddin, K., Ali, L., Alauddin, S., Rabbani, M., et al. (2009). *The Probable Impacts of Climate Change on Poverty and Economic Growth and the Options of Coping with Adverse Effect of Climate Change in Bangladesh, Support to Monitoring PRS and MDGs in Bangladesh.* Planning Commission, Government of the Peoples's Republic of Bangladesh & UNDP Bangladesh.

Raphael, A. (2024). Understanding soil organic carbon formation for climate-smart agriculture. https://medriva.com/health/environmental-health/understanding-soil-organic-carbon-formation-for-climate-smart-agriculture/

Rushton, K.R., Zaman, M.A., Hasan, M. (2020). Monitoring groundwater heads and estimating recharge in multi-aquifer systems illustrated by an irrigated area in northwest Bangladesh. *Sustainable Water Resources Management*, *6*: 22.

Sarker, M.A.R., Alam, K., Gow, J. (2012). Exploring the relationship between climate change and rice yield in Bangladesh: An analysis of time series data. *Agricultural Systems*, *112*: 11–16.

Sarker, R.R., Rashid, M.H., Islam, M.R., Jahiruddin, M., Islam, K.R., Jahangir, M.M.R. (2023). Conservation agriculture's impact on total and labile organic carbon pools in calcareous and non-calcareous floodplain soils under a subtropical rice-based system. *PLOS One*. https://doi.org/10.1371/journal.pone.0293257.

Saseendran, S., Singh, K., Rathore, L., Singh, S., Sinha, S. (2000). Effects of climate change on rice production in the tropical humid climate of Kerala, India. *Climatic Change*, *44*: 495–14.

Schmidhuber, J., Tubiello, F.N. (2007). Global food security under climate change. *PNAS*, *104*: 19703–19708.

Siderius, C., Hellegers, P., Mishra, A., van Ierland, E., Kabat, P. (2014). Sensitivity of the agroecosystem in the Ganges basin to inter-annual rainfall variability and associated changes in land use. *International Journal of Climatology*, *34*: 3066–77.

Sinha, S., Swaminathan, M. (1991). Deforestation, climate change, and sustainable nutrition security: A case study of India. *In*: Myers, N. (Ed.), *Tropical Forests and Climate*. pp. 201–209. Cham: Springer.

Smith, P., Gregory, P.J. (2013). Climate change and sustainable food production. *Proceedings of the Nutrition Society*, *72*: 21–28.

Spijkers, M. Ad. (2011). Implications of Climate Change on Agriculture and Food Security in South Asia. *In*: Lal, R., Sivakumar, M.V.K., Faiz, S.M.A., Rahman, A.H.M.M., Islam, K.R. (Eds.), *Climate Change and Food Security in South Asia*, 217–28. Springer Dordrecht Heidelberg, London. Sultana, H., Ali, N. (2006). Vulnerability of wheat production in different climatic zones of Pakistan under climate change scenarios using CSM-CERES-Wheat Model. *In*: *2nd International Young Scientists' Global Change Conference* (pp. 7–9). Beijing.

Sultana, H., Ali, N., Iqbal, M. M., Khan, A.M. (2009). Vulnerability and adaptability of wheat production in different climatic zones of Pakistan under climate change scenarios. *Climatic Change*, *94*: 123–42.

Sundermeier, A.P., Islam, R., Yogendra Raut, Y., Randall Reeder, R., Dick, W.A. (2011). Continuous no-till impacts on soil biophysical carbon sequestration. *Soil Science Society of America Journal*, *75*: 1779–88.

Tesfaye, L.D. (2017). Evaluation and screening of durum wheat germplasm to stem rust races at seedling stage. *International J Agriculture and Environmental Research*, *3*: 3113–32.

UN-Water (2013). *Climate Change* (WWW document).

USDA (2020). *Summary Report: 2017*. National Resources Inventory, Natural Resources Conservation Service, Washington, DC, and Center for Survey Statistics and Methodology, Iowa State University, Ames, Iowa. https://www.nrcs.usda.gov/ wps/portal/nrcs/main/national/technical/nra/nri/results.

USEPA (2023). *Overview of Greenhouse Gases*. United States Environmental Protection Agency, (https://www.epa.gov/ghgemissions/overview-greenhouse-gases#:~:text=Concentration% 20%2C%20or%20abundance%2C%20is%20the).

Vidanage, S., Abeygunawardane, P. (1994). An economic assessment of global warming on agriculture: The case of paddy production in Sri Lanka. *Marga*, *13*: 33–44.

Vinke, K., Martin, M.A., Adams, S., Baarsch, F., Bondeau, A., Coumou, D., Donner, R.V., Menon, A. (2017). Climatic risks and impacts in South Asia: Extremes of water scarcity and excess. *Regional Environmental Change*, *17*: 1569–83.

Voet, D., Voet, J.G. (2010). *Biochemistry*. (4th Edition). Wiley, Hoboken. ISBN: 978-0-470-57095-1.

Wallace, A. (1994). Strategies to Avoid Global Greenhouse Warming: Stashing Carbon Away in Soil is One of the Best. *Commun. Soil Sci. Plant Anal., 25*: 37–44.

Wheeler, T., von Braun, J. (2013). Climate Change Impacts on Global Food Security. *Science, 341*: 508–513.

Wright, J., Kenner, S., Lingwall, S. (2022). Utilization of Compost as a Soil Amendment to Increase Soil Health and to Improve Crop Yields. *Open J. Soil Science, 12*: 216–24.

Yakhin, O.I., Lubyanov, A.A., Yakhin, I.A., Brown, P.H. (2017). Biostimulants in Plant Science: A Global Perspective. *Frontiers in Plant Science, 7*: 2049.

Yelikbayev, B.K., Pagano, M.C., Mamedov, A.I., Islam, K.R. (2023). Alfalfa–organic amendments impact soil carbon sequestration and its lability in reclaimed loess. *Land Degradation and Development, 35*: 88–101.

Yousaf, M., Li, J., Lu, J., Ren, T. (2017). Effects of fertilization on crop production and nutrient-supplying capacity under rice-oilseed rape rotation system. *Scientific Reports, 7*: 1270.

Zahid, A., Ali, S., Ahmed, M., Iqbal, N. (2020). Improvement of Soil Health through Residue Management and Conservation Tillage in Rice-Wheat Cropping System of Punjab, Pakistan. *Agronomy, 10*: 1844.

Zhai, F., Zhuang, J. (2009). *Agricultural Impact of Climate Change: A General Equilibrium Analysis with Special Reference to Southeast Asia*. ADBI Working Paper 131. Tokyo: Asian Development Bank Institute. http://www.adbi.org/workingpaper/2009/02/23/2887.

Zhao, J., Liu, D., Huang R. (2023). A Review of Climate-Smart Agriculture: Recent Advancements, Challenges, and Future Directions. *Sustainability, 15*: 3404.

Ziska, L.H., Bunce, J.A., Shimono, H., Gealy, D.R., Baker, J.T., Newton, P.C.D., Reynolds, M.P., Jagadish, K.S.V., Zhu, C., Howden, M., Wilson, L.T. (2012). Food security and climate change: On the potential to adapt global crop production by active selection to rising atmospheric carbon dioxide. *Proc. Royal Soc. (B), 279*: 4097–4105.

Ziska, L.H. (2014). Increasing Minimum Daily Temperatures are Associated with Enhanced Pesticide Use in Cultivated Soybean along a Latitudinal Gradient in the Mid-Western United States. *PLOS One, 9*: e98516.

Chapter 2

Stepping Towards a Climate Smart India: Impacts and Initiatives

Sheshakumar Goroshi[1*], *S.C. Bhan*[1], *Ramaraj Palanisamy*[1], *K.K. Singh*[1] and *Kamaljit Ray*[2]

Introduction

Climate change is a real and serious threat, causing dangerous and widespread disruption in nature and affecting the lives of billions of people (IPCC, 2022) around the world. Climate change is unequivocal and impacts every country on earth and sector. India is both a major greenhouse gas emitter and one of the most vulnerable countries in the world to projected climate change. Being an agrarian economy with 55% of the country's net sown area under rainfed and 61% of India's farmer population, Indian agriculture remains as the most vulnerable sector to climate change. As in previous reports (TAR, FAR, and AR5), IPCC in its 6th Assessment Report reiterated the wide consequences of climate change on agriculture (IPCC, 2021). Increase in atmospheric temperature can have direct impacts in reducing crop duration, photosynthesis, and crop yield. It can affect the survival and distributions of pest populations, increase soil nutrient mineralization, decrease nutrient use efficiency, and increase water loss resulting in increasing demand of irrigation water and plant nutrients. Climate change has significant indirect effects on agriculture with its effects on irrigation water availability, droughts and floods, soil fertility, and erosion (Pathak, 2023).

Impact of climate change on agriculture is studied systemically since 1980s, resulting in a large body of research that has been well summarized in several national and international reviews and reports. Most studies indicate increasing impacts of climate change on agriculture with time, with the largest impacts affecting low-latitude (tropical), lower-income countries of South Asia, Latin America, and Sub-

1 India Meteorological Department.
2 Ministry of Earth Sciences, New Delhi, India.
* Corresponding author: sheshakumar.goroshi@imd.gov.in

Saharan Africa that have limited adaptive capacity (Aggarwal et al., 2019). This study presents the impact of climate change on agriculture and India's initiatives to tackle it through review of peer reviewed literature.

Overall Framework Representing the Impacts and Their Management

The framework followed in this review on climate change impacts on India and initiatives to tackle it is presented here diagrammatically. (Fig.1)

Climate Change and Indian Agriculture: Transitioning from 'Vulnerable' to 'Climate Smart'

Numerous studies highlight the impact of climate change on Indian agriculture, emphasizing the need for adaptation and mitigation strategies. Consolidating past research, focusing on key areas of improvement, and addressing gaps in adaptation and planning at all levels are essential for making agriculture climate smart. This chapter aims to address these concerns.

Projected Changes in Climate

The climate change projections for the Sixth Assessment Report (AR6) of Intergovernmental Panel on Climate Change (IPCC) are made using the Shared Socioeconomic Pathway (SSP) scenarios under the Coupled Model Inter-comparison Project 6 (CMIP6). IPCC has produced scenario narratives and projections from its Third Assessment Report (TAR) till AR6. A compilation all the scenarios and projections for India captured from peer reviewed literature is presented in Table 1.

The IPCC, in the 2000s developed the emission pathways in its special report on emissions scenarios (SRES) and is reported in the Third Assessment Report (TAR) (Nakicenovic et al., 2000). It developed four families of emission pathways, namely A1, B1, A2, and B2 based on different socio-economic development assumptions under the Coupled Model Inter-comparison Project 3 (CMIP3) (http://www. ipcc-data.org/). IPCC in the Assessment Report 5 (AR5) developed a set of new-emission scenarios (IPCC, 2013) termed as representative concentration pathways (RCPs) under the CMIP5. The scenario assumptions and projections are evolving and the recent is the AR6 (IPCC, 2021). In AR6, CMIP6 have included four illustrative Shared Socioeconomic Pathway (SSP) scenarios, to explore a set of climate futures. The low-emissions' scenario SSP1-2.6; one that is mid-range, SSP2-4.5; and two that are high-emissions' scenarios, SSP3-7.0 and SSP5-8.5. In the scenario designation, SSP refers to the Shared Socioeconomic Pathway assumptions underlying the scenario, while the second number is the scenarios' radiative forcing at 2100.

All the scenarios and story lines project an increase in temperature and rainfall for India, CMIP3 based models project an increase in the temperature by 3.19°C through the most preferred business as-usual A1B scenario while highest increase of 4.10°C is through the most pessimistic A2 scenario. CMIP5 models project 1.50°C–4.30°C increase in temperature based on RCP2.5 and RCP 8.5 scenarios, respectively. The most preferred business as-usual scenario is RCP 4.5 projected an increase of 2.40°C.

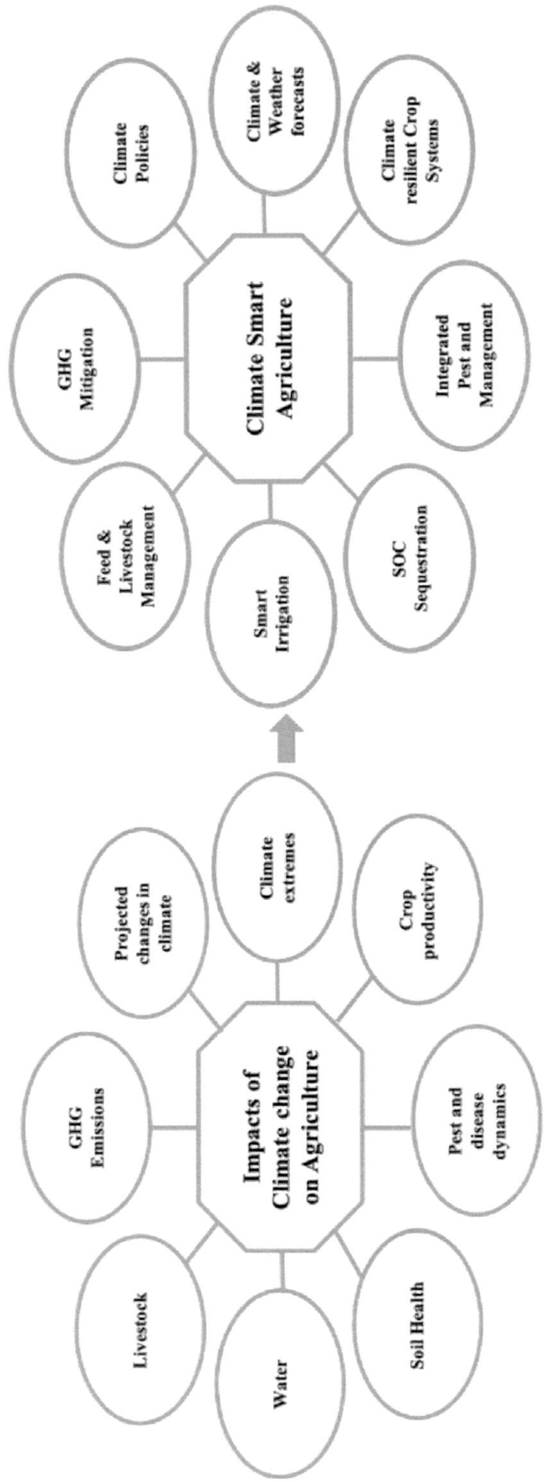

Fig. 1 Diagrammatic representation of the cliamte change impacts and management.

Table 1 Projected changes in temperature and rainfall for India

References	*Scenario*	*Change in Temperature (°C)*	*Change in Rainfall (%)*
Third Assessment Report (TAR)			
Rupakumar et al., 2006	A2	4.1	23
	B2	2.9	18
Chaturvedi et al., 2012 Bal et al., 2016	A1B	3.19	24
Assessment Report 5 (AR5)			
Chaturvedi et al., 2012	RCP 2.6	1.5	6
	RCP 4.5	2.4	10
	RCP 6.0	2.8	9
	RCP 8.5	4.3	14
Assessment Report 6 (AR6)			
Almazroui, et al., 2020	SSP1-2.6	1.0	9.7
	SSP2-4.5	2.1	14.4
	SSP5-8.5	4.2	27.3

Note: *All the projections are for the 30-year average of 2071–2100 relative to the 1961–1990 baselines. However, CMIP3 projections are for the period 2067–2097, CMIP5 projections are for the period 2070–2099, and CMIP6 projections are for the period 2081–2100.

CMIP6 models projects increase in temperature from 2.10°C–4.20°C through the SSP2-4.5 and SSP5-8.5 scenarios, respectively. All the assessments and scenarios showed an increase in Indian rainfall. CMIP3 projection for India shows an increase of 18, 23, and 24% for A2, B2, and A1B scenarios, respectively. CMIP5 projections show an increase of 6% and 14% for RCP 2.6 to RCP 4.5 scenarios. Similarly, CMIP6 projections show an increase of 14.4–27.3 % for the SSP2-4.5 and SSP5-8.5 scenarios, respectively. It is evident from these scenario projections that, in the most preferred business-as-usual scenarios (A1B, RCP 4.5, and SSP4.5) temperature is expected to increase 2.1°C or above, which is well above the 2°C target. Similarly, rainfall is also projected to increase indicating the expected increase in extreme rainfall events. These increase in temperature and rainfall will impact the agriculture production with varying degrees of magnitude depending on the crop, soil and adaptive capacity.

Extreme Events

Globally and more specifically in India, rainfall variability and intensity are increasing, seasons are shifting; sea level is rising, and extreme events such as cyclones, heat waves, heavy rain, floods, and droughts are becoming intense and more frequent (MoEFCC, 2021). Extreme weather events (floods, droughts, heat and cold waves, flash floods, cyclones, and hail-storms) very severely affect agricultural production (Aggarwal et al., 2021). Utilizing the CMIP6 scenarios, Das and Umamahesh (2019), used Heat Wave Magnitude Index (HWMI) to characterize the future heat wave conditions over India. Findings from the study indicate that the TMax anomaly is likely to increase (SSP126 < SSP245 < SSP370 < SSP585) in future causing more intense, high frequency, and long duration heat wave events over India. The study also highlighted that, currently unaffected large regions of southern, northeast, and western parts are expected to be severely affected by heat wave events. More intense heat wave events are likely to occur under SSP585 scenario. Reddy and Saravanan (2023) studied extreme rainfall events, utilising CMIP6 scenarios. They used the indices, viz., RX1DAY, RX5DAY, R10MM, R20MM, and CWD for the assessment. The results reveal an increase in these indices over the Western Ghats and north-eastern regions of India. The temporal variation for the SSP585 scenario predicts significant increases of about 45.41%, 149.40%, 52.26%, and 45.92% in R10MM, R20MM, RX1DAY, and RX5DAY over the current climate, respectively. These findings reaffirm the paramount necessity for adaptation strategies to address the adverse consequences of foreseen heat wave and extreme rainfall events.

Projected Impacts on Crop Productivity

Climate change affects all the plant processes including photosynthesis, respiration, water relations, nutrient uptake, and phenology affecting crop yield and quality. Indian agriculture has been facing climate change induced stresses and related losses already. Recent studies indicate that terminal heat stress has been affecting the wheat yield in Eastern and Central India. Similarly, Cold wave (December 2002–January 2003) has caused considerable damage to horticultural crops such as mango, guava, papaya, brinjal, tomato, potato. The apple production zones have shifted up to 2500 m above mean sea level from 1250 m AMSL. This is partially attributed to non-fulfilment of chilling requirement because of increase in temperatures in Himachal Pradesh.

Elevated temperature affects most crops and result in early maturity, shortened duration and enhanced respiration, which will cause decline in grain and biomass yields. Also, decline in grain quality is observed with increased night temperature (Anand et al., 2012; Chakrabarti et al., 2012). Rice, wheat, and maize are the major crops of India and are risk-prone as well, productivity of rice and wheat in the Indo-Gangetic Plains (IGP) is projected to reduce and similarly, maize in the mid-IGP and Southern Plateau is also projected to reduce. It is estimated that crop yield in India will reduce due to climate change by 4.5–9.0% (Naresh Kumar et al., 2020). Impact of climate change on major crops of India, viz., rice, wheat, and maize revealed varying levels of yield penalty for each degree of increase in temperature. For an increase in temperature less < 2°C, rice yield is projected to decrease by 6% (Saseendran et al., 2000) and Pathak et al. (2021), and wheat yield is projected to decrease by 5.2% (Gupta et al., 2017). For a projected increase in temperature between 2–3°C, rice yield was projected to decrease up to 8.4% (Mathauda et al., 2000). For maize, the yield decreases for temperature increase of 3°C and is projected to be between 10–30% (Kalra et al., 2007).

To minimise crop losses, several adaptation options are researched and suggested. Agronomic adaptations such as crop diversification help to minimise the adverse effects of elevated temperature, saving irrigation water, improved nutrient management, reducing water logging, and better yield. Conservation agriculture (CA) with zero or minimum tillage with direct seeding in rice, residue cover and crop rotations have potential to provide climate resilience (Pathak et al., 2021). Co-benefits of CA in rain-fed agriculture is reduced nutrient loss along eroded soil with high intensity rainy events (Prasad et al., 2015). Intercropping/mixed cropping is one of promising adaptation option, growing more than one crop to increase productivity, nutrient use efficiency, and income. Similarly, Integrated farming system that encompasses crop, livestock, and fishery is a climate smart practice to improve livelihood of the vulnerable farmers (Sutton et al., 2013).

Pest Dynamics in a Changing Climate

An estimated annual loss of 13.6% globally, 17.5% in India is caused by Insect pest (Dhaliwal et al., 2010), climate change is likely to increase the pest-associated losses further. Several insect pests, which are the minor pests, are likely to become more devastating with climate change (Sharma, 2012). For all the insect species, higher temperatures will result in faster development, resulting in rapid increase of pest populations as reproductive cycle gets shortened. Host plant resistance to insects is one of the most environmentally friendly components of pest management. However, climate change may alter the interactions between the insect pests and their host plants (Sharma, 2012). Temperature not only affects the rate of insect development, but also has a profound effect on fecundity and sex ratio of parasitoids (Dhillon and Sharma, 2009) making natural pest control less effective. The rising temperature is projected to shift a range of crop pests and diseases to higher latitudes (Rosenzweig et al., 2001).

Climate change makes the already complex pest epidemiology and its dynamics more complex. Integrated pest management should be given more emphasis to tackle the increasing emergence of pests and diseases (Gautam et al., 2017). The interactions between insect pests and their natural enemies need to be studied carefully to devise appropriate methods for using predators and parasitoids in pest management. Studies on pesticide efficacy to understand its persistence in the environment is needed. Therefore, there is need to develop pesticide formulations and the application schedules that will be least affected by climate change.

Soil Health

As soil upholds all the minerals that are required for crop growth, soil management is one of the most crucial measures for climate change adaptation (Bedano et al., 2016; Chen et al., 2017; He et al., 2018). Elevated temperature is projected to impact soil process, it is anticipated to decline soil organic carbon, increased nitrogen mineralization in the short-term, increased de-nitrification, volatilization losses of nitrogen causing negative impacts on soil fertility and impact the mineral nutrition of crops (St. Clair and Lynch, 2010; Chakrabarti et al., 2012). Soil biodiversity is directly and indirectly related to several soil bio-physical processes and properties; any impact of soil biodiversity will impact the fertility. Increase in soil temperature due to climate change will alter microbial mediated nitrification and denitrification dynamics in soil environment due to shift of nitrifiers and denitrifies population. This affects nutrient cycling and organic carbon accumulation (Dijkstra and Cheng, 2007).

Heavy rainfall and high winds can accelerate the process of soil erosion. To prevent wind-induced soil erosion, tree planting and hedgerow planting are used in semiarid areas, while vegetation cover, contour ploughing, and contour hedgerows are common in humid and coastal areas. In mountains, mini-irrigation facilities, water harvesting, and terrace gardening help control soil erosion. Changing tillage practice and shifting to zero tillage with residue retention help cropping system to adapt to water stress, excess water due to untimely rainfall and hot temperature. Sapkota et al. (2015) found that the change in tillage practices moderates the effect of hot temperature (reduced canopy temperature by 1–4°C) and increases irrigation water productivity by 66–100% compared to traditional production systems, thus adapting to water and heat stress situations. Sequestration of soil organic carbon (SOC) is one of the important strategies not only to mitigate climate change but also to improve soil quality. Even a minor increase in SOC can have positive effects on a range of soil physical properties and thus potentially improve the resilience of soil to stress and contribute to climate change adaptation (Chakraborty et al., 2014; Powlson et al., 2016).

Water Management

In India, the agricultural sector is the largest consumer of water, accounting for approximately 83% of the available water (Gupta et al., 2022). It is crucial to accurately assess future changes in irrigation water usage for effective water and agricultural management (Foster et al., 2020). Predictions suggest that increasing temperatures could result in a reduction in irrigation water availability from both surface sources and groundwater. This decrease is projected to occur in the Indus, Ganges, and Brahmaputra River basins, exacerbating the existing decline in groundwater levels. Therefore, it is essential to implement the "per drop more crop" principle, particularly for water-intensive crops such as rice, wheat, and sugarcane, which collectively contribute to 90% of India's crop production (D'Souza, 2022).

Water harvesting presents a potential solution for managing irrigation water deficits throughout the seasons (Mandal et al., 2020). Laser land levelling has gained popularity as a method to enhance water-use efficiency (Jat et al., 2020). Studies have shown significant water savings in rice (26–30%), wheat (26–33%), maize (22–33%), and cotton (26–43%) through laser land levelling (Jat et al., 2020). Additionally, the System of Rice Intensification (SRI) has been reported to increase crop yield by more than 10% while using 25–47% less water compared to conventional flooded systems in India (Barah, 2009). Similarly, the adoption of micro-irrigation systems, such as sprinklers and drip irrigation, can lead to water savings ranging from 12–84%, depending on location and crop (Kumar, 2016).

At the policy level, the Ministry of Jal Shakti was established to consolidate efforts aimed at sustainable water management. The ministry has devised schemes to improve irrigation efficiency, expand outreach, and optimize water resource utilization through innovative measures. These programs include the Pradhan Mantri Krishi Sinchayee Yojana (PMKSY), Atal Bhujal Yojana, Micro Irrigation Fund, NABARD and Sahi Fasal Campaign, and National Water Mission.

Livestock Management

Livestock plays a crucial role in the agricultural system of India, with small and marginal farmers and landless labourers owning around 70% of the livestock. These livestock owners, who have limited resources, are particularly susceptible to the effects of climate change, placing them at greater risk (Tripathy, 2022). Research conducted at NDRI as part of NICRA has shown that a temperature increases of over 4°C above normal during summer and a temperature decrease of over 3°C below normal during winter adversely affect milk production in crossbred cattle and buffaloes. The decline in milk yield ranged from 10 to 30 percent during the first lactation and 5–20% during the second and third lactations. Heat waves and cold waves both have cumulative effects on milk production in cattle and buffaloes over the short to long term. For India, the leading global milk producer in 2005, it is projected that milk production per adult animal in the grassland-based, arid,semi-arid system will decrease by 25% by the end of the century compared to 2005 levels (Thornton et al., 2022).

The shortage of feed and fodder, including 40% dry fodder, 36% green fodder, and 52% concentrate on a dry matter basis (DAHDF, 2014), along with the inferior quality of roughage, primarily crop residues, already poses a major constraint on achieving the full genetic potential in livestock production in India. Considering the current availability of feed and fodder and the adverse impact of climate change on crop production, the gap between the supply and demand for feed and fodder is expected to widen. This situation is further exacerbated. Poultry birds are also highly vulnerable to climate change due to their limited tolerance for temperature fluctuations (Behura and Samal, 2022). Heat stress during the growth period of broilers not only results in lower body weight but also quality losses. In layers, heat stress negatively impacts egg production and quality (Wasti et al., 2020).

Implementing modifications in nutritional management can help reduce the internal heat load on animals. Balanced feeding and effective feeding management improve the conversion of feed nutrients into animal products while reducing nitrogen and methane emissions. Methane production tends to be lower when forages are ensiled rather than dried. Grinding and pelletizing forages enhance feed utilization and decrease methane emissions (Valli, 2020). Likewise, agronomic approaches such as integrated crop-livestock systems and mixed forage systems offer opportunities for climate change adaptation and environmental benefits (Sekaran et al., 2021).

Climate Services

Climate services in India have become increasingly important in the face of climate change. As the impacts of climate change continue to intensify, accurate and timely information on weather patterns and climate forecasts is crucial for understanding and mitigating its effects. Climate services play a vital role in providing data and analysis that help policymakers and stakeholders make informed decisions and develop effective adaptation strategies. To provide valuable assistance to farmers in India, the India Meteorological Department (IMD) in collaboration with the Ministry of Earth Sciences (MoES), Indian Council of Agriculture Research (ICAR), State Agriculture Universities (SAUs), Indian Institute of Technology (IITs), and other organizations have established a weather forecast-based District Level Agrometeorological Advisory Service (AAS) known as Gramin Krishi Mausam Sewa (GKMS). GKMS project delivers advanced weather information and crop-specific agrometeorological advisories to farmers across the country. By utilizing innovative instruments and technology, the GKMS project enables efficient dissemination of information, empowering farmers to make informed decisions at the farm level. Having successfully implemented the GKMS at the district level, the focus now shifts towards expanding its reach to the block level, considering the vast number of districts (764) and blocks (7,233) in India (source: https://lgdirectory.gov.in/). To further enhance the effectiveness of the system, IMD collaborated with the Regional Integrated Multi-Hazard Early Warning System (RIMES) in Thailand to develop the Agromet-Decision Support System. This system integrates various activities into a single platform, facilitating comprehensive decision-making for agricultural purposes (Singh et al., 2023). The advisories provide timely and actionable crop information from sowing to harvesting, livestock management, and disseminating information on agriculture-related aspects to farmers (Nain et al., 2019). This project currently provides services to over 2.5 million farmers in India and has an estimated economic impact of about USD 10 billion. Capacity building and support of local government is crucial in better uptake by farmers and adopt to the changing climate.

GHG Emissions and Mitigation in Agriculture

Agriculture accounts for 18% of total greenhouse gas (GHG) emissions in India and is the second major emitter of greenhouse gases (GHGs) after the energy sector (71%). The majority of agricultural GHG emissions occur at the primary production stage and are generated through livestock rearing (enteric fermentation and manure management) (59% of agricultural GHGs), the use of chemical fertilisers and their associated impacts on soils (21%), paddy rice cultivation (18.3%), as well as residue management practices (1.7%). GHG emissions per kg of product were greater for livestock products than for crops, except for rice (Ministry of Environment, Forest and Climate Change, 2015). India at the 26th session of the Conference of the Parties (COP26) to the United Nations Framework Convention on Climate Change (UNFCCC) expressed to intensify its climate action by presenting the five nectar elements (Panchamrit) of India's climate action. As per the updated NDC, India now stands committed to reduce emissions Intensity of its GDP by 45% by 2030, from the 2005 level. This update is a step towards achieving long term goal of reaching net-zero by 2070 (Government of India, MoEFCC, 2023). To achieve this commitment, multi-sectoral efforts are vital and reduction from agriculture one of the key aspects.

Mitigation of GHG emission from agriculture can be achieved by changing land-use management practices and enhancing input-use efficiency. Experiments in India showed that methane emission from lowland rice fields can be reduced by 40–50% with alternate wetting and drying (AWD), growing shorter duration varieties, and using neem-coated urea according to soil health card (SHC) and leaf colour chart (LCC). Dry direct-seeding of rice, which does not require continuous soil submergence, can reduce methane emission by 70–75%. Sequestration of carbon in agricultural soil can be promoted with the application of organic manure, crop residues, and balanced nutrients. India has taken several proactive steps for addressing the issues of climate change in agriculture (Pathak, 2022).

Initiatives for Climate Smart Agriculture in India

This section discusses the initiatives of government to offset the impacts of climate change through various Missions and Policy level interventions. Also discusses the available proven other management options in agriculture.

National Missions to Tackle Climate Change

Government of India has launched key missions, programmes, and policies to tackle the issue of climate change and make agriculture more resilient (Table 2).

The Prime Minister's National Action Plan on Climate Change has identified agriculture as one of the eight national missions. Focused studies on climate change impacts on agriculture were conducted since 2004 by Indian Council of Agricultural Research (ICAR) and Ministry of Agriculture and Farmers Welfare. Under major projects such as National Network Project on Climate Change (from 2004–2013) and National Initiative on Climate Resilient Agriculture (NICRA from 2011–2014), and National Innovations in Climate Resilient Agriculture (NICRA, 2015 onwards). The flagship project NICRA is aimed to enhance the resilience of Indian agriculture, encompassing crops, livestock, and fisheries to variable and changing climate through development risk management technologies; to demonstrate site specific technology packages on farmers' fields for adapting to current climate risks; and to enhance the capacity of scientists and other stakeholders in climate resilient agricultural research and its application (NICRA Report, 2016).

In 2010, Ministry of Agriculture and Farmers' Welfare, launched the National Mission for Sustainable Agriculture (NMSA) to support and develop sustainable agriculture and devise appropriate adaptation strategies for ensuring food security, enhanced livelihood opportunities and economic stability. The mission focuses on 10 key dimensions including improved crop seeds, livestock and fish cultures; water use efficiency; pest management; improved farm practices; nutrient management; agricultural insurance; credit support; markets; access to information; and livelihood diversification (https://nmsa.dac.gov.in). There are many programs, policies that support climate change adaptation and mitigation directly and indirectly (Table 2).

Conclusions

Climate change presents significant challenges to agriculture, impacting food and livelihood security. To ensure agricultural sustainability, adaptation to climate change is crucial. This involves adopting technologies, sustainable land management practices, leveraging local knowledge, and establishing supportive policies and institutions. Adaptation options vary by location and context. In India, national-level policies address climate change adaptation. Collaboration between international and national institutions is key to effectively addressing the multifaceted challenges of climate change and securing a sustainable future for agriculture. Efforts for continuous

Table 2 National Initiatives by Government of India to tackle climate change.

National Missions	*Programmes/Policy*	*Initiatives Supporting Climate Smart Agriculture*
1. National Solar Mission 2. National Mission for Enhanced Energy Efficiency 3. National Mission on Sustainable Habitat 4. National Water Mission 5. National Mission for Sustaining the Himalayan Eco-system. 6. National Mission for a Green India 7. National Mission for Sustainable Agriculture 8. National Mission on Strategic Knowledge for Climate Change	1. The flagship project National Initiative on Climate Resilient Agriculture (NICRA) (http://nicra-icar.in). 2. National Food Security Mission (NFSM) (https:/nfsm.gov.in). 3. Mission for Integrated Development of Horticulture (MIDH) (https://midh.gov.in). 4. National Mission on Oilseed and Oil Palm (NMOOP) (https://nmoop.gov.in). 5. Pradhan Mantri Krishi Sinchayee Yojana (PMKSY) (https://pmksy.gov.in). 6. Rashtriya Krishi Vikas Yojana (RKVY) (https://rkvy.nic.in). 7. Paramparagat Krishi Vikas Yojana (PKVY) (https://pgsindia-ncof.gov.in). 8. The National Agroforestry Policy (NAP).	1. Kisan Credit Card (KCC) Scheme 2. Pashu Kisan Credit Card Scheme 3. Pradhan Mantri Kisan Maandhan Yojana 4. National Agriculture Market (e-NAM) 5. Soil Health Card Scheme 6. Pradhan Mantri Annadata Aay Sanrakshan Yojna (PM-AASHA) 7. Formation and Promotion of 10,000 Farmer Producer Organizations (FPOs) 8. Agriculture Infrastructure Fund (AIF) 9. Rainfed Area Development and Climate Change 10. National Project on Agro-Forestry 11. National Mission on Horticulture 12. National Project on Organic Farming 13. Micro Irrigation Fund Scheme 14. Gramin Bhandaran Yojana 15. Pradhan Mantri Kisan Samman Nidhi (PM-KISAN) 16. Pradhan Mantri Urja Suraksha Evam Utthaan Mahabhiyaan (PM-KUSUM) 17. National Action Plan on Climate Change (NAPCC) 18. National Mission for "Green India"

advancement of alternative adaptation measures in agriculture are ongoing, such as the exploration of stress-tolerant crop varieties, resilient cropping systems, farming techniques etc. through research. India's policies and institutions are increasingly responsive to climatic risks. Adapting agriculture to climate change involves utilizing scientific measures and local knowledge. Coordination among ministries and institutions is vital for allocating resources and implementing Climate-Smart Agriculture to mitigate risks in Indian agriculture amidst an uncertain and changing climate.

Acknowledgment

The authors express their heartfelt gratitude to Dr. Mrutyunjay Mohapatra, Director General of Meteorology, India Meteorological Department for his encouragement, guidance, and continued support.

References

Aggarwal, P.K., Roy, J., Pathak, H., Kumar, S.N., Venkateswarlu, B., Ghosh, A., Ghosh, D. (2021). Managing climatic risks in agriculture. *In*: *Discussion Paper Indian Agriculture Towards 2030*. Food and Agricultural Organization, Rome.

Aggarwal, P., Vyas, S., Thornton, P., Campbell, B.M. (2019). How much does climate change add to the challenge of feeding the planet this century? *Environmental Research Letters*, *14*(4): 043001.

Almazroui, M., Saeed, S., Saeed, F., et al. (2020). Projections of precipitation and temperature over the South Asian Countries in CMIP6. *Earth System and Environment*, *4*: 297–20. https://doi.org/10.1007/s41748-020-00157-7.

Anand, A., Singh M.P., Nagarajan, S. (2012). Quantification of climate change impacts on crops: Field experimentation. *In*: H. Pathak, P.K. Aggarwal, S.D. Singh (Eds.), *Climate Change Impact, Adaptation, and Mitigation in Agriculture: Methodology for Assessment and Applications*, xix + 302. Indian Agricultural Research Institute.

Bal, P.K., Ramachandran, A., Palanivelu, K., et al. (2016). Climate change projections over India by a downscaling approach using PRECIS. *Asia-Pacific Journal of Atmospheric Sciences*, *52*: 353–69. https://doi.org/10.1007/s13143-016-0004-1.

Barah, B.C. (2009). Economic and ecological benefits of System of Rice Intensification (SRI) in Tamil Nadu. *Agricultural Economics Research Review*, *22*: 209–14.

Bedano, J.C., Domínguez, A., Arolfo, R., Wall, L.G. (2016). Effect of good agricultural practices under no-till on litter and soil invertebrates in areas with different soil types. *Soil and Tillage Research*, *158*: 100–109.

Behura, N.C., Samal, L. (2022). Effect of Climate Change on Sustainable Backyard Poultry Production. *In*: *Impact of Climate Change on Livestock Health and Production* (pp. 203–212). CRC Press.

Chakrabarti, B., Jain, N., Bhatia, A., Gupta, S.K., Pathak, H. (2012). Impact of climate change on soil fertility. *In*: H. Pathak, P.K. Aggarwal, S.D. Singh (Eds.), *Climate Change Impact, Adaptation, and Mitigation in Agriculture: Methodology for Assessment and Applications*, xix + 302. Indian Agricultural Research Institute.

Chakraborty, D., Watts, C.W., Powlson, D.S., Macdonald, A.J., Ashton, R.W., White, R.P., et al. (2014). Triaxial testing to determine the effect of soil type and organic carbon content on soil consolidation and shear deformation characteristics. *Soil Science Society of America Journal*, *78*: 1192–1200.

Chaturvedi, R.K., Joshi, J., Jayaraman, M., Bala, G., Ravindranath, N.H. (2012). Multi-model climate change projections for India under representative concentration pathways. *Current Science*, 791–802.

Chen, Z., Wang, H., Liu, X., Zhao, X., Lu, D., Zhou, J., et al. (2017). Changes in soil microbial community and organic carbon fractions under short-term straw return in a rice–wheat cropping system. *Soil and Tillage Research*, *165*: 121–27.

DAHDF (Department of Animal Husbandry and Dairying). (2014). *National Action Plan for Dairy Development* [PDF file]. Retrieved from https://www.dahd.nic.in/ sites/default/files/final%20NAP_0.pdf.

Das, J., Umamahesh, N.V. (2019). A review of impacts of climate change on food and nutrition security. *Intern. J. Climate Change Strategies Management*, *11*: 670–91.

Dhaliwal, G.S., Jindal, V. and Dhawan, A.K. (2010) Insect pest problems and crop losses: changing trends. Ind. J. Ecol., 37: 1–7.

Dhillon, M.K. and Sharma, H.C. (2009) Influence of temperature and Helicoverpa armigera food on survival and development of the parasitoid, Campoletis chlorideae. Ind. J. Plant Prot., 36: 240–244.

Dijkstra, F.A., Cheng, W.X. (2007). Moisture modulates rhizosphere effects on C decomposition in two different soil types. *Soil Biol. Biochem.*, *39*: 2264–74.

Foster, T., Mieno, T., Brozović, N. (2020). Satellite-based monitoring of irrigation water use: Assessing measurement errors and their implications for agricultural water management policy. *Water Resources Research*, *56*(11): e2020WR028378.

Gautam, S., Schreinemachers, P., Uddin, M.N., Srinivasan, R. (2017). Impact of training vegetable farmers in Bangladesh in integrated pest management (IPM). *Crop Protection*, *102*: 161–69.

Government of India, MoEFCC. (2023). India stands committed to reducing Emissions Intensity of its GDP by 45 percent by 2030, from 2005 level [Press release]. Retrieved from https://pib.gov.in/PressReleasePage.aspx?PRID=1885731.

Gupta, H., Reddy, K.K., Gandla, V., Paridula, L., Chiluka, M., Vashisth, B. (2022). Freshwater discharge from the large and coastal peninsular rivers of India: A reassessment for sustainable water management. *Environ. Sci. Pollution Res.*, 1–18.

Gupta, R., Somanathan, E., Dey, S. (2017). Global warming and local air pollution have reduced wheat yields in India. *Climatic Change*, *140*: 593–604.

He, Y., He, X., Xu, M., Zhang, W., Yang, X., Huang, S. (2018). Long-term fertilization increases soil organic carbon and alters its chemical composition in three wheat–maize cropping sites across central and south China. *Soil and Tillage Research*, *177*: 79–87.

IPCC. (2013). *Climate Change 2013: The Physical Science Basis*. Contribution of Working Group I to the Fifth Assessment Report of the Intergovernmental Panel on Climate Change [Eds., Stocker, T.F., D. Qin, G.-K. Plattner, M. Tignor, S.K. Allen, J. Boschung, A. Nauels, Y. Xia, V. Bex, P.M. Midgley]. Cambridge University Press.

IPCC. (2021a). *Climate Change 2021: The Physical Science Basis*. Contribution of Working Group I to the Sixth Assessment Report of the Intergovernmental Panel on Climate Change (Eds., Masson-Delmotte, V., P. Zhai, A. Pirani, S.L. Connors, C. Péan, S. Berger, N. Caud, Y. Chen, L. Goldfarb, M.I. Gomis, M. Huang, K. Leitzell, E. Lonnoy, J.B.R. Matthews, T.K. Maycock, T. Waterfield, O. Yelekçi, R. Yu, B. Zhou). Cambridge University Press.

IPCC. (2021b). Summary for Policymakers. *In*: *Climate Change 2021: The Physical Science Basis*. Contribution of Working Group I to the Sixth Assessment Report of the Intergovernmental Panel on Climate Change (Eds., Masson-Delmotte, V., P. Zhai, A. Pirani, S.L. Connors, C. Péan, S. Berger, N. Caud, Y. Chen, L. Goldfarb, M.I. Gomis, M. Huang, K. Leitzell, E. Lonnoy, J.B.R. Matthews, T.K. Maycock, T. Waterfield, O. Yelekçi, R. Yu, and B. Zhou). Cambridge University Press.

IPCC. (2022). *Climate Change 2022: Impacts, Adaptation, and Vulnerability*. Contribution of Working Group II to the Sixth Assessment Report of the Intergovernmental Panel on Climate Change (Eds., H.-O. Pörtner, D.C. Roberts, M. Tignor, E.S. Poloczanska, K. Mintenbeck, A. Alegría, M. Craig, S. Langsdorf, S. Löschke, V. Möller, A. Okem, B. Rama]. Cambridge University Press.

Jat, M. L., Chakraborty, D., Ladha, J. K., Rana, D. S., Gathala, M. K., McDonald, A., Gerard, B. (2020). Conservation agriculture for sustainable intensification in South Asia. *Nature Sustainability*, *3*: 336–43.

Kalra, N., Chander, S., Pathak, H., Aggarwal, P., Gupta, N., Sehgal, M., et al. (2007). Impacts of climate change on agriculture. *Outlook on Agriculture*, *36*: 109–18.

Kumar, M. D. (2016). Water saving and yield enhancing micro irrigation technologies in India: Theory and practice. *In*: M. Dinesh Kumar (Ed.), *Micro Irrigation Systems in India* (pp. 13–36). Berlin, Germany: Springer.

Mandal, S., Vema, V.K., Kurian, C., Sudheer, K.P. (2020). Improving the crop productivity in rainfed areas with water harvesting structures and deficit irrigation strategies. *Journal of Hydrology*, *586*: 124818.

Mathauda, S., Mavi, H., Bhangoo, B., Dhaliwal, B. (2000). Impact of projected climate change on rice production in Punjab (India). *Tropical Ecology*, *41*: 95–98.

MoEFCC. (2021). *India: Third Biennial Update Report to the United Nations Framework Convention on Climate Change*. Ministry of Environment, Forest and Climate Change, Government of India.

Nain, M.S., Singh, R., Mishra, J.R. (2019). Social networking of innovative farmers through WhatsApp messenger for learning exchange: A study of content sharing. *Indian Journal of Agricultural Sciences*, *89*(3): 556–58.

Nakicenovic, N., Alcamo, J., Davis, G., Vries, B.D., Fenhann, J., Gaffin, S., Zhou, D. (2000). *Special report on Emissions Scenarios*. IPCC.

Naresh Kumar, S., Chander, S., Sinha, P., Padaria, R.N. (2020). *Climate Change Impact, Vulnerability, and Adaptation: Agriculture*. A report for India's Third National Communication to UNFCCC, ICAR-Indian Agricultural Research Institute, New Delhi, p. 21.

Pathak, H. (2022). Making Indian agriculture climate smart. *Indian J. Fertilisers*, *18*: 396–403.

Pathak, H. (2023). Impact, adaptation, and mitigation of climate change in Indian agriculture. *Environmental Monitoring and Assessment*, *195*: 52.

Pathak, H., Srinivasarao, C.H., Jat, M.L. (2021). Conservation agriculture for climate change adaptation and mitigation in India. *J. of Agricultural Physics*, *21*: 182–96.

Powlson, D.S., Stirling, C.M., Thierfelder, C., White, R.P., Jat, M.L. (2016). Does conservation agriculture deliver climate change mitigation through soil carbon sequestration in tropical agro-ecosystems? *Agric. Ecosys. Environ.*, *220*: 164–74.

Prasad, Y.G., Srinivasa Rao, C.H., Prasad, J.V.N.S., Rao, K.V., Ramana, D.B.V., Gopinath, K.A., Srinivas, I., Reddy, B.S., Adake, R., Rao, V.U.M., Maheswari, M., Singh, A.K., Sikka, A.K. (2015). Technology demonstrations: enhancing resilience and adaptive capacity of farmers to climate variability. *National Innovations in Climate Resilient Agriculture Project*. ICAR-Central Research Institute for Dryland Agric.

Ram Fishman, Xavier Giné, Hanan G. Jacoby, "Does a drop per crop help groundwater extraction to stop?" The website of India water portal, October 4, 2021. https://www.indiawaterportal.org/articles/does-drop-crop-help-groundwater-extraction-stop; M.G. Chandrakanth, "Farm subsidies account for 21% farm income per hectare with continuing patronage of governments," *The Times of India*, February 27, 2021, https://timesofindia.indiatimes.com/blogs/economic-policy/farm-subsidies-account-for-21-farm-income-per-hectare-with-continuing-patronage-of-governments/

Reddy, N.M., Saravanan, S. (2023). Extreme precipitation indices over India using CMIP6: A special emphasis on the SSP585 scenario. *Environ. Sci. Pollut. Res.*, *30*: 47119–47143.

Rosenzweig, C., Iglesius, A., Yang, X.B., Epstein, P.R., Chivian, E. (2001). Climate change and extreme weather events - Implications for food production, plant diseases, and pests. NASA Publications 24. https://digitalcommons.unl.edu/nasapub/24.

Rupa Kumar, K., Sahai, A.K., Kumar, K.K., Patwardhan, S.K., Mishra, P.K., Revadekar, J.V., Pant, G.B. (2006). High-resolution climate change scenarios for India for the 21st century. *Current Science*, 334–45.

Sapkota, T.B., Jat, M., Aryal, J.P., Jat, R., Khatri-Chhetri, A. (2015). Climate change adaptation, greenhouse gas mitigation, and economic profitability of conservation agriculture: Some examples from cereal systems of Indo-Gangetic Plains. *Journal of Integrative Agriculture*, *14*: 1524–33.

Saseendran, S., Singh, K., Rathore, L., Singh, S., Sinha, S. (2000). Effects of climate change on rice production in the tropical humid climate of Kerala, India. *Climatic Change*, *44*: 495–14.

Sekaran, U., Lai, L., Ussiri, D.A., Kumar, S., Clay, S. (2021). Role of integrated crop-livestock systems in improving agriculture production and addressing food security: A review. *Journal of Agriculture and Food Research*, *5*: 100190.

Sharma, H.C. (2012). Effect of global warming on insect-host plant-environment interactions. *In*: *24th International Congress of Entomology*, 19–24 Aug., Daegu, South Korea.

Singh, K., Ghosh, K., Bhan, S., Singh, P., Vishnoi, L., Balasubramanian, R., Singh, R. (2023). Decision support system for digitally climate informed services to farmers in India. *Journal of Agrometeorology*, *25*: 205–14.

St. Clair, S.B., Lynch, J.P. (2010). The opening of Pandora's Box: Climate change impacts on soil fertility and crop nutrition in developing countries. *Plant Soil*, *335*: 101–15.

Sutton, M.A., Bleeker, A., Howard, C.M., Bekunda, M., Grizzetti, B. (2013). *Our Nutrient World: The Challenge to Produce More Food and Energy with Less Pollution*, 128. Global Overview of Nutrient Management. Centre for Ecology and Hydrology, Edinburgh on behalf of the Global Partnership on Nutrient Management and the International Nitrogen Initiative.

Thornton, P., Nelson, G., Mayberry, D., Herrero, M. (2022). Impacts of heat stress on global cattle production during the 21st century: A modelling study. *The Lancet Planetary Health*, *6*: e192-e201.

Tripathy, G.K. (2022). Climate Change and Policy Interventions for Livestock Sector. *In*: *Impact of Climate Change on Livestock Health and Production* (pp. 289–95). CRC Press.

Valli, C. (2020). Mitigating enteric methane emission from livestock through farmer-friendly practices. *Global Climate Change and Environmental Policy: Agric. Persp.*, 257–73.

Wasti, S., Sah, N., Mishra, B. (2020). Impact of heat stress on poultry health and performances, and potential mitigation strategies. *Animals*, *10*: 1266.

Chapter 3

Climate Change Impacts on Food Security in Nepal

Jaya K. Gurung[1], *Divas B. Basnyat*[1], *Srijana Thapa*[1] and *Rojina Haiju*[1]

Introduction

Nepal's climate is highly variable and influenced by seasonal and temporal variations along with the east-to-west progressing summer monsoon. Precipitation is driven by monsoon between June to September when about 80% of the rainfall occurs. While the spatial variability of rainfall across the country ranges between less than 200 mm to above 5,000 mm per year, the average annual temperature ranges from 25°C in the lowlands (terai plain) to –1°C in high altitudes (USAID, 2017). Nepal also possesses physio-climatic characteristics such as rugged terrain, steep topography, and fragile geology, which result in intense monsoon precipitation and very low rainfall in the rest of the months (NDRI, 2013). As a result, Nepal is highly susceptible to disasters, experiencing recurring natural hazards such as flash floods in the mountains and hills, riverine floods, and inundation in the southern lowlands (MoFE, 2019a, b)

The livelihood security and socio-economic development of Nepal are significantly influenced by the agriculture sector (MoFE, 2021). It is the main source of income and the country's food security. About 75% of agricultural production in Nepal is monsoon dependent (USAID, 2017). The intensity of the summer monsoon and the onset and offset dates of the monsoon are crucial factors to influence agriculture (Nayava, 2017); however, the variability of monsoon rains directly impacts crop productivity in Nepal (Malla, 2008). Furthermore, climate change poses a significant threat in Nepal. As Nepal is one of the most climate-vulnerable countries in the world, the country's condition is exacerbated by the climate change impacts. Nepal has already witnessed the consequences of climate change, with temperature rise and shifts in precipitation patterns (NAP, 2019). Also, agriculture is highly sensitive to short-term weather changes. Climate change impacts such as intense rainfall in a short period hampers groundwater recharge and affect in crop productivity, and ultimately in the country's economy (Malla, 2008).

[1] Nepal Development Research Institute, Manbhawan, Lalitpur, Nepal.

* Corresponding author: info@ndri.org.np

According to the study of the National Adaptation Plan (NAP), the temperature in Nepal has increased in all zones, from the lowlands to the high mountains. The future climate change projection predicts that Nepal's temperature will rise further for all seasons. Also, the climate change scenario for near future (2016–2045), mid-future (2036–2065), and far future (2071–2100) shows that the climate will be more warmer and wetter in the future, and there will be decrease in precipitation during the monsoon (NAP, 2019).

Science-based knowledge and innovative technologies can play a crucial role in addressing the concerns related to climate change effects in agriculture for food security. The development of information technology in agriculture sector has helped to combat the threats of climate change and improve crop productivity (UN, 2017). The availability of information via mobile phones and mobile internet as part of Agro-met service is assisting farmers to effectively manage the risks and mitigate vulnerability caused by climate change effects (Baumueller, 2013). Hence, information communication technology (ICT), including mobile phones, are considered powerful digital platforms that are capable to reach large groups of farmers simultaneously in rural areas (Santosham and Lindsey, 2015; WB, 2019).

Future Climate Projections

The Intergovernmental Panel on Climate Change (IPCC) Sixth Assessment Report on the physical basis of climate change concluded that it is the human influence that is warming the global climate and also has contributed to many observed changes in weather and climate extremes (IPCC, 2021). Approximately 80% of Nepal's population lives in remote areas, making them vulnerable to various natural hazards, such as earthquakes, droughts, landslides, and extreme weather events (MoHA, 2017).

As temperatures rise, in Nepal, climate hazards such as floods, landslides, drought are expected to increase in both intensity and frequency (MoFE, 2021). Due to climate hazards, millions of people in Nepal are at risk, including reductions in agricultural production, food insecurity, water stress, loss of forests and biodiversity, and damage to infrastructures. Consequently, this will lead to an increase in food insecurity (MoSTE, 2014).

The climate change projections using the global climate models indicate that climate change in the Himalayan region is rapidly accelerating and is expected to further increase in all seasons. Additionally, it is projected that the climate in Nepal will be significantly warmer and wetter in the future, but will be a decrease in precipitation during the pre-monsoon season (NAP, 2019). To select the climate models and to understand the projected future changes in Nepal for the NAP purpose, two Representative Concentration Pathways (RCPs), RCP4.5 and RCP8.5, with the uncertainties of the projected future changes, were considered (MoFE, 2019).

Note: The Fifth Assessment Report of IPCC has put forward four representative concentration pathways (RCPs) to project the future climate scenarios. These four RCPs represent the potential trajectories of greenhouse gas concentration on the different levels of emissions in the future. The four RCPs (RCP 2.6, RCP4.5, RCP6.0, RCP8.5) span the range of different radiative forcing values for the year 2100 ranging from 2.6–8.5 W/m^2 (MoFE, 2019).

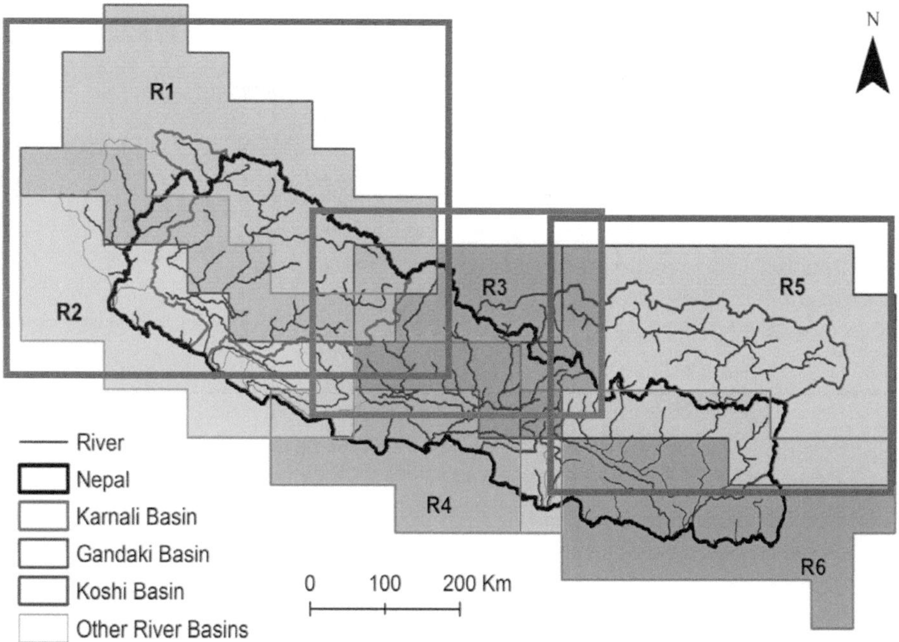

Fig. 1 Regions of Nepal used for Climate Change Assessment. (*Source:* NDRI)

Figure 1 shows regions 1, 2, and 3 that are the upper regions of Nepal, spanning from West to East. It is used to assess the climate change projections. The climate is characterized by two main periods: the monsoon period (JJAS) from June to September and the winter season (DJFM) from December to March.

Figure 2 displays changes in precipitation from the historical to the future periods, 2040–2059, based on various General Circulation Models (GCMs). The X-axis represents changes in precipitation, while the Y-axis represents changes in temperature. The figures demonstrate the considerable uncertainty in the future climate change projections, indicating the need for a more flexible adaptation approach. In summary, it shows that most models project higher monsoon precipitation, but the winter precipitation remains more uncertain. The future is expected to be warmer, although the magnitude of the temperature change varies between the model performance.

Projection of temperature changes

- Warmer winter temperatures projected for the future compared to monsoon (JJAS).
- For the upper region, temperature increases during the monsoon months and are expected to be around 2°C in the RCP4.5 scenario and 2.6°C in the RCP8.5 scenario.
- For all upper regions, in winter months (DJFM), the model averages predict temperature increase of about 2.7°C and 3.4°C in the RCP4.5 and RCP8.5 scenarios, respectively.

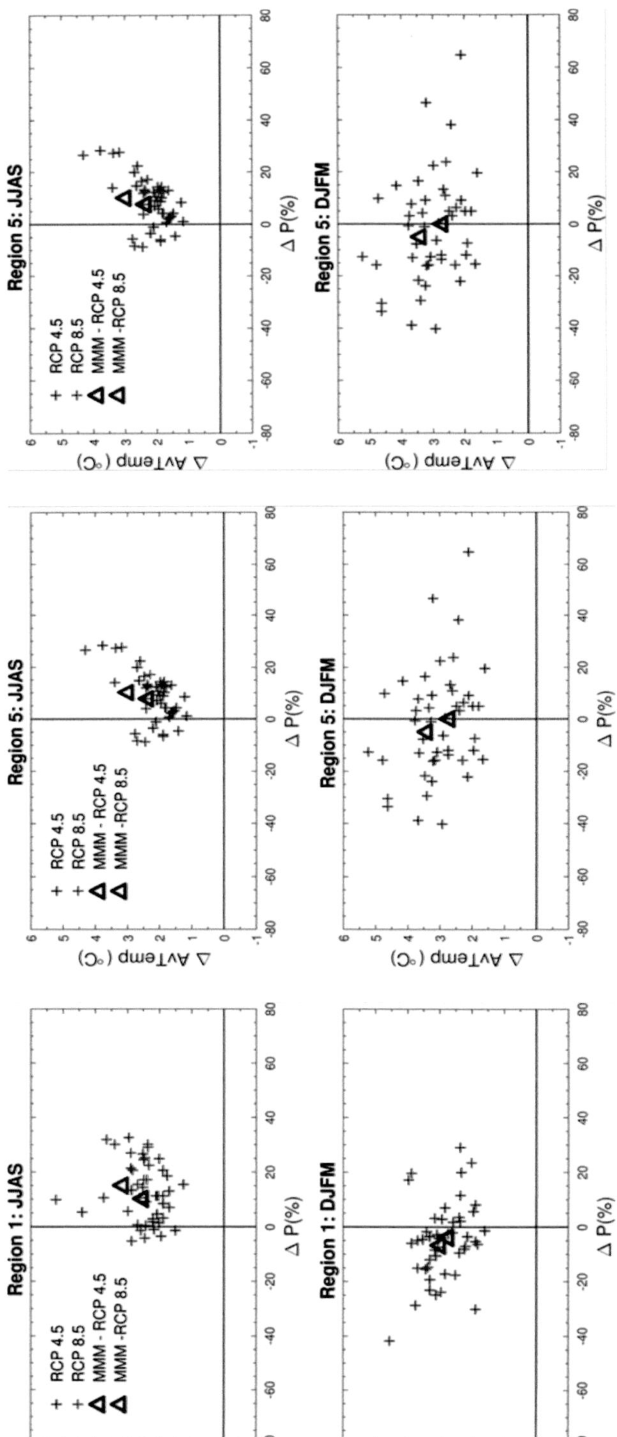

Fig. 2 Projected Temperature and Precipitation Changes in the 2040–2059 periods.

- Under the RCP4.5 scenario, selected CMIP 5 models project temperature rises from approximately 1.2°C–4.4°C for Nepal. In the RCP8.5 scenario, the models project temperature increase from 1.6°C–5.2°C.

Projection of precipitation changes

- Projections for precipitation show wide variation and are less distinct compared to temperature projections.
- The average increase in mean monsoon precipitation for upper regions 1, 3, and 5 is 10.32%, 8.64%, and 7.59%, respectively, in the RCP4.5 scenario. Under the RCP8.5 scenario, the increase in precipitation projections is 15.2%, 11.79%, and 9.98% for the upper regions 1, 3, and 5. However, some models indicate a decrease in precipitation as well.
- In the RCP4.5 scenario, precipitation decreases for upper regions 1, 3, and 5 are 3.91%, 3.89%, and 0.03%, respectively. In the RCP8.5 scenario, the winter precipitation decreases by 5.19%, 6.30%, and 4.82% for the upper 3 regions.

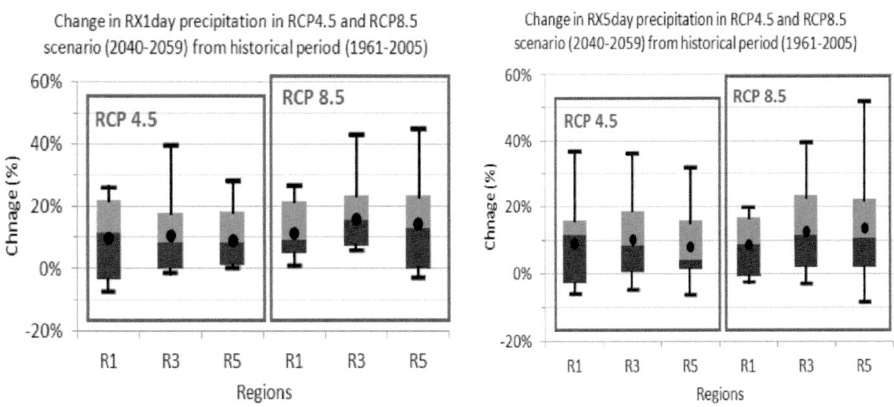

Fig. 3 Projected Changes of RX1day Precipitation.

Figure 3 shows the projected changes in RX1day (1-day Maximum Annual rainfall) and RX5day (Maximum Annual 5-day rainfall) for different selected GCM models. Higher extreme precipitation means higher flood and related sediment load. GCM models project an increase in 1-day and 5-day maximum precipitation.

Impacts on Agriculture and Food Security

About 43% of Nepal's land is forest and 24% is designated as agricultural land. Two-thirds of the country's population depends on agriculture, which contributes 26% of the national gross domestic production (GDP) of Nepal (NPC, 2019). The land distribution per person varies across the country, and over half of all farmers cultivate less than a ha of land (MoLCPA, 2019; NPC, 2019).

Figure 4 shows the land use map of Nepal developed by the Department of Irrigation for Irrigation Master Plan (IMP), 2018 considering irrigation development status of Nepal.

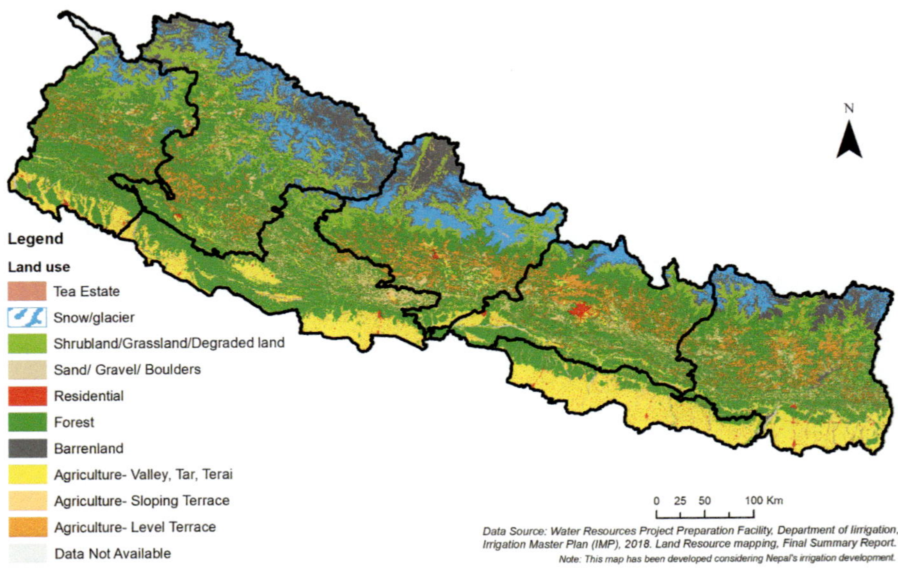

Fig. 4 Land Use Map of Nepal (NPC, 2019).

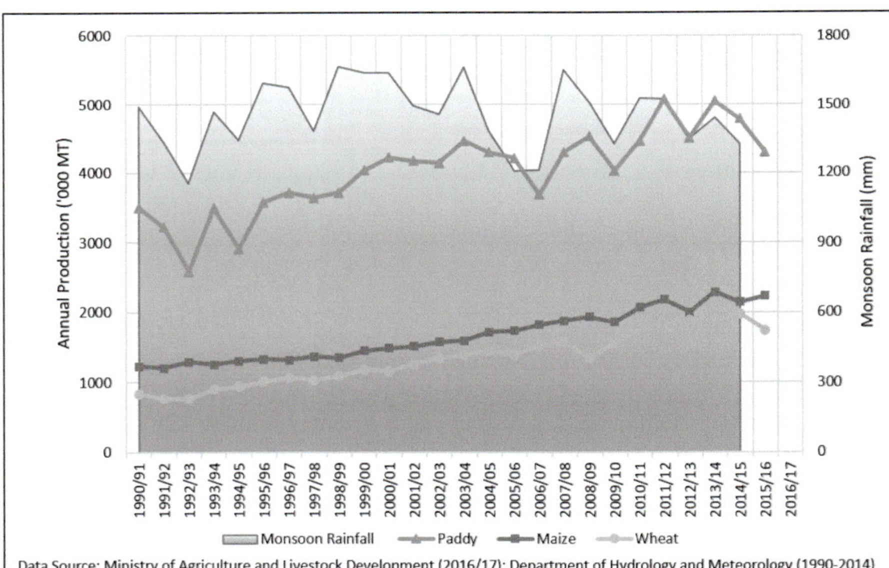

Fig. 5 Annual Cereal Crop Production and Monsoon (June–Sept.) Precipitation.

Figure 5 demonstrates that the cereal crop production in Nepal has shown an increasing trend over the years. However, paddy production is susceptible to variations in monsoon rainfall. When there is less than normal monsoon rainfall, paddy production experiences a decline. This indicates that paddy production in

Nepal is still heavily dependent on rainfall, as the country lacks adequate irrigation infrastructure.

In the absence of adequate irrigation facilities, farmers heavily rely on the monsoon season for water supply, making paddy cultivation vulnerable to climatic variability and rainfall patterns. To increase paddy production and reduce the vulnerability to weather fluctuations, investing in irrigation infrastructure and adopting climate-smart agriculture become essential strategies.

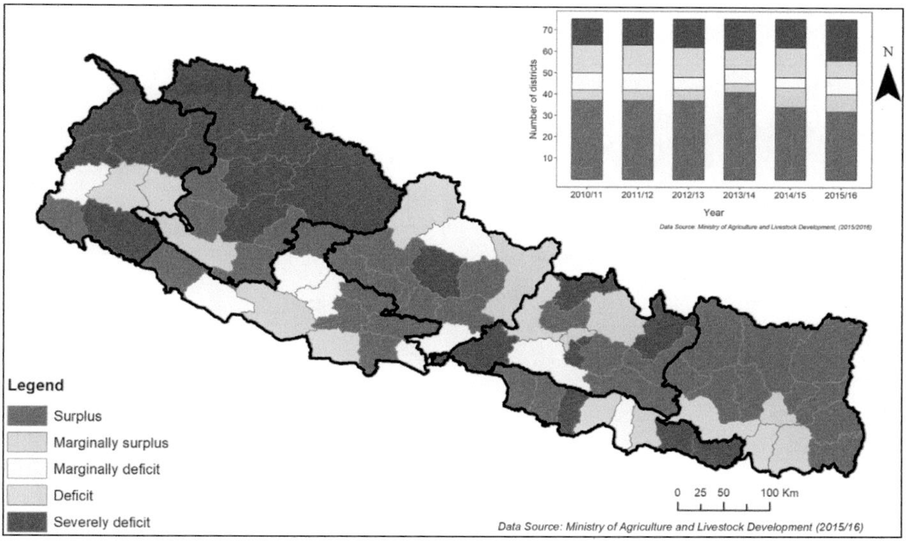

Fig. 6 Food Self-Sufficiency Status (NPC, 2019).

Food demand in Nepal has been increasing over time. The unplanned and haphazard urbanization of agricultural land, particularly in the southern part of the country, has escalated pressure on the food supply. As a consequence, food imports in Nepal have reached unpreceded levels since 2011. Figure 6 shows the food self-sufficiency status of Nepal in 2015/16. This situation has underscored significance of domestic agriculture ensuring the country's food security status (NPC, 2019).

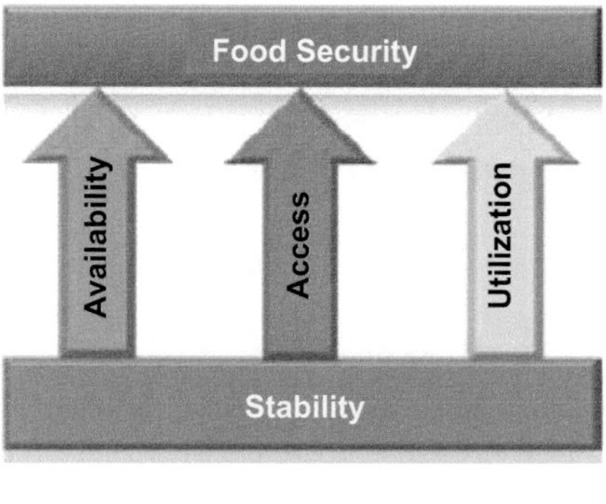

Fig. 7 Food Security: 4 Dimensional Framework (NPC, 2019).

Nepal is ranked in 82nd out of 113 countries in the Global Food Security Index. The country faces a notable disparity in availability of food across different regions. Exposure to extreme climate events, particularly droughts in the mid-hills and mountains, along with complex geographical conditions and a poor road network, leads to a high dependency of food aid in Nepal. These challenges significantly impact the country's food security situation. Figure 7 above shows the four-dimensional framework for food security and stability.

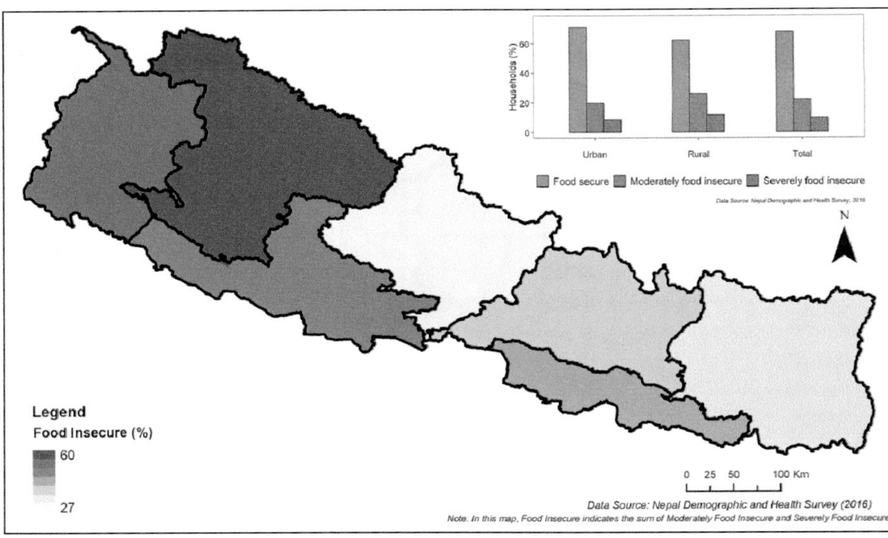

Fig. 8 Food security map of Nepal as per province (NPC, 2019).

Figure 8 indicates Nepal's progress in food security; however, the accompanying map shows that chronic food insecurity is concentrated mainly in two parts of the country: Mid- and Far-Western regions and as well as the central mountain regions. Additionally, the rural-urban disparities further influence the geographical dispersion of hunger, with food insecurity being more prevalent in hard-to-access regions of the country (NPC, 2019).

Figure 9 illustrates a distinct seasonal relationship as the vulnerable population relies on rain-fed agricultural practices. As a result, the crop cycle and weather conditions play a crucial role in determining the food security of the region (NPC, 2019).

Climate change in Nepal is already affecting food security through temperatures, changing rainfall patterns, and an increase in the frequency of extreme weather events (GoN, 2019). The GCMs project a rise in temperature in the future; whereas in the case of precipitation, the models project increase in winter precipitation. However, some models also project a decrease in winter precipitation (NPC, 2019). Figure 10 (A and B) show that average changes in temperature is high in Western Nepal. In Fig. 10B, the annual mean temperature changes for further three decades (2036–2065) can be noticed high in Western Nepal.

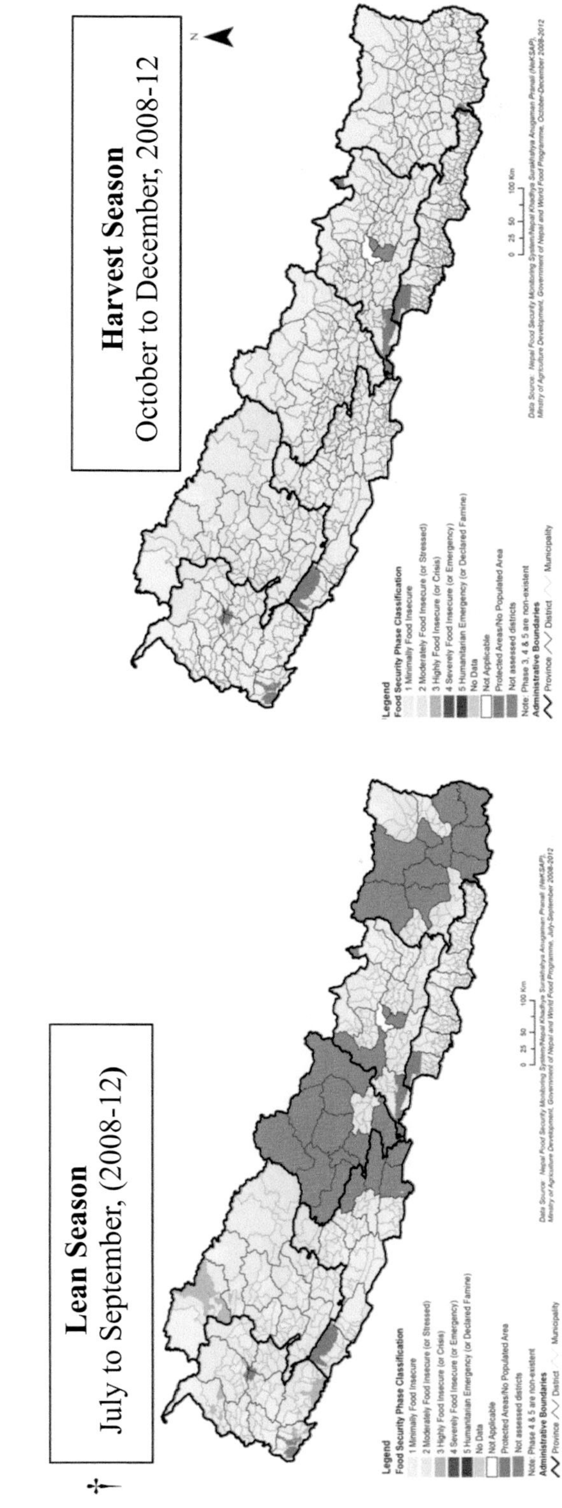

Fig. 9 Food Security Phase Classification (NPC, 2019).

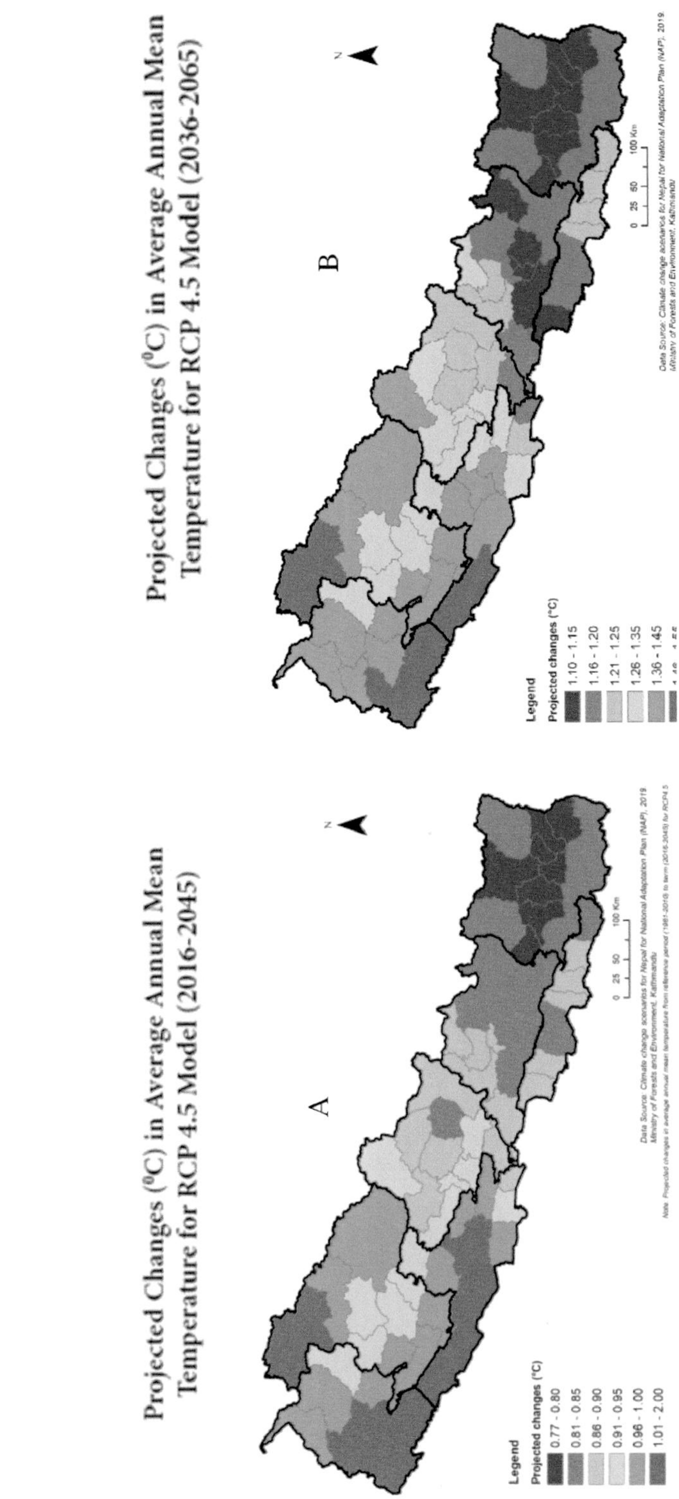

Fig. 10 Projected range of average annual mean temperature (°C) change in Nepal (NPC, 2019).

The annual change in temperature in future is projected high in western Nepal in contrast to eastern Nepal, as shown in Fig. 10 (A and B).

Figure 11 (C and D) shows the map of projected changes in percentage average annual precipitation for the selected scenario, RCP 4.5 in the year 2016–2045 and in the year 2036–2065. It is predicted that in the future (2036–2065) the precipitation will be less in the east compared to the west in contrast to the projection in the period 2016–2045.

Initiatives/Measures on Adaptation and Mitigation

Climate Smart Agriculture (CSA) technologies and practices offer promising opportunities for addressing climate-related challenges and fostering economic growth and development. In the country, modern Information and Communication Technologies (ICT)-based approaches such as mobile phone applications have been utilized in pilot projects to provide agro-advisories and weather forecasts across different regions (WB, 2019).

These initiatives aim to empower farmers with timely and relevant information, enabling them to make informed decisions about their agricultural practices, optimize resource use, and enhance their resilience to climate impacts. By integrating technology and climate-smart practices, Nepal seeks to enhance agricultural productivity and food security while adapting to the changing climate conditions.

Examples of CSA Practices in Nepal

- *Nutrient management* – e.g., Using leaf color charts, improving the timing, etc.
- *Improved Water and Irrigation Management* – e.g., Using waste-water collection and rainwater harvesting techniques.
- *Soil conservation techniques* – Such as zero tillage.
- *Crop intensification techniques* – e.g., Inter-cropping.
- *Use of drought tolerant and high-yielding varieties* – Such as lentils.
- *Infrastructure development for risk reduction and effective response* – Establishment of gene banks, check dams, embankment of river channels.
- *Livestock and animal husbandry* – Diet quality management, local breeds, species, enhance fodder management.
- *Fishery and aquaculture* – complementary adaptive response, practices of growth of local species, and adoption of climate resilient and stress-tolerant fish species.

Agro-Met Services in Nepal

It is clear that agriculture is highly affected by climate change and its variability. The great diversity of crops in Nepal is attributed to the presence of microclimatic zones. Moreover, most of the indigenous crops are sensitive to the climate conditions (Shrestha, 2012). Climate change is rapidly threatening the climatic conditions of each microclimatic zone, causing stress to these crops. Nepalese agriculture is heavily relying on monsoon, and the monsoon pattern is likely to be influenced by climate change. Over the past two decades, the early pre-monsoon rains, occurring around April and May after the long dry winter in Nepal, have become increasingly erratic. The onset of monsoon has also shifted by 1–2 weeks compared to the past and the

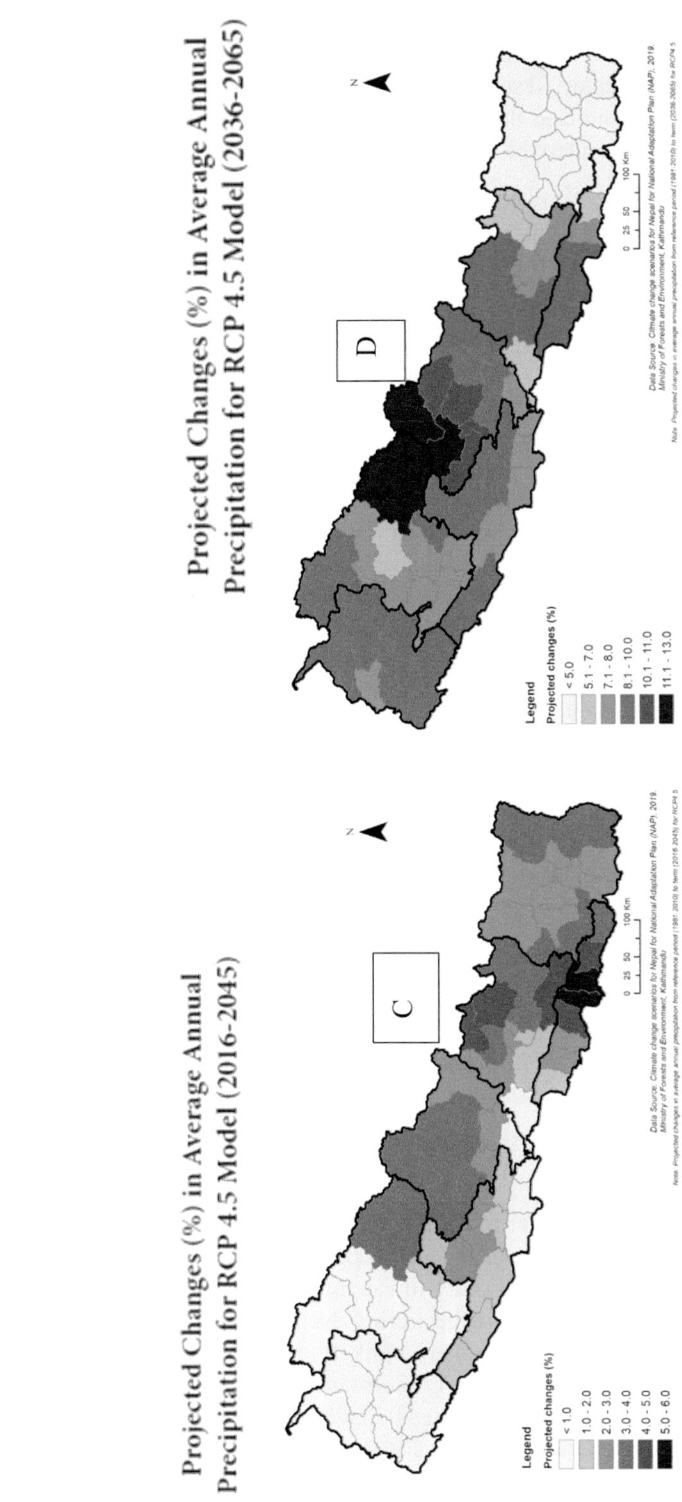

Fig. 11 Projected range of average annual precipitation (%) change in Nepal (NPC, 2019).

exit of monsoon has changed accordingly, affecting water availability. Consequently, the timing of sowing of rice, the stable food crop in Nepal, has been altered, leading to a significant drop in production. Similarly, unprecedented rains have become more common in recent times. For instance, in October 2021, a nationwide downpour of over 500 mm in 24 hr. was recorded (DHM, 2021) occurring after the monsoon season had officially exited. This unusually heavy rain led to the loss of millions of tons of rice ready for harvest, significantly impacting food security in Nepal. However, the rapid advancement of information technology could serve as a valuable tool to protect people from the high risk of livelihood and food insecurity.

Information Technology Penetration in Nepal

The percentage of mobile (Smart phone) penetration in Nepal stands at an impressive 73% or more, according to Nepal Telecom 2018. Moreover, internet coverage reaches significant portion of population, 91% of the total, and this coverage is continuously expanding as per the Nepal Telecom Authority in 2021. The rapid growth in mobile and internet usage indicates a substantial advancement in technology adoption within the country.

Initially, the primary use of this technology was for socialization, staying connected with family and friends, and entertainment purposes. However, as technology continues to permeate various aspects of society, it is increasingly being leveraged for professional development and economic activities, including the farming systems of Nepal.

The widespread use of IT is opening up new opportunities for farmers in Nepal. They can access agricultural information, weather forecasts, maker prices, and other crucial data through mobile apps and the internet. These tools enable farmers to make more informed decisions about their farming practices, optimize resource use, and enhance productivity. Integration of IT into agriculture is helping to bridge information gaps and improve overall agricultural efficiency and sustainability in Nepal.

Agro-Met Service Pioneer Initiatives

Development agencies such as Climate Change Agriculture and Food Security (CCAFS), Food and Agriculture Organization (FAO), and World Food Program (WFP) collaborated with Government agencies of Nepal, including the Ministry of Agriculture and Livestock Development, the Department of Hydrology and Meteorology (DHM), and the Nepal Agriculture Research Council (NARC). In addition, think-tank research institutes like Nepal Development Research Institute (NDRI) and Practical Action also played instrumental roles on these pioneer initiatives of the Agro-Met Service between 2012–2015. During that time, DHM lacked adequate infrastructures for weather forecasting. Therefore, weather data from the World Meteorological Organization (WMO), and the International Crop Research Institute for the Semi-Arid Tropics (ICRISAT)/GCIAR were used. This data was then interpolated and verified by DHM.

Later, the World Bank-supported program called Pilot Project for Climate Resilient (PPCR) was implemented and scaled up, including the Agriculture

Advisory Program. This initiative significantly strengthened DHM by providing necessary infrastructure for weather forecasting. Consequently, DHM started offering weekly weather information and forecasts, encompassing data on rainfall, maximum temperature, minimum temperature, relative humidity, and the wind/cloud conditions.

The District Agriculture Development Office (DADO) and Livestock Development Office (DLSO) provided crop status update, including crop stage, pest and diseases, crop cover area, and livestock health. The Agriculture Management Information System (AMIS), under the Ministry of Agriculture Development (MoAD) in coordination with NARC and DHM, prepared agro-advisory bulletins. These bulletins were disseminated to all the extension offices of the Ministry of Agriculture and Livestock Development through various channels such as websites, Short Message Services (SMS), Agriculture-Application, Newspapers, and FM Radio. The communication between the service providers and the beneficiaries including farmers and stakeholders, has been significantly augmented by utilizing the advanced information technology. This includes the establishment of Automatic Weather Station and providing smart mobile phones equipped with farmer SIM cards.

Agro-Met Services at Present

At present, with the success of some pilot programs, the Government of Nepal has taken ownership and institutionalized the Agro-Met Service. The weekly bulletin is now published by the Agro-Meteorological Division of the DHM under the Ministry of Energy, Water Resource, and Irrigation (MoEWRI).

The weekly weather forecast is province-specific, covering regions from east to west, including mountain-hills and Terai areas across the country. The weekly weather forecast is provided to NARC, the main agency of Nepal for agriculture research. Subject matter specialists at NARC prepare the advisory messages based on forecast, which are then published regularly. These advisory messages are disseminated through various means such as website, SMS, FM radio, and news bulletins, to effectively reach farmers and other key stakeholders.

Opportunities and Challenges

The approach of the Agro-Met service has fostered collaboration between experts in agriculture and meteorology, enhancing their communication skills. Meteorologists have developed their capacities to provide the forecast specifically tailored to the agriculture sector. As a result, beneficiaries such as farmers and stakeholders are well-informed about weather forecasting mechanisms and are developing their capacities to risks, using weather forecasts. Despite the substantial evolution in ICT use in the agro-meteorological sector, several challenges persist. It remains difficult to prepare district/local level bulletins tailored to the specific demands of users. On the other hand, farmers lack the necessary skills to extract information from national bulletins that are not customized to their districts. Moreover, there is a significant opportunity to make the existing agro-met service more interactive and user-friendly. Service-providing agencies need to enhance their capacity to provide accurate and customized information, while beneficiaries must build their capacity to adopt to climate-responsive measures. The Agro-Met Advisory Services have proven successful in optimizing agricultural production and easing the stress caused by weather (Timilsina et al., 2019).

Prospects of Regional Cooperation

Weather and climate change are often transboundary problems that lead to interdependence between countries, making transboundary cooperation essential. The Indo-Gangetic Basin, the largest watershed in the world, is shared by several Asian countries including China, Pakistan, Nepal, India, Bhutan, and Bangladesh, which are adjacent to one another. Given this shared resource, the exchange of hydrological-meteorological data and information between upstream basins and downstream regions has become crucial for the prudent utilization water resources and the protection of life and property from water-related disasters.

According to Prof. Brahma Chellaney (Challaney, 2011) many of Asia's water sources cross national boundaries, and as water availability diminishes, international tensions are likely to escalate. The economies of most Asian countries heavily rely on agriculture, which is increasingly affected by climate change. In this context, adopting an integrated approach that spans upstream and downstream communities, irrespective of country borders, present an opportunity to effectively address climate-related risks.

References

Baumueller, H. (2013). *Mobile Technology Trends and Their Potential for Agriculture Development* (ZEF Working Paper Series). Center for Development Research, University of Bonn.

Challaney, B. (2011). *Water: Asia's New Battleground.* Georgetown University Press. ISBN, 1589017986, 9781589017986.

DHM, (2021). *Post Monsoon Precipitation Monitoring.* http://old.dhm.gov.np/uploads/climatic/1267328706Post%20Monsoon_Precipitation_monitoring_30th%20November%202021.pdf.

GoN, (2019). Government of Nepal National Position Paper on Disaster Risk Reduction and Management. *Global Platform for Disaster Risk Reduction, 2019.* http://drrportal.gov.np/uploads/document/1514.pdf.

IPCC, (2021). *Climate Change 2021* (The Physical Science Basis, Summary for Policymakers) [Working Group]. Intergovernmental Panel on Climate change. https://www.ipcc.ch/report/ar6/wg1/downloads/report/IPCC_AR6_WGI_SPM_final.pdf

Malla, G. (2008). Climate Change and Its Impact on Nepalese Agriculture. *The Journal of Agriculture and Environment, 9.* https://www.climatenepal.org.np/sites/default/files/doc_resources/gpnpy6vdd7i_2.pdf.

MoFE, (2019a). *Climate Change Scenarios for Nepal* (National Adaptation Plan). Ministry of Forestry and Environment, Government of Nepal.

MoFE, (2019b). *Climate Change Scenarios for Nepal for National Adaptation Plan (NAP).*

MoFE, (2019c). *Climate Change Scenarios for Nepal for National Adaptation Plan (NAP).*

MoFE, (2021). *Vulnerability and Risk Assessment and Identifying Adaptation Options* (Summary for Policy Makers). Ministry of Forestry and Environment, Government of Nepal. https://www.mofe.gov.np/uploads/documents/vulnerability-repnew1630571413pdf-2940-766-1658827788.pdf.

MoHA, (2017). *Nepal Disaster Report.* Ministry of Home Affairs, Government of Nepal. http://drrportal.gov.np/uploads/document/1321.pdf.

MoLCPA, (2019). *Government of Nepal, Ministry of Land Management, Cooperatives, and Poverty Alleviation*. https://molcpa.gov.np/

MoSTE, (2014). *Economic Impact Assessment of Climate Change in Key Sectors in Nepal* (Summary Report). Ministry of Science Technology and Environment, Government of Nepal. https://reliefweb.int/report/nepal/economic-impact-assessment-climate-change-key-sectors-nepal.

NAP, (2019). *National Adaptation Plan 2021–2050* (Summary for Policy Makers). National Adaptation Plan.

Nayava, J.L. (2017). Monsoon rainfall and its impact on rice production in Nepal. *Rice Science and Technology in Nepal*, 497–506.

NDRI, (2013). *Disaster Risk Reduction and Climate Change Adaptation in the Koshi River Basin, Nepal*. Nepal Development Research Institute.

NPC, (2019). *The Food Security Atlas of Nepal (NPC)*. Government of Nepal, National Planning Commission.

Santosham, S. and Lindsey, D. (2015). *Bridging the Gender Gap: Mobile Access and Usage in Low- and Middle-income Countries* (Connected Women).

Shrestha, D. (2012). Indigenous Vegetables of Nepal for Biodiversity and Food Security. *International Journal of Biodiversity Advanced Conservation*, *5*: 98–108. https://doi.org/10.5897/IJBC11.124.

Timilsina, A., Shrestha, A., Gautam, A.K., Gaire, A., Malla, G., Paudel, B.P., Rimal, R., Upadhyaya, K.P. and Bhandari, H.L. (2019). A practice of agro-met advisory service in Nepal. *Journal of Bioscience and Agriculture Research*, *21*(02): 1778–85.

UN. (2017). *The Role of Science, Technology and Innovation in Ensuring Food Security by 2030*. United Nations Conference on Trade and Development, New York, and Geneva.

USAID. (2017). *Climate Risk Profile Nepal*. https://www.climatelinks.org/resources/climate-risk-profile-nepal.

World Bank (2019). *Climate Smart Agriculture Profile Nepal*. World Bank. https://climateknowledgeportal.worldbank.org/sites/default/files/2019-06/CSA_Profile_Nepal.pdf.

Chapter 4

Climate Change
Causes, Effects, and Solutions for Agriculture and Food Security of Pakistan

Sajid Masood[1], *Niaz Ahmed*[1]* and *Muhammad Baqir Hussain*[2]

Introduction

Pakistan, positioned at the convergence of the Middle East and South Asia (30°N latitude and 70°E longitude), spans a total area of 796,095 km^2, comprising 770,875 km^2 of land and 25,220 km^2 of water. Its diverse geography includes the flat Indus plain in the east, imposing mountains in the north and northwest, and the Baluchistan Plateau in the west. With land boundaries with Afghanistan (2,430 km), China (523 km), India (2,912 km), and Iran (909 km), Pakistan enjoys both geographical and cultural connections. The 1,046 km coastline along the Arabian Sea facilitates essential maritime activities. Varied terrain contributes to diverse ecosystems and abundant natural resources. K2 (Mt. Godwin-Austen), the highest point, stands at 8,611 m elevation, while the lowest point lies at sea level along the Indian Ocean. Arable land resources, with approximately 77,181 square miles of irrigated land, support crucial agricultural activities.

Pakistan's climate is characterized as dry and extreme, with exceptionally hot summers and extremely cold winters, minimal rainfall, and four distinct seasons: Summer (May to mid-September), Winter (November to February), Autumn (September to November), and Spring (March and April). The country exhibits diverse climates across four zones:

North and North-Western Mountainous Area: Characterized by extremely harsh winters and pleasantly moderate summers.

Upper Indus Plain: Experiences scorching summers, minimal rainfall, and short but enjoyable winters.

Coastal Areas and the Lower Indus Valley: Stable temperatures due to sea-land breezes, minimal rainfall, and a pleasant climate with low temperature variations.

[1] Department of Soil Science, Bahauddin Zakariya University, Multan, Pakistan.
[2] Department of Soil and Environmental Sciences, MNS University of Agriculture, Multan, Pakistan.
* Corresponding author: niaz.ahmed@bzu.edu.pk

Plateau of Baluchistan and the Thar Desert: Elevated summer temperatures, dry and hot conditions, and severe winters with occasional snowfall.

The 6th national population and housing census marked a significant event, revealing a 57% increase in the total population to 207.8 million. Urban areas experienced a higher growth rate of 2.70%, posing challenges such as low man-land ratio and land degradation. Assessing changes in Land Use and Land Cover (LULC) through geospatial techniques is crucial for effective planning and sustainable resource management. Agriculture, the largest employer, faces challenges from population and urbanization pressures, requiring increased food production and adaptation to changing consumption patterns. Despite public spending and support, agricultural growth has slowed, emphasizing the need for a transition to more sustainable practices.

The Pakistani government actively oversees important crops, formulating policies to ensure a continuous supply of essential food items at reasonable prices. Emphasizing financial inclusion in the agriculture sector, the government aims to amplify productivity, exports, and economic growth driven by rural development. The World Bank supports Pakistan in reshaping its agri-food system through Advisory Services and Analytics (ASA), developing knowledge-based products to guide policy formulation, investments, and collaborative discussions. The topics covered range from research fostering agricultural productivity to addressing diverse food and nutrition needs of households.

Causes and Impacts of Climate Change on the Environment

Temperature

Climate change is primarily propelled by the rise in greenhouse gas concentrations in the Earth's atmosphere, such as carbon dioxide (CO_2), methane (CH_4), nitrous oxide (N_2O), and other gases (IPCC, 2021). These gases trap heat from the sun, creating a natural greenhouse effect that warms the planet. Human activities, notably the burning of fossil fuels for energy, industrial processes, and deforestation and agriculture, have significantly increased greenhouse gas concentrations, leading to global warming (IPCC, 2021). Consequently, climate change is causing an overall increase in Earth's temperatures. The warming planet results in more frequent and intense heatwaves, shifts in weather patterns, and disruptions to ecosystems. The scientific consensus, supported by extensive research and observations, underscores the role of human activities in driving temperature increases and emphasizes the need for mitigation efforts to address the impacts of climate change on ecosystem services (IPCC, 2021).

Heatwaves

In the 2015 Paris Agreement, discussions centered on the alarming global temperature rise of 1.5–2.0°C from pre-industrial levels. The 2021 IPCC report attributes this phenomenon to human activities, particularly the combustion of fossil fuels trapping heat and greenhouse gases, especially carbon dioxide, in the atmosphere. Projections for 2081–2100 outlined in the IPCC report indicate a potentially substantial temperature increase ranging from 3.3°C–5.7°C (Warner and Laczko, 2006). Pakistan, committed to the Paris Agreement, faces a staggering 300% increase in carbon emissions by

2030 (Faisal et al., 2019). The country grapples with frequent, intense, and prolonged droughts and heatwaves, exemplified by the devastating 2015 heatwave claiming over 2,000 lives. In 2022, several deadly heatwaves and prolonged droughts impacted agriculture, leading to crop failures, livestock losses, food and freshwater scarcity, and economic challenges (Prescia, 2021). Addressing and mitigating these climate-induced challenges is crucial for both the environment and millions of people.

Glaciers Melting

Climate change poses a formidable threat to the world's glaciers, intensifying the process of glacial melting through interconnected mechanisms. As global temperatures rise due to greenhouse gas accumulation, the elevated atmospheric warmth directly impacts glacial regions (IPCC, 2021). Warmer temperatures accelerate glacier melting, particularly at lower elevations and during extended warmth periods (Hock et al., 2019). The albedo effect exacerbates this phenomenon, as reduced reflective ice exposes darker surfaces, absorbing more solar radiation and amplifying the melting process (Pelto et al., 2020). Changes in precipitation patterns, a hallmark of climate change, also contribute to more rapid glacial melting. The warming atmosphere and oceans create feedback loops, perpetuating the cycle of melting, releasing freshwater into oceans, influencing currents, and impacting regional climate systems (Zemp et al., 2019). In Pakistan's Karakoram mountains, the Siachen Glaciers, one of the world's largest, are rapidly melting due to global warming, posing a significant threat to water resources and increasing flood risks (Khan et al., 2015).

Extreme Floods

Pakistan has faced devastating flash floods due to the abrupt release of water from melting glacial lakes. Severe instances occurred in 2010 and 2011, causing extensive damage, displacement, and loss of life. The 2010 floods destroyed or damaged approximately 1.1 million homes, marking one of Pakistan's most significant humanitarian crises. In 2020, monsoon rains led to urban flooding, causing significant damage in Sindh alone, impacting one-third of Pakistan and claiming lives, displacing millions, and damaging infrastructure (Hussain et al., 2023). Urgent reconstruction efforts are crucial for reviving agriculture, ensuring food security, and addressing critical infrastructure needs, requiring a substantial financial commitment of PKR 854 billion (Ministry of Planning Development and Special Initiative, 2022).

Water Scarcity and Crop Production

Climate change influences precipitation patterns, leading to more intense and erratic rainfall, impacting water availability for crops worldwide (IPCC, 2021). Rising temperatures contribute to increased evaporation, loss of soil moisture, and reduced water bodies, diminishing the water supply for crops (IPCC, 2021). Changes in hydrological cycles and more frequent droughts deplete soil moisture, reduce groundwater levels, and limit water access for sustained crop growth (Dai, 2011). Rising sea levels contribute to saltwater intrusion into coastal aquifers, rendering freshwater unsuitable for agriculture (Rahmstorf et al., 2012). Extreme weather events, like storms and hurricanes, compound challenges by causing flooding, soil erosion, and disrupting water retention, exacerbating water scarcity for agriculture (IPCC, 2021). Addressing these challenges requires sustainable water management practices, enhanced irrigation efficiency, drought-resistant crop varieties, and global efforts to mitigate climate change (Barnett et al., 2005; Dai, 2011; Rahmstorf et al., 2012; IPCC, 2021).

Storms and Cyclones

As global surface temperatures rise, cyclones and intense storms increase. Higher temperatures contribute to higher wind speeds in tropical storms. Rising sea levels bring erosive forces to higher locations, exposing them to impacts not typically seen. Cyclones, fueled by warmer oceans, intensify, posing increased threats. Human-induced climate change plays a significant role in influencing cyclones through mechanisms like warming the air and oceans and inducing a rise in sea levels. Climate change establishes conditions conducive to the formation, intensification, and persistence of more potent storms reaching land, carrying increased moisture and causing destructive floods and mudslides (World Weather Attribution).

Land Degradation

Anthropogenic climate change significantly contributes to land degradation, defined as the diminishing capacity of land to yield benefits under specific land use and management approaches (Barrow, 1992). This phenomenon compromises soil biological, chemical, and physical attributes, threatening agricultural production (Lal, 2010; Blum and Nortcliff, 2013). Given the pivotal role of soils in ensuring food security, addressing climate change's impact on soil health is crucial for global agricultural sustainability (Douglas, 1994).

Table 1 The effect of time scale effect of climate change on soil parameters, properties, and characteristics. Climate change impact will be become serious on soil health with the passage of time. Adapted and modified from Karmakar et al. (2016).

Time scale and effect on Soil Parameters

Temperature, moisture contents, bulk density, total porosity, infiltration rate, permeability, composition of soil air and nitrate contents	Total water capacity, field capacity, hydraulic conductivity, microbiota, pH, nutrients status and composition of soil solution.	Wilting percentage, soil acidity, cation exchange capacity and exchangeable cations.	Specific surface, clay mineral association and organic matter contents	Primary mineral composition and chemical composition of mineral part	Texture, particle-size distribution and particle density
10^{-0}	$<10^{-1}-10^{0}$	$10^{0}-10^{1}$	$10^{1}-10^{2}$	$10^{2}-10^{3}$	$>10^{3}$
Compaction, drainage and workability			Tree roots soil biota, sialic, calcareous, sodic and vertic properties	Trees root and Colour (yellowish/reddish), iron concentrations, soil depth cracking, soft powdered lime and indurated sub soil.	Parent material, depth and abrupt textural change

Time scale and effect on soil properties and characteristics

Soil Salinity

Soil salinity refers to the accumulation of salts in the soil, a process driven by factors such as the deposition of minerals with salt content, capillary rise of salt-laden groundwater, unsustainable agricultural practices, and industrial activities. Measured in deci-siemens per meter (dS/m), soil salinity indicates a high concentration of soluble salts and low exchangeable sodium ions [$EC > 4$ dS/m; $ESP < 15\%$; $pH < 8.5$; $SAR < 13$]. This salinity level significantly impacts soil quality, vegetation growth, and crop yields (Wang et al., 2018). Among the numerous factors contributing to soil salinity, climate change exacerbates the situation. Drivers include sea-level rise and intrusion (Hu and Lindo-Atichati, 2019), increasing surface temperatures, irrigation and drainage practices (Singh et al., 2018), altered rainfall patterns leading to reduced leaching and salt accumulation in the plant root zone, and the influx of highly saline underground water from aquifers, whether confined or unconfined affected by seawater inundation through cyclones or intentional flooding for aquaculture, which further contributes to salt buildup (Xie et al., 2020).

Globally, salt-affected soils cover an extensive area of approximately 1,060.1 million hectares, found in North and Central Asia (211.7 M ha), Africa (209.6 M ha), and South America (129.3 M ha) (FAO-ITPS-GSP, 2015). According to the FAO report on the Status of the World's Soil Resources in 2015, the distribution of salt-affected soils, categorized as saline and sodic soils, varies by region. North and Central Asia, with the highest area of salt-affected soils, has a greater presence of sodic soils (120.2 M ha) than saline soils (91.5 M ha). In South Asia, the area of saline soils (82.3 M ha) exceeds that of sodic soils (1.8 M ha). Saline soils are most prevalent in regions such as Sinkiang, Qinghai, Shaanxi, and Inner Mongolia, covering 2% of China's total land area (Wan et al., 2019). Africa, as a continent, holds almost 63% of the world's total saline soils (Kahime et al., 2018). Approximately 20% of the world's irrigated lands across 120 countries are affected by salinization, and this issue is on the rise due to climate change and unsustainable irrigation management (Meena et al., 2019). Irrigation-induced soil salinization emerges as a leading cause of land degradation in arid and semi-arid regions.

Erosion

Wind and water erosion emerge as the most prevalent forms of land degradation. In arid and semi-arid regions, the risk of wind erosion is particularly severe, occurring when soil is left exposed due to cultivation and overgrazing from excessive livestock. Wind action results in the displacement and deposition of soil particles from the topsoil surface, posing a threat to structures, machinery, and land as undesirable soil accumulates (Andrien, 1994). Water erosion encompasses various processes such as splash erosion, rill and gully erosion, and sheet erosion (Douglas, 1994). Splash erosion initiates water erosion as raindrops impact the bare soil, breaking up surface soil aggregates and propelling particles into the air. The soil particles filling the gaps between surface aggregates reduce water infiltration, leading to increased runoff. Water flowing above the soil surface can pick up soil particles released by splash erosion and detach particles from the soil surface, resulting in the consistent removal of soil particles, known as sheet erosion. Rill and gully erosion occur when runoff becomes concentrated into channels. Rills are small streamlets or head cuts that can be controlled through tilling. Gullies, on the other hand, erode substantial amounts of soil during heavy rainfall, being larger versions of rills. Gullies are not easily mitigated through regular tilling practices.

Drought

Climate change and drought exert both direct and indirect effects on land degradation in both forested and agricultural settings. These effects include potential rises in heat stress on vegetation, alterations in soil moisture availability, heightened soil erosion from wind and rain, nutrient loss in the soil, soil salinization, and an overall decline in vegetation and biomass (Li and Fang, 2016). Such impacts can intensify existing instances of land degradation and trigger it in new regions. In turn, land degradation can contribute to climate change by diminishing carbon sequestration in the affected areas' soils and vegetation.

Droughts and extreme heat events wreak havoc on grazing lands and diminish yields of feed crops, leading to increased livestock mortality rates and reduced income from livestock sales. This, in turn, results in decreased household incomes and heightened food insecurity. In food systems primarily focused on crop farming, the combination of land degradation and gradual environmental changes depletes soil moisture and/or nutrients, leading to diminished crop yields. The specific impacts and outcomes hinge on the resilience and adaptive capacity of the particular farming system. Crucial determinants include the nature of the farming economy (subsistence versus market-oriented, rain-fed versus irrigated, livestock versus field crops), institutional engagement, and the available adaptation options for households. These options may encompass technological solutions (such as irrigation), managerial adjustments (like altering crop choices), or economic diversification (for example, seeking off-farm employment) (Webb et al., 2017). In the realm of household-level adaptation to drought and land degradation, shifts in migration patterns may emerge. The potential or inclination for migration is influenced by the range of adaptation alternatives available to households, shaped by various socio-economic factors operating at multiple scales (Mapedza et al., 2019).

Organic Matter Depletion, and Soil Health

Organic matter stands out as a crucial indicator of soil health, contributing to soil biology, chemical buffering, structural stability, water and oxygen retention, and nutrient storage. It serves as a habitat for a diverse range of soil microorganisms. The sensitivity of organic matter to climate changes is evident in its decomposition rate, which accelerates with rising temperatures. The expansion and intensification of agriculture have led to a decline in soil organic matter (SOM) levels. A significant portion of SOM is released as carbon dioxide into the atmosphere, underscoring the substantial contribution of intensively cultivated lands to greenhouse gas emissions (Tate, 1992). The increased atmospheric concentration of CO_2 has resulted in enhanced plant growth, indicating a heightened utilization of SOM. Consequently, the SOM content is profoundly affected by climate change.

Pests and Diseases

Global climate changes exert significant impacts on agriculture, extending their influence on insect pests and diseases. Both agricultural crops and their corresponding pests experience direct and indirect effects as a result of climate change. Direct impacts manifest in the reproduction, development, survival, and dispersal of pests, while indirect effects alter the relationships between pests, their environment, and other insect species such as natural enemies, competitors, vectors, and mutualists (Prakash et al., 2014). Insects, being poikilothermic organisms, have their body temperature dependent on the environmental temperature. Consequently, temperature emerges as the most crucial environmental factor influencing insect behavior, distribution, development, and reproduction (Kocmankova et al., 2010). Therefore, it is highly likely that the primary drivers of climate change, including increased atmospheric CO_2, elevated temperature, and decreased soil moisture, could significantly impact the population dynamics of insect pests and, consequently, the extent of crop losses (Fand et al., 2012).

Climate change creates novel ecological niches, providing opportunities for insect pests to establish and spread in new geographic regions while shifting from one area to another (FAO, 2020). The intricate physiological effects resulting from rising temperatures and increased CO_2 levels can profoundly influence interactions between agricultural crops and insect pests (Caulfield and Bunce, 1994). Consequently, farmers should anticipate encountering new and intensified pest challenges in the coming years due to the evolving climate. The spread of crop pests across physical and political boundaries poses a threat to food security, constituting it a global issue by affecting all countries and regions (FAO, 2020).

Fisheries and Aquaculture

In the pursuit of sustainability in aquatic production, aquaculture has emerged as an important practice with increased economic viability. Given the ongoing unsustainable harvests from capture fisheries, the aquaculture sector is recognized as a pivotal solution to meet the escalating global demand for aquatic products (Askarysary et al., 2012; FAO, 2020). According to FAO (2020), the contribution of aquaculture to global fish production has consistently increased, reaching 82.1 million tons (46%) out of the estimated 179 million tons of global production. Moreover, projections indicate that the share of aquaculture production in global fish production is anticipated to grow from the current 46% to 53% by 2030 (FAO, 2020).

In the realm of aquaculture, a substantial body of recent literature highlights observable shifts in climate, such as rising temperatures, altered precipitation patterns, and an increased frequency of certain extreme events, impacting water resources. While some changes are already evident, others are still unfolding (Fleming et al., 2014). The effects of climate change on aquaculture sustainability have garnered significant attention lately, given the sector's crucial role in global food security, nutrition, and livelihoods (Blanchard et al., 2017; FAO, 2020). Despite some uncertainties, various projections indicate that the entire aquaculture value chain is susceptible to the impacts of climate change (Bueno and Soto, 2017; Barange et al., 2018).

Livestock Production

Climate change poses a significant challenge to current livestock systems on a global scale. The rising temperatures and associated shifts in mean climate variables and climate variability have repercussions on feed and water resources, as well as on animal health and production. Additionally, climate change impacts various stages of the livestock product supply chain, including processing, storage, transport, retailing, and consumption. The ability of existing livestock systems to sustain livelihoods and meet the growing demand for livestock products is consequently under threat.

The livestock sector holds a pivotal role in global food supply and food security. Livestock products, encompassing meat, milk, and eggs, contribute to 15% and 31% of the global per capita calorie and protein supply, with regional variations. Approximately 30% and 6% of global ruminant meat and milk production originate from grazing systems, often situated on land unsuitable for cropping (Herrero et al., 2013). Beyond food production, livestock serves various other functions, such as providing draught power, transportation, nutrients for poor soils, income generation and diversification, and acting as a form of financial capital. These diverse roles contribute to the overall well-being and resilience of many communities (CIRAD, 2016). More than 844 million people worldwide derive income from agriculture, with the livestock sector contributing about 40% of agricultural value-added (The World Bank, 2020). Although the full extent and magnitude of the impacts remain uncertain, climate change will inevitably affect livestock contributions to food security and other dimensions of sustainability.

Climate Change, Food Security, and Population

Climate Change and Food Security

"Food security is attained when individuals, at all times, have the resources to acquire an adequate, safe, and nutritious food supply that aligns with their dietary needs and preferences, fostering an active and healthy life" (World Food Summit, 1996). This definition encapsulates four fundamental dimensions of food security: the availability of food, economic and physical access, efficient utilization, and the sustainability of these three dimensions over time (Fig. 1).

Despite substantial progress in reducing hunger in recent decades, 800 million people still grapple with chronic undernourishment, as reported in 2015. Approximately 161 million children under the age of five are stunted, and 500 million individuals contend with obesity. Moreover, two billion people lack essential micronutrients for a healthy life. The Food and Agriculture Organization (FAO) predicts a 60% increase in food production by 2050 to meet growing demand. However, global food production alone is inadequate to eradicate hunger, as almost 800 million people still experience food insecurity. The key lies in ensuring universal access to a sufficient quantity and quality of food at all times.

According to the United Nations, 836 million people were living in extreme poverty (earning less than US $1.25 per day) in 2015. The International Fund for Agricultural Development (IFAD) notes that at least 70% of the extremely poor reside

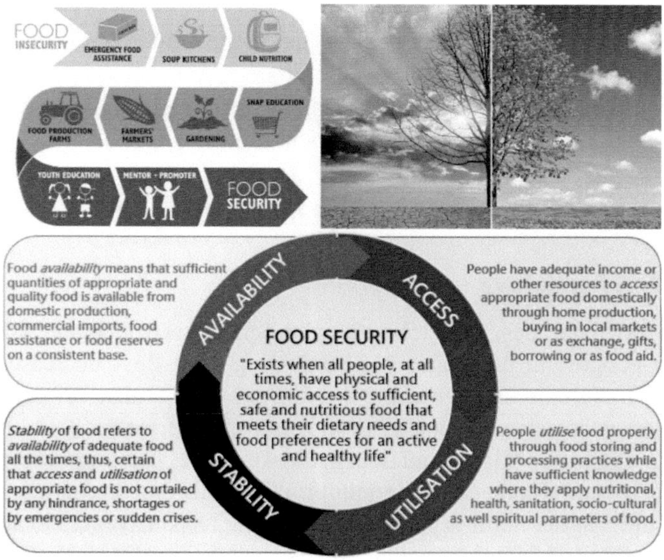

Fig. 1 Key factors contribute to food security and insecurity of the world.

in rural areas, relying on agriculture for their livelihoods, either partially or entirely. Around 80% of the food consumed in Asia and sub-Saharan Africa is produced by an estimated 500 million smallholder farms, supporting almost 2 billion people in the developing world. The latest report from the IPCC emphasizes that climate change exacerbates risks to food security for the most vulnerable countries and populations. IPCC AR5 identifies four out of eight key risks induced by climate change with direct consequences for food security: (i) Loss of rural livelihoods and income, (ii) Loss of marine and coastal ecosystems and livelihoods, (iii) Loss of terrestrial and inland water ecosystems and livelihoods, (iv) Food insecurity and breakdown of food systems.

Pakistan emerges as a significant emerging economy in South Asia, grappling with the dual challenges of poverty and hunger. With a substantial reliance on the agriculture sector, contributing to 21% of the GDP (Zhou et al., 2019), and employing 43.5% of the total workforce, the country's per capita income remains low at $1512. Despite a 2% growth rate in the agriculture sector, Pakistan envisions achieving food security by ensuring both economic and physical access to nutritious food for its population by 2025 (Pakistan Economic Survey, 2015).

The agricultural sector serves as the backbone for a majority of the population, and the country has committed to addressing food security in alignment with the United Nations Sustainable Development Goals (SDGs), aiming for zero hunger by 2030. Despite these efforts, challenges persist, exacerbated by rising average temperatures and climate-related events. Regrettably, Pakistan ranks as the 11th most food insecure country globally, with a notable increase in food insecure districts from 38 in 2003 to 102 in recent years.

This alarming situation is primarily attributed to the adverse impacts of climate change, marking it as a pivotal factor in Pakistan's food security. The five key dimensions of food security—production, access, distribution, utilization, and cleanliness—are significantly influenced by the changing climate (FAO, 2020). Recognizing this, Pakistan must continue to address these challenges through comprehensive policies and actions to secure a sustainable and resilient future.

Climate Change Mitigation and Adaptation Strategies

Climate-smart Agriculture

Over the past two decades, significant global progress has been made, lifting around 200 million people out of hunger and witnessing a 40–26% decrease in chronic malnutrition in children. However, the World Bank reports that 702 million people still endure an extremely poor quality of life, and approximately 793 million people suffer from malnutrition (Hertel and Lobell, 2014). Addressing food security has been a global priority, and climate-smart agriculture (CSA) has emerged as a solution. CSA aims to integrate agricultural expansion and climate responsiveness to achieve food security and growth goals amid changing weather patterns and increasing food demand. CSA initiatives focus on sustainable production, resilience-building, greenhouse gas reduction, and managing trade-offs and synergies between productivity, adaptation, and mitigation (Kench et al., 2018).

In the context of Pakistan, where agriculture holds social, economic, and cultural significance, CSA presents promising opportunities. Climate change poses a looming threat to millions of Pakistani families that are dependent on agriculture, risking economic setbacks. Existing CSA measures across the nation can be enhanced by leveraging current financing, exploring additional funding sources, promoting farmer-friendly policies, and empowering institutions to take action. Innovations such as laser land leveling and solar-powered irrigation systems, along with management adjustments like crop diversification, suitable cropping models, and improved planting dates, can contribute to a more productive agricultural system (Awan and Yaseen, 2017).

Pakistan's 10 Agro-ecological Zones (AEZs) are crucial for understanding the diverse factors affecting agriculture, including soil composition, climate, land, and physiographic elements. Government initiatives should focus on investing in research to develop heat-resistant, drought-tolerant, and pest-resistant crop and livestock varieties. Establishing data-driven models to simulate the potential effects of climate change on agricultural production across AEZs is vital for formulating accurate adaptation and mitigation policies (Khan et al., 2021).

To mitigate the impacts of climate change, a comprehensive plan incorporating information technology, GIS techniques, and remote sensing for precise farming methods is essential. This approach enhances input use efficiency and provides valuable data on soil, crop, and climate-related issues to determine optimal cropping patterns for diverse AEZs. Ensuring widespread adoption of CSA requires not only technological advancements but also financial support from national and international sources, especially for smallholders. In Pakistan, fostering CSA adoption involves crafting climate-resilient policies, integrating climate change threats into national planning agendas, and educating farmers about the latest technologies and practices through farmer programs (Awan and Yaseen, 2017).

Cropping Systems

Climate change is anticipated to bring warmer conditions (Battisti and Naylor, 2009), marked by an increased frequency of events such as anomalous warm seasons (Weisheimer and Palmer, 2005), dry spells, and heightened intensity of precipitations (Seneviratne et al., 2012). While elevated atmospheric CO_2 concentration can increase photosynthesis but the positive effects saturate at higher concentrations (Long et al., 2006), and the negative impacts of warmer conditions and irregular events often outweigh this benefit. While some positive effects of climate change may be observed in temperate and boreal regions, where longer growing seasons could potentially benefit agriculture (Altieri et al., 2015). Overall, it is predicted that climate change may have adverse effects on crop yields and yield stability globally, particularly in tropical regions (Challinor et al., 2014). It would be worth noting that yield instability can be considered a desirable quality if driven by positive deviations from the average yield. Additionally, climate change is expected to increase the occurrence of pathogens, pests, and weeds (Altieri et al., 2015), introducing a biotic dimension to the disturbances induced by climate.

Efficient carbon sequestration in agriculture relies heavily on the management of cropping systems, including strategies such as crop rotation, intercropping, cover cropping, and more. These systems play a crucial role in determining optimal yields, the total increase in carbon sequestered through biomass, and the amount retained in the soil. Soil carbon sequestration serves as a multi-purpose strategy. It not only restores degraded soils but also enhances land productivity, promotes biodiversity, safeguards the environment, and curbs the CO_2 enrichment. This, in turn, contributes to a shift in greenhouse gas emissions and aids in mitigating the impacts of climate change.

Farmers lack control over climatic conditions, but they do possess the ability to avoid, adapt to, or even capitalize on local conditions to a certain extent. The goal is to minimize yield losses under unfavorable conditions and maintain or increase maximum yields under favorable conditions. However, climate change may lead to cropping systems producing lower and more unstable yields (Ray et al., 2015). Hence, it is essential to address climate change through both mitigation and adaptation strategies.

Mitigation involves directly reducing the environmental impact of cropping systems, such as curbing climate change-inducing emissions or enhancing sinks of greenhouse gases through sustainable practices like using organic fertilizers and natural pest control. On the other hand, adaptation focuses on lessening the negative impact of current or anticipated climatic conditions on production and exploiting the positive aspects. This is achieved by adopting specific management practices, for instance, adjusting sowing dates and selecting crops that can tolerate specific climatic conditions. While most research concentrates on the climate mitigation aspect of management practices, there is limited understanding of their effectiveness from a climate adaptation perspective.

Irrigation Management

Irrigated agriculture, though occupying only 20% of the total cultivated land, is characterized by more intensive cropping management practices. On average, irrigated agriculture employs large quantities of chemical fertilizers and other reactive agrochemicals compared to most rain-fed systems. The irrigated lands contribute to carbon storage in soils and increased yields through practices like using crop residues as soil cover. However, irrigation also has implications for energy consumption, particularly in the pumping and treatment of water.

With water scarcity and irrigation development, there is an expanding reliance on groundwater resources, both in absolute terms and relative to other sources (Siebert et al., 2010). Water scarcity prompts the agricultural sector to explore non-traditional means like desalination and wastewater reuse, all of which have high energy requirements, often met through the burning of fossil fuels, thereby emitting CO_2. Groundwater, covering 38% of all irrigated lands, contributes to energy consumption and CO_2 emissions, as seen in China and India (Shah, 2009).

Modern irrigation technologies, including drip irrigation, further escalate energy demands. In Spain, irrigation modernization led to a 21% reduction in water consumption between 1950 and 2008, but energy demand surged by 657%. The use of solar energy for irrigation emerges as a potential option to reduce GHG emissions associated with irrigation. Solar-based irrigation systems, offering a reliable and low-cost clean-energy solution, gain importance in areas with high solar radiation. In regions lacking reliable access to electricity or where diesel fuel is expensive, solar-based irrigation initiatives can provide broader access to energy for agricultural purposes (FAO and GIZ, 2015). Several countries are integrating solar-based irrigation systems into their national action plans on climate change to reduce agricultural GHG emissions. However, the cost-effectiveness of solar-based irrigation systems raises concerns about potential over-exploitation of groundwater. Therefore, the implementation of appropriate policies and regulations becomes crucial to control water use.

In conclusion, the options for directly mitigating climate change through irrigation align with those for agriculture as a whole. The mitigation potential is higher in areas with intensive groundwater irrigation, determined primarily by increased irrigation intensity, allowing for greater soil organic C (SOC) sequestration potential in tropical conditions and increased crop productivity. However, the benefits of climate change mitigation may be counteracted by more intensive use of inputs (Turral et al., 2011).

Land Reclamation and Soil Productivity

Soil productivity refers to the ability of the soil to yield a specific crop or plant under a defined set of agricultural management practices. It encompasses both soil fertility and various inherent and management-related factors influencing plant growth and development, typically gauged in terms of inputs such as water and/or nutrient input versus crop yield. The pivotal soil functions influencing crop productivity include providing physical support, serving as a rooting medium with plant-available water, facilitating air exchange for respiration, and supplying essential nutrients. Human activities can exert substantial impacts on soil productivity by influencing dynamic soil properties. Agricultural management decisions, including choices related to tillage, fertilization, crop rotation, irrigation, and drainage, play important roles in shaping soil productivity to address food security.

Safeguarding soil quality and productivity while utilizing chemical fertilizers necessitates a collective global effort, given their potential to impact soil characteristics (Abinandan et al., 2019). Excessive chemical fertilization, particularly nitrogen (N), often leads to soil acidification and degradation, resulting in diminished soil fertility and crop yield (Shah, 2009). The combination of organic and chemical fertilization increases SOC and nitrogen (TN) levels, influencing the composition of bacterial communities. Soil microbes, by participating in organic matter breakdown, nutrient cycling, and bioremediation, play a crucial role in maintaining soil functional properties.

Organic amendments and chemical fertilization contribute to soil nutrient enrichment to support crop productivity. For instance, soil when amended with biochar (@ 30 ton/ha) has shown an increase in SOM (0.30%) after 5 months in acid soils compared to calcareous soils, which experienced a more significant loss of carbon due to mineralization. Biochar increased ammonium, phosphorus, and potassium availability in both soils but did not affect the total nitrogen content. Recognized for its higher cation-exchange capacity and surface area, biochar is known for its ability to retain macronutrients like phosphorus and nitrogen in the soil (Lehmann, 2007). Plant growth-promoting bacteria, as influenced by organic amendments, ecologically contribute to enhanced phosphate solubilization, improved moisture retention, and effective water repellency, leading to improved plant growth. In the rhizosphere, biochar significantly enhanced phosphorus content, with the impact of inorganic phosphate-solubilizing bacteria (iPSB) inoculation proving more significant. This suggests that the iPSB community inoculation had a greater influence on soil phosphorus release than that of biochar (Zhou et al., 2019).

Reforestation and Afforestation

Afforestation and reforestation stand as effective strategies for mitigating the impacts of climate change by fixing CO_2 via photosynthesis. Trees play a crucial role by capturing CO_2 from the atmosphere during their growth, storing carbon in living biomass, dead organic matter, and soils, making forestation a negative emissions method. Afforestation involves establishing new forests, while reforestation focuses on re-establishing previous forest areas that have undergone deforestation or degradation.

Carbon dioxide uptake during forest growth varies from 20–100 years, depending on the tree species, until the trees reach maturity. Once mature, C sequestration rates slow down significantly, and at this stage, forest products can be harvested and utilized. However, it is essential to carefully plan forest management practices and activities to ensure the maximization of co-benefits associated with afforestation and reforestation (Royal Society, 2018). These co-benefits encompass improved biodiversity, flood control, and enhanced soil, water, and air quality (Harper et al., 2018).

Carbon storage in forests proves to be a highly effective method for reducing GHG emissions, although permanent storage faces vulnerabilities from both natural and human disturbances. Forests are susceptible to risks such as natural disasters like fire, droughts, and diseases, as well as human-induced deforestation activities. These factors can compromise the integrity of C storage in forests. Unlike storage in geological formations, biogenic storage has a shorter lifespan, underscoring the importance of protecting forests and managing them sustainably (Fuss et al., 2018).

Additionally, forestation projects demand substantial land resources, potentially competing with other land uses (Royal Society, 2018). There are two major types of reforestation, i.e., urban and rural:

"Urban reforestation" focuses on thoughtfully reintroducing trees into cities, strategically creating shade, wildlife habitats, rainfall catchments, and green spaces. This approach aims to enhance the quality of life for both residents and animals in urban environments.

"Rural reforestation" aims to establish C sinks in areas previously subjected to clear-cutting, selected felling, or possessing the space and favorable climate for rapid tree growth. Responsible rural reforestation involves careful consideration of appropriate tree species, local water sources, and the potential impact on nearby communities. When executed appropriately, this type of reforestation holds significant potential to directly influence global CO_2 levels.

Reforestation has several benefits to ecosystems and communities

"Trees are nature's carbon processors": Human reliance on burning fossil fuels has resulted in an excess of the greenhouse gas, especially CO_2. Through photosynthesis, trees engage in C sequestration, naturally absorbing and processing CO_2.

"Trees serve as vital wildlife habitats": A wide array of wildlife, from insects to larger mammals, depends on trees for sustenance and shelter. The loss of habitat poses a significant threat to various species facing extinction.

"Trees act as natural barriers against erosion and flooding": The roots of trees play a crucial role in anchoring soil, preventing erosion, and absorbing runoff. Additionally, the canopies of leaves contribute to better rainfall distribution and increased evaporation. Deforestation heightens the vulnerability of both urban and rural areas to flooding, particularly as extreme climate events become more frequent.

"Tree planting initiatives generate employment opportunities": Thoughtful reforestation projects can provide a substantial economic boost to local communities by creating jobs for individuals involved in planting and maintaining these burgeoning forests.

Conclusions

Climate change is an obvious factor influencing agriculture and food security worldwide, including Pakistan. Recent events such as drought, erratic rainfall, floods, heatwaves, chilling, disease, and insect outbreaks in Pakistan during the last three years have drastically damaged the economy and led to food insecurity in 20% of Pakistan's population. However, these issues can be addressed through establishing policies and adopting strategies for reducing and ameliorating the climate change impacts on agriculture and food security in Pakistan. Forestation, efficient irrigation systems, climate-smart agriculture practices, resource-efficient crop varieties, livestock, fisheries, renewable energy sources, and crop and food diversification are strategies for mitigating the impacts of climate change on agriculture and food security globally, especially in developing countries like Pakistan.

References

Abinandan, S., Subashchandrabose, S.R. and Venkateswarlu, K. (2019). Soil microalgae and cyanobacteria: The biotechnological potential in the maintenance of soil fertility and health. *Critical Reviews in Biotechnology*, *39*: 981–98.

Altieri, M.A., Nicholls, C.I., Henao, A. and Lana, M.A. (2015). Agroecology and the design of climate change-resilient farming systems. *Agronomy for Sustainable Development*, *35*: 869–90.

Andrien, M., 1994. Social communication in nutrition: A methodology for intervention. M-67, ISBN 92-5-103367-6. FAO, Rome. *12*: 24–36.

Askarysary, A., Velayatzadeh, M. and Karimisary, V. (2012). Proximate composition of farmed fish *Oncorhynchus mykiss and Cyprinus carpio* from Iran. *Adv. Environ Biol.*, *6*: 2841–45.

Awan, A.G. and Yaseen, G. (2017). Global climate change and its impact on agriculture sector in Pakistan. *American Journal of Trade and Policy*, *4*: 41–48.

Barange, M., Bahri, T., Beveridge, M.C.M., Cochrane, A.L., Funge-Smith, S. and Paulain, F. 2018. *Impacts of Climate Change on Fisheries and Aquaculture, Synthesis of Current knowledge, Adaptation and Mitigation Options*. Rome: FAO.

Barnett, T.P., Adam, J.C. and Lettenmaier, D.P. (2005). Potential impacts of a warming climate on water availability in snow-dominated regions. *Nature*, *438*: 303–309.

Barrow, C.J. (1992). *World Atlas of Desertification* (UNEP). London.

Battisti, D.S. and Naylor, R.L. (2009). Historical warnings of future food insecurity with unprecedented seasonal heat. *Science*, *323*: 240–44.

Blanchard, J.L., Watson, R.A., Fulton, E.A., Cottrell, R.S., Nash, K.L. and Bryndum-Buchholz, A. (2017). Linked sustainability challenges and trade-offs among fisheries, aquaculture and agriculture. *Nature Ecology and Evolution*, *1*: 1240–49.

Blum, W.E.H. and Nortcliff, S. (2013). Soils and Food Security. *In*: Brevick, E.C. and Burgess, L.C. (Eds.), *Soils and Human Health*. CRC Press, Boca Raton.*12*: 299–321.

Bueno, P.B. and Soto, D. 2017. *Adaptation Strategies of the Aquaculture Sector to the Impacts of Climate Change*. Rome: FAO.

Caulfield, F. and Bunce, J.A.(1994). Elevated atmospheric carbon dioxide concentration affects interactions between *Spodoptera exigua* (Lepidoptera: Noctuidae) larvae and two host plant species outdoors. *Environmental Entomology*, *23*: 999–1005.

Challinor, A.J., Watson, J., Lobell, D.B., Howden, S.M., Smith, D.R. and Chhetri, N. (2014). A meta-analysis of crop yield under climate change and adaptation. *Nat. Clim. Change*, *4*: 287–91.

CIRAD. (2016). Livestock Farming and Local Development. *In*: *Agri-Chains and Sustainable Development*. Montpellier, 12–16 December.

Dai, A. 2011. Drought under global warming: A review. *Wiley Interdisciplinary Reviews: Climate Change*, *2*: 45–65.

Douglas, A.E. (1994). Symbiotic interactions. Oxford University Press, London. *Earth Science Reviews*, *163*: 94–17.

Faisal, N., Khattak, M.J.K., Sathar, Z., Khalil, M. and Khan, K. (2019). Climate, population, and vulnerability in Pakistan: Exploring evidence of linkages for adaptation. Islamabad Population Council, *22*: 112–16.

Fand, B.B., Kamble, A.L. and Kumar, M. (2012). Will climate change pose serious threat to crop pest management: A critical review? *International Journal of Scientific Research*, *2*: 1–14.

FAO. (2020). *The State of World Fisheries and Aquaculture. Sustainability in Action*. Rome: FAO.

FAO. (2020). *Climate Related Transboundary Pests and Diseases*.

FAO and German Agency for International Cooperation (GIZ). (2015). International Workshop: *Prospects for Solar-powered Irrigation Systems (SPIS) in Developing Countries*. Final Report. Rome.

FAO-ITPS-GSP. (2015). *Status of the World's Soil Resources*. FAO-ITPS-GSP Main Report, FAO, Rome, Italy. *12*: 125–27.

Fleming, A., Hobday, A.J., Farmery, A., Van Putten, E.I., Pecl, G.T. and Green, B.S. (2014). Climate change risks and adaptation options across Australian seafood supply chains: A preliminary assessment. *Climate Risk Management*, *1*: 39–50.

Fuss, S., Lamb, W.F., Callaghan, M.W., Hilaire, J. and Creutzig, F. (2018). Negative emissions—part 2: costs, potentials and side effects. *Environmental Research Letters*, *13*: 063002

Gilani, H., Shrestha, H.L., Murthy, M.S.R., Phuntso, P., Pradhan, S., Bajracharya, B. and Shrestha, B. (2015). Decadal land cover change dynamics in Bhutan. *J. Environ. Manag.*, *148*: 91–100.

Herrero, M., Havlik, P., Valin, H., Notenbaert, A.M., Rufino, M.C., Thornton, P.K., Blummel, M., Weiss, F., Grace, D., Obersteiner, M., Havlikb, P., Valin, H., Notenbaert, A.M., Rufino, M.C., Thornton, P.K., Blummel, M., Weiss, F., Grace, D. and Obersteiner, M. (2013). Biomass use, production, feed efficiencies, and greenhouse gas emissions from global livestock systems. *Proceedings of the National Academy of Sciences U.S.A.*, *110*: 20888.

Hertel, T.W. and Lobell, D.B. 2014. Agricultural adaptation to climate change in rich and poor countries: Current modeling practice and potential for empirical contributions. *Energy Economics*, *46*: 562–75.

Hock, R., Bliss, A., Marzeion, B., Giesen, R.H., Hirabayashi, Y., Huss, M. and Zemp, M. (2019). "Glacier MIP – A model inter-comparison of global-scale glacier mass-balance models and projections." *Journal of Glaciology*, *65*: 453–67.

Hu, Y. and Lindo-Atichati, D. (2019). Experimental equations of seawater salinity and desalination capacity to assess seawater irrigation. *Science of the Total Environment*, *651*: 807–12.

Hussain, M.A., Shuai, Z., Moawwez, M.A., Umar, T., Iqbal, M.R., Kamran, M. and Muneer, M. (2023). A review of spatial variations of multiple natural hazards and risk management strategies in Pakistan. *Water*, *15*: 2–32.

IPCC (Intergovernmenal Panel on Climate Change). (2021). *Climate Change 2021: The Physical Science Basis*." Contribution of Working Group I to the Sixth Assessment Report. Cambridge University Press.

Iqbal, M.F. and Khan, I.A. (2014). Spatiotemporal land use land cover change analysis and erosion risk mapping of Azad Jammu and Kashmir, Pakistan. *Egypt J. Rem Sens Spa Sci.*, *17*: 209–29.

Kahime, K., Ben Salem, A., El hidan, M.A., Messouli, M. and Chakhchar, A. (2018). Vulnerability and adaptation strategies to climate change on water resources and agriculture in Morocco: Focus on Marrakech-tansifet-al haouz. *International Journal of Agricultural and Environmental Research*, *4*: 58–77.

Karmakar, R., Das, I., Dutta, D. and Rakshit, A. (2016) Potential Effects of Climate Change on Soil Properties: A Review. Sci. Int., *4*: 51–73.

Kench, P.S., Ryan, E.J., Owen, S., Bell, R., Lawrence, J., Glavovic, B., Blackett, P., Becker, J., Schneider, P. and Allis, M. (2018). Co-creating resilience solutions to coastal hazards through an interdisciplinary research project in New Zealand. *J. Coastal Res.*, *12*: 1496–1500.

Khan, M., Ejaz, N., Amin, M., Punthakey, J., Mitchell, M., Allan, C. and Hassan, G.Z. (2021). *Improving Groundwater Management to Enhance Agriculture and Farming livelihoods: Integrating Web and Mobile-based Applications for Groundwater Management*. Institute for Land, Water and Society, Charles Sturt University, Albury, NSW. *11*: 2640.

Khan, M.S., Gul, B. and Khan, H. (2015). *Impact of Climatic Change on Flora of High Altitudes in Pakistan*. Springer E Books. *22*: 361–81.

Kocmankova, E., Trnka, M., Juroch, J., Dubrovsky, M., Semeradova, D., Mozny, M., Zalud, Z., Pokorny, R. and Lebeda, A. (2010). Impact of climate change on the occurrence and activity of harmful organisms. *Plant Protection. Science*, *45*: 48–52.

Lal, R. (2010). Managing soils and ecosystems for mitigating anthropogenic carbon emissions and advancing global food security. *Journal of Biosciences*, *60*: 708–21.

Lehmann, J. (2007) Bio-energy in the black. Frontiers in Ecology and the Environment., *5*: 381–387.

Li, Z. and Fang, H. (2016). Impacts of climate change on water erosion: A review. *Earth-Science Reviews*, *163*: 94–17.

Long, S.P., Ainsworth, E.A., Leakey, A.D. B., Nosbsrger, J. and Ort, D.R. (2006). Food for thought: Lower-than-expected crop yield stimulation with rising CO_2 concentrations. *Science*, *312*: 1918–21.

Mapedza, E. and McLeman, R. (2019). Chapter 1 - Drought risks in developing regions: Challenges and opportunities. *In*: Mapedza, Everisto; Tsegai, D.; Bruntrup, M.; McLeman, R. (Eds.), *Drought Challenges: Policy Options for Developing Countries*. Amsterdam, Netherlands: Elsevier. *Current Directions in Water Scarcity*, *2*: 1–14.

Meena, M.D., Yadav, R.K., Narjary, B., Yadav, G., Jat, H.S., Sheoran, P., Meena, M.K., Antil, R.S., Meena, B.L. and Singh, H.V. (2019). Municipal solid waste (MSW): strategies to improve salt affected soil sustainability: A review. *Journal of Waste Management*, *84*: 38–53.

Ministry of Planning Development and Special Initiative. (2022). *Pakistan Floods 2022 Post-Disaster Needs Assessment*.

Pal, S. and Ziaul, S.K. (2016). Detection of land use and land cover change and land surface temperature in English Bazar urban center. *The Egyptian Journal of Remote Sensing and Space Science*, *13*: 1–24.

Pelto, M., Maurer, M. and Zemp, M. (2020). Annual balance of North Cascade glaciers from 1984 to 2020. *Northwest Science*, *94*: 27–40.

Prakash A., Rao J., Mukherjee A.K. and Jeyaveeran, B. (2014). *Climate Change: Impact on Crop Pests*. Applied Zoologists Research Association (AZRA), Central Rice Research Institute; Odisha, India. *81*: 2–19.

Prescia, M. (2021). The immediate threat of climate change in Pakistan. *Clim. Refugees*, *12*: 24–35.

Rahmstorf, S., Cazenave, A., Church, J.A., Hansen, J.E., Keeling, R.F., Parker, D.E. and Somerville, R.C.J. (2012). Sea level rise: A review of recent scientific literature. *Rev. Geophy.*, *50*: 1–23.

Ray, D.K., Gerber, J.S., Macdonald, G.K. and West, P.C. (2015). Climate variation explains a third of global crop yield variability. *Nature Communications*, *6*: 1–9.

Royal Society. 2018. *Greenhouse Gas Removal*. Accessed. *28*: 25–38.

Seneviratne, S.I., Nicholls, N., Easterling, D., Goodess, C.M., Kanae, S., Kossin, J., Luo, Y., Marengo, J., Mc Innes, K., Rahimi, M., Reichstein, M., Sorteberg, A., Vera, C., Zhang, X., Rusticucci, M., Semenov, V., Alexander, L. V., Allen, S., Benito, G. and Zwiers, F.W. (2012). *Changes in Climate Extremes and Their Impacts on the Natural Physical Environment. Managing the Risks of Extreme Events and Disasters to Advance Climate Change Adaptation*. Special Report of the Intergovernmental Panel on Climate Change, *12*: 109–30.

Shah, T. (2009). Climate change and groundwater: India's opportunities for mitigation and adaptation. *Environmental Research Letters*, *4*: 1–25.

Siebert, S., Burke, J., Faures, J.M., Frenken, K., Hoogeveen, J., Doll, P. and Portmann, F.T. (2010). Groundwater use for irrigation: A global inventory. *Hydro Earth Sys Sci.*, *14*: 1863–80.

Singh, S., Ghosh, N.C., Gurjar, S., Krishan, G., Kumar, S. and Berwal, P. (2018). Index-based assessment of suitability of water quality for irrigation purpose under Indian conditions. *Environmental Monitoring and Assessement*,*190*: 29–36.

Tate, K.R. (1992). Assessment, based on a climo-sequence of soils in tussock grasslands, of soil carbon storage and release in response to global warming. *J. Soil Science*, *43*: 697–707.

Turral, H., Burke, J. and Faures, J.M. (2011). *Climate Change, Water, and Food Security*. FAO Water Report No. 36. Rome, FAO.

Wan, X., Gong, F., Qu, M., Qiu, E. and Zhong, C. (2019). Experimental study of the salt transfer in a cold sodium sulfate soil. *KSCE Journal of Civil Engineering*, *23*: 1573–85.

Wang, Y., Deng, C., Yan, L. and Ziru, N. (2018). Identifying change in spatial accumulation of soil salinity in an inland river watershed, China. *Science Total Environ.*, *621*: 177–85.

Warner, K. and Laczko, F. (2008). *Migration, Environment and Development: New Directions for Research*. Center for Migration Studies. *21*: 235–51.

Webb, N.P., Marshall, N.A., Stringer, L.C., Reed, M.S., Chappell, A. and Herrick, J.E. (2017). Land degradation and climate change: Building climate resilience in agriculture. *ESA*, *15*(8): 450–59.

Weisheimer, A. and Palmer, T.N. (2005). Changing frequency of occurrence of extreme seasonal temperatures under global warming. *Geophysical Research Letters*, *32*: 1–5.

World Bank. (2020). *World Development Indicators*. https://databank.worldbank.org/source/world-development-indicators

World Food Summit. (1996). *Rome Declaration on World Food Security*. Rome, FAO.

Xie, W., Chen, Q., Wu, L., Yang, H., Xu, J. and Zhang, Y. (2020). Coastal saline soil aggregate formation and salt distribution are affected by straw and nitrogen application: A 4-year field study. *Soil and Tillage Research*, *198*: 102–16.

Zemp, M., Huss, M., Thibert, E., Eckert, N., McNabb, R., Huber, J. and Paul, F. (2019). Global glacier mass changes and their contributions to sea-level rise from 1961 to 2016. *Nature*, *568*: 382–86.

Zhou, D., Shah, T., Ali, S., Ahmad, W., Din, I.U. and Ilyas, A. (2019). Factors affecting household food security in rural northern hinterland of Pakistan. *J. Saudi Soc. Agric. Sci.*, *18*: 201–10.

Chapter 5

Food Security and Atmospheric CO_2 Mitigation through Increasing Soil Carbon Storage

The 4 per 1000 Initiative and the Importance of Photosynthetically Active Plants

Thierry HEULIN[1]*

Introduction

Climate Change: Consensus on the Facts....and the End of Sceptical Attitude?

More than just a concept, climate change is a scientific theory. It is a well-documented explanation, supported by numerous scientific papers and scientific consensus, attributing observed changes in the Earth's climate to human activities since the beginning of the industrial revolution, primarily due to the increase in greenhouse gas emissions. To date, significant research has focused on the mechanisms, processes, and consequences of climate change.

Scientist (and suffragette) Eunice Newton Foote was the first to publish an article suggesting that an accumulation of carbon dioxide in the Earth's atmosphere could lead to an increase in surface temperatures (Foote, 1856). In the mid-20th century, evidence of the impact of human activities on climate change was attributed to the work of Guy Stewart Callendar (1938), who established a connection between rising carbon dioxide (CO_2) levels and global warming, albeit without providing direct experiments, but based on observational data and calculations. Today, a century and

[1] Lab Microbial Ecology of the Rhizosphere (LEMiRE), Aix Marseille Univ, CEA, CNRS, UMR 7265 BIAM, F-13115 Saint Paul-Lez-Durance, France.

* Corresponding author: thierry.heulin@cea.fr.

a half after Foote's work, there is irrefutable scientific evidence that human activities are the primary driver of climate change.

In recent years, while the scientific consensus on human-induced climate change remains strong, certain aspects of the concept have become the subject of debates and controversies. It is important to note that these controversies often come from increasingly minority scientists or individuals who question scientific consensus in general (is the Earth truly round/spherical?).

Predicting Climate Change (IPCC_AR6_2022): Different Scenarios

The report by the UN's leading climate experts (Intergovernmental Panel on Climate Change, IPCC, Report 2022) sounds the alarm. Different scenarios are possible depending on our capacity to reduce CO_2 emissions (Fig. 1).

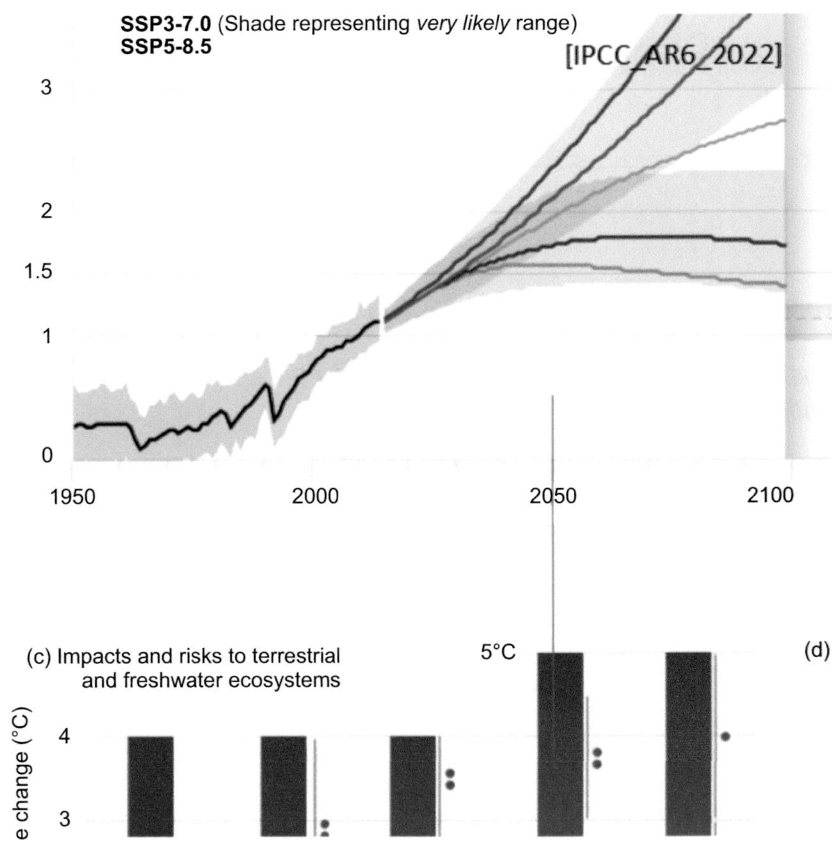

Fig. 1 Scenarios of global surface temperature evolution (from IPCC Report 2022).

Even with global warming limited to +2°C compared to the pre-industrial era, by 2050, the consequences will be cataclysmic. The figures they put forward are very worrying:

The central figure in this report is global warming compared to the pre-industrial period (1850–1900). The IPCC uses two assumptions: +2°C by 2050 is the ceiling set by the Paris Climate Agreement, but +1.5°C is the target set by the countries that signed the Paris Agreement in 2015. The average increase observed is currently +1.1°C.

The IPCC also estimates that with a warming of more than 2°C, the melting of the Greenland and West Antarctic ice sheets could be a point of no return. These ice sheets alone contain enough water to cause a 13 m rise in sea level. In a previous report, the IPCC estimated that sea levels would rise by 1.1 meter by the end of the 21st century, since the total melting of these glaciers seems unlikely.

Furthermore, the IPCC report points to a 4 to 10% drop in the production of the main crops over the last 10 years. As a result, at this rate, 80 million more people could go hungry by 2050 with a global warming of +2°C.

How to Capture CO_2 to Save the Climate?

Several negative emission technologies are known (Minx et al., 2018), and mainly:

(i) The combination of Bioenergy (BE: 3rd generation fuels, biogas/methanization) with Carbon Capture and Storage (CCS) techniques (BECCS),
(ii) Soil C storage (optimizing the transfer of biomass from agrosystems into soils in the form of stable organic matter), and
(iii) Afforestation, reforestation.

The main challenge will to be to assess, in an interdisciplinary manner, the effectiveness of different negative emission technologies (NET) and their co-benefits with other services such as storage potential, costs, energy consumption, long-term efficiency, competition for water and nutrients (Smith et al., 2015), land competition (Graves et al., 2016), biodiversity impacts, consequences in terms of overall health, nutrition, organization of technical and economic sectors, and social change.

One of the main conclusions is that C storage in soils has the greatest potential to remove CO_2 from the atmosphere, along with the most diverse co-benefits and at the lowest cost per tonne of CO_2.

What is the "4 per 1000" Initiative?

The "4 per 1000" initiative is a global network of international cooperation aimed at accelerating the adoption of best practices in the fields of agriculture, forestry, and other land uses (AFOLU).

Why Storing C in Soils?

- To mitigate climate change
  Attenuate GHG emissions. Organic carbon can be stored in soils for decades to millennia.
  Low-cost negative emission technology
- To adapt to climate change
  Soil organic matter increases water retention capacity by soils.

Decreases sensitivity to erosion.

- To contribute to food security and restore degraded soils.
  Major role of soil organic matter in soil fertility
  Yields stability.
  Better food quality
  Restoring degraded soils.

In summary, it is obvious that the "4 per 1000" initiative is an effective climate change adaptation strategy (Abbadie et al., 2017).

The "4 per 1000" Initiative: Where Does It Come From?

An annual increase of "4 per 1000" in C stocks at the global soil surface (860*0.004) would almost compensate the annual increase in CO_2 in the atmosphere (Fig. 2).

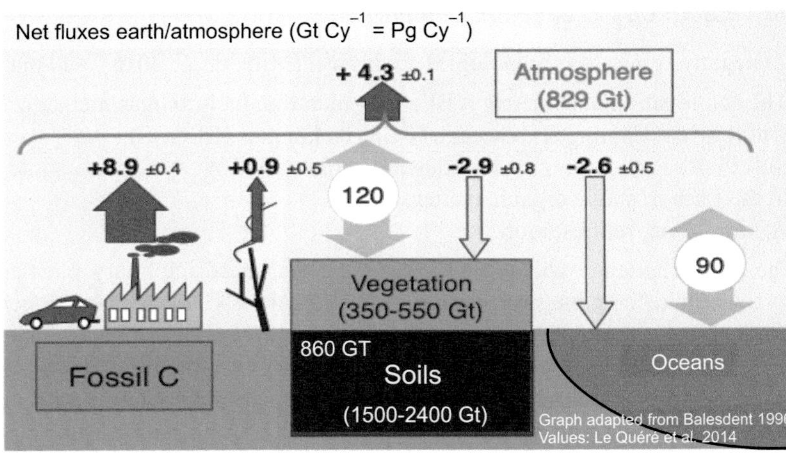

Fig. 2 Graph of C net fluxes between earth surface and atmosphere, adapted from Balesdent (1996), updated with values from Le Quéré et al. (2014).

What are the Goals of the "4 per 1000" Initiative?

The main goal is to increase carbon sequestration in soils in the form of organic matter to:

- Enhance food security,
- Adapt agriculture to climatic change, and
- Mitigate climate change,

while also pursuing the sustainable development goals adopted by the United Nations (Fig. 3).

Who are the Partners and Members of the "4 per 1000" Initiative (Feb. 2020)?

For an overview of Partners and Members: www.4p1000.org

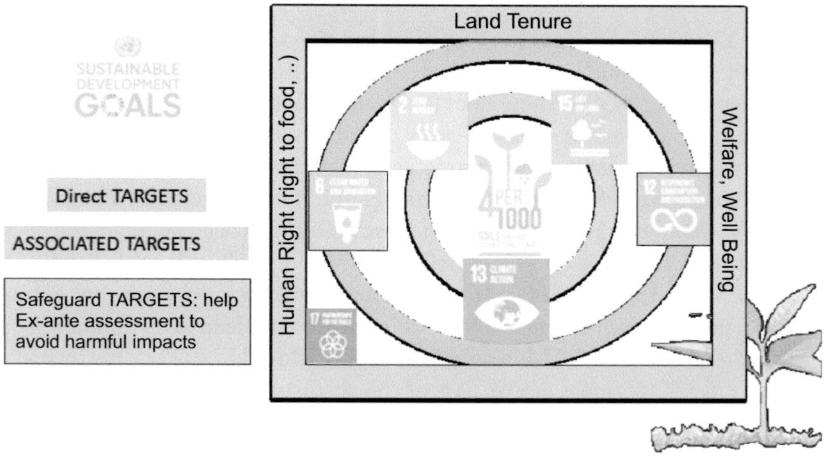

Fig. 3 Targets of 4‰ and SDGs (Chotte, J.L., personal communication).

FORUM partners

448 signatories of the Paris Declaration

Including 46 countries & provinces, 13 international organizations, 11 foundations & development banks, 144 NGOs, 89 research & education bodies & Universities, 47 Farmers organizations and 98 enterprises

CONSORTIUM Members

Already 214 signatories of the Declaration of Intention n°2

Chair : **Dr. Ibrahim MAYAKI** (ES NEPAD)

Vice-Chair : **Mr. Stephane LE FOLL** (ex - Min. Agriculture – France)

Why is the Priority to Increase C Storage in Cultivated Soils, Including Deeper Layers of Soils?

Beyond the urgency of mitigating atmospheric CO_2, it is crucial to increase the C content in soils, not only in the topsoil but also in deeper layers, to prevent soil degradation and enhance soil fertility.

Why Choose Photosynthetically Active Plants over Crop Residues?

While crop residues are important for maintaining soil carbon content, they can potentially increase emissions of N_2O, a major greenhouse gas, from the soil. This call is for innovative approaches to mitigate these emissions (Abalos et al., 2022). Furthermore, crops residues could be utilized for purposes other than their incorporation into the soil.

Therefore, we have considered the possibility of searching for a plant phenotype that shows a positive correlation with sustainable carbon storage in the soil. This phenotype serves as an integrative proxy for the plant's adaptation to its soil environment: a significant part of the plant's photosynthate is exuded into the soil (10–20%, Nguyen et al., 2003) and nourishes the microbiota associated with the rhizosphere and root surface. Among the wide diversity of microbial functions stimulated by root exudates (mainly carbohydrates, organic acids and amino acids), the production of exopolysaccharides (EPS) by various rhizobacterial species, including *Paenibacillus*, *Pantoea*, *Burkholderia*, and non-symbiotic *Rhizobium*, has been documented over the last 30 years (Gouzou et al., 1993; Amellal et al., 1998; Achouak et al., 1999; Alami et al., 2000; Bezzate et al., 2000) (Fig. 4). The interactions between root exudates, EPS-producing rhizobacteria, and soil particles/aggregates lead to the formation of the rhizosheath (George et al., 2014; Ndour et al., 2020; Mo et al., 2023). This rhizosheath can be considered the most active part of the rhizosphere, a soil fraction which is physically, chemically, and biologically influenced by photosynthetically active plants through root exudates) and easily sampled (Gouzou et al., 1993).

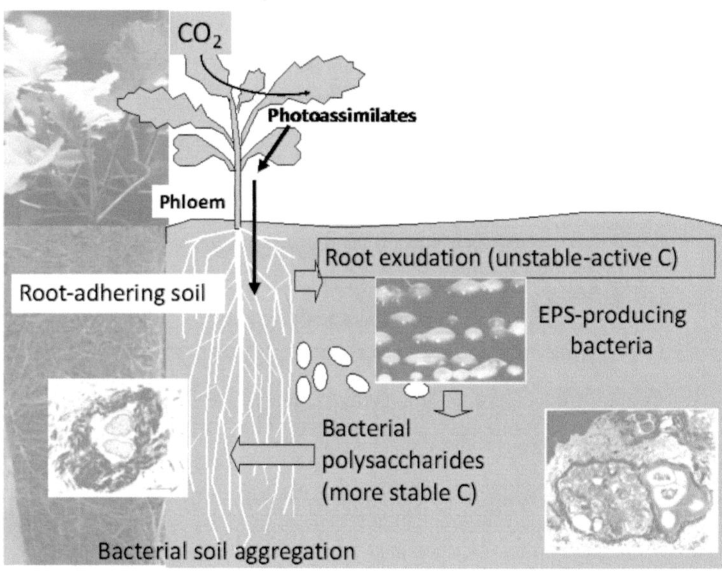

Fig. 4 Schematic representation of 'plant-soil' interactions involving EPS-producing bacteria.

For this purpose, since 2000, we have been conducting a multidisciplinary study (plant genetics, plant physiology, microbial ecology, and soil science) on pearl millet, a tropical crop primarily cultivated in dry and poor soils (Africa and India) as a rain-fed crop, which can be considered as a proof of concept. The main results of this study are as follows:

- In an initial study, the rhizosheath was characterized in a collection of 86 pearl millet lines (*Pennisetum glaucum*) (Photo 1) using the ratio of soil dry mass adhering to the roots to the dry mass of root tissues (RAS/RT) (Ndour et al., 2017). This ratio showed significant variation among the lines, with a range 2–3 times higher between the lowest and highest average values (Fig. 5). For nine lines with contrasting aggregation properties, we then compared the bacterial diversity and composition in the rhizosheath. Bacterial alpha-diversity increased with the "RAS/RT ratio". Regarding taxonomic composition, Rhizobiales were stimulated in lines with high aggregation level, while Bacillales were stimulated in lines with a low ratio. These results illustrate the impact of pearl millet's genetic variation on soil physical properties and microbial diversity (Ndour et al., 2017).

Photo 1 Rhizosheath of pearl millet grown on arenosol (Senegal) ©Marcel Nahim-Diouf/IRD-UCAD.

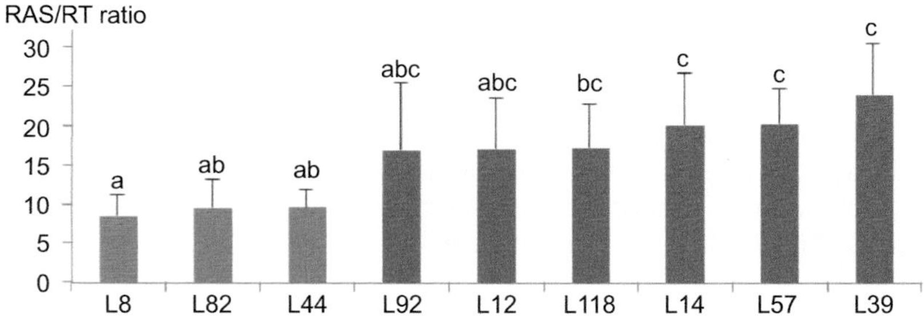

Fig. 5 Comparison of the root-adhering soil (dm) per root biomass (dm) (RAS/RT ratio) of nine selected pearl millet inbred lines (n = 10 per line) (Ndour et al., 2017).

In the second work, we confirmed the findings of the first study by evaluating the impact of genetic variability in pearl millet at the rhizosheath level. To do this, we phenotyped a collection of 181 pearl millet lines based on the rhizosheath size (RAS/RT ratio), and then on a subsample of 13 contrasting genotypes, we measured RAS/RT ratio (Fig. 6) as well as microbial biomass, pH, mineral nitrogen content, and six enzymatic activities involved in major nutrient cycles. Additionally, metabarcoding of 16S rDNA and ITS metabarcoding was performed to estimate the diversity of fungal bacterial populations (Ndour et al., 2021). We showed that certain enzyme activities (chitinase, acid phosphomonoesterase, FDA-hydrolysis, and β-glucosidase) were higher in the rhizosheath of lines with characterized high value of rhizosheath. Certain bacterial genera such as *Bradyrhizobium* and *Mesorhizobium* were associated with the rhizosheath of the lines characterized for their high RAS/RT value. This study demonstrated that rhizosheath size is correlated to soil nutrient dynamics, certain enzymatic activities involved in the carbon and nitrogen cycling, and microbiota diversity (Ndour et al., 2021).

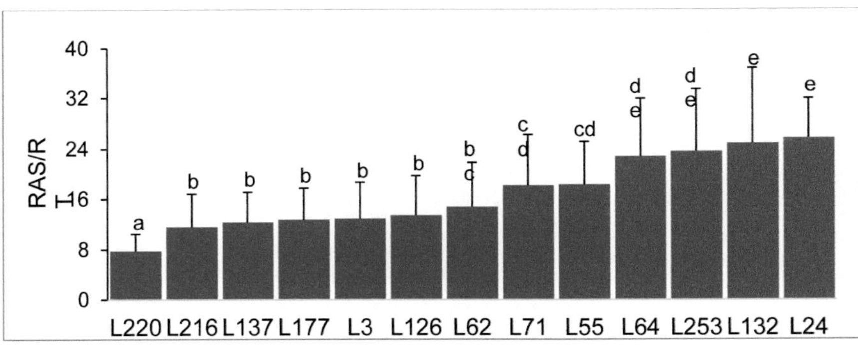

Fig. 6 Phenotype verification of the 13-pearl millet contrasting lines (Ndour et al., 2021).

- To evaluate rhizodeposition of C in soil, we grew four pearl millet lines (C4 plant: $\delta^{13}C$ of −12.8 ‰, $F^{14}C = 1.012$), with contrasting rhizosheath size (RAS/RT), in a C3 soil type (organic matter dominated by C3 plant residues with $\delta^{13}C$ of −22.3 ‰, $F^{14}C = 1.045$) (Ndour et al., 2022). The results showed a significant increase in $\delta^{13}C$ in the rhizosheath of the four pearl millet lines compared to the control soil, suggesting that this approach detects plant C input into the soil, at an early stage of pearl millet growth. The concentration of plant C in the rhizosheath did not vary significantly among the pearl millet lines, but the absolute amount of plant C varied significantly when considering the rhizosheath of these different lines (Fig. 7). Using a conceptual model and data from two carbon isotope measurements, we found a priming effect for all pearl millet lines. It is important to note that the magnitude of this priming effect (C_{lost}/C_{new} ratio) is higher for the line with less rhizosheath (low aggregation) than for the line with a higher rhizosheath value (high aggregation), indicating a greater carbon sequestration potential for the latter (Ndour et al., 2022).

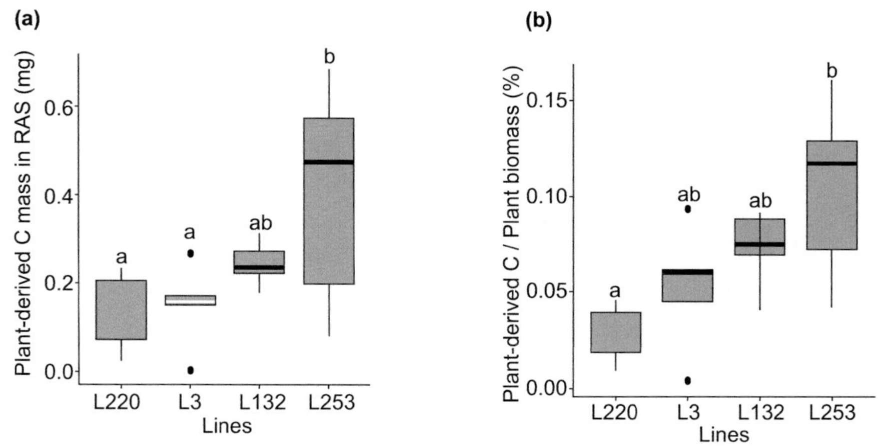

Fig. 7 Mass of plant-derived carbon deposited (PDCD in mg C) in the RAS of the four pearl millet lines (B). Amount of plant-derived carbon per plant biomass (in %) produced by the four pearl millet lines (C) (Ndour et al., 2022).

In these three published studies, we established the link between the variability of pearl millet genotypes to a phenotype (rhizosheath) that integrates plant activity (photosynthesis, translocation, root exudation) and soil characteristics including physical properties (texture, structure), chemical properties (pH, CEC...) and the diversity and activities of the microbiota, particularly the EPS-producing bacterial populations that structure and 'glue' the soil around the roots, resulting in rhizosheath formation.

It remained to be verified whether this rhizosheath formation was under genetic control in pearl millet and primarily regulated by root exudation.

Is Plant Breeding a Means to Increase C Storage in Soil?

We have demonstrated that the rhizosheath (estimated using root-adhering soil mass) is a heritable trait in pearl millet and is correlated with changes in microbiota structure and functions. We have studied the correlation between rhizosheath and root hair development, root architecture and symbiosis with arbuscular mycorrhizal fungi, and analysed the genetic control of this trait using genome-wide association sampling (GWAS) combined with bulk segregant analysis and gene expression studies (de la Fuente et al., 2022) (Fig. 8). Twelve QTLs for rhizosheath formation have been identified through GWAS. Validation of five of these QTLs Bulk was achieved through segregant analysis in a biparental population. Combining genetics with a comparison of global gene expression in the root tip of contrasting inbred lines revealed candidate genes that may control rhizosheath formation in pearl millet. Our study indicates that rhizosheath formation is subject to complex genetic control in pearl millet (involving multiple genes, each with a minimal impact) and suggests that it is primarily regulated by root exudation (de la Fuente et al., 2022).

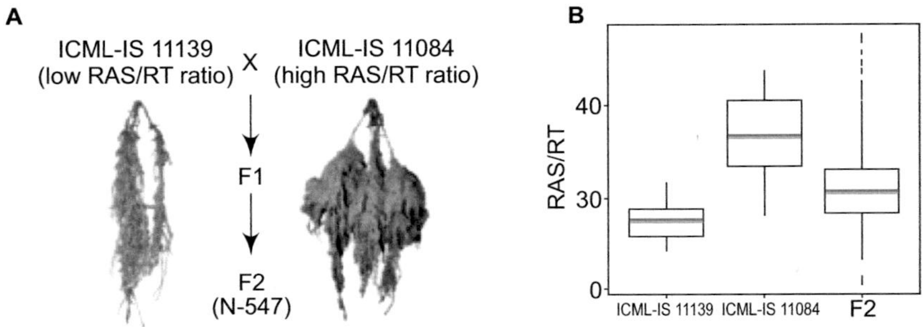

Fig. 8 Genetic dissection of root soil aggregation in pearl millet by Bulk Segregant Analysis (BSA). (A) Cross established for Bulk Segregant Analysis (BSA) between two pearl millet inbred lines with contrasted rhizosheath phenotype. (B) Boxplot showing the distribution of RAS/RT ratio in line ICML-IS 11139, ICML-IS 11084, and F2 population.

In a recent-review-article, we have examined studies on the biological factors involved in rhizosheath formation, the methods used to quantify it, and its importance in plant nutrition. We have gathered evidence from the literature and from our own work, particularly on pearl millet, showing that transformation of root exudates by EPS-producing bacteria effectively contributes to soil aggregation and thus increases the rhizosheath volume. This, in turn, improves water uptake and plant nutrition. The growing interest in this process has led to the genetic mapping of potential QTLs controlling this phenotype (de la Fuente et al., 2022), enabling the development of new tools for selecting plant varieties that are not only tolerant to abiotic stresses but also capable of long-term C storage in soils (Ndour et al., 2020).

Based on this proof of concept achieved with pearl millet, we can therefore conclude from all these studies that plant breeding aimed at promoting "plant-microbiota-soil" interaction phenotype is not only possible but also desirable for the purpose of long-term C storage in soils.

Conclusions

Beyond the impact of this approach, advocating the use of crop plants for soil C storage, on atmospheric CO_2 mitigation, which will only be measurable in the medium to long term, its impact on soil fertility and food security could have beneficial effects in the short to medium term. The Bangladesh is concerned about the effects of climate change, particularly sea level rise, as well as food security. In the specific context of agriculture in Bangladesh, the low C content of soils (1.0–1.3%) necessitates considering this approach of plant breeding, particularly for rice.

References

Abalos, D., Recous, S., Butterbach-Bahl, K., De Notaris, C., Rittl, T.F., Topp, C.F.E., Petersen, S.O., Hansen, S., Bleken, M.A., Rees, R.M. and Olesen, J.E. (2022). A review and meta-analysis of mitigation measures for nitrous oxide emissions from crop residues. *Sci. Total. Environ., 828*: 154388. https://doi.org/10.1016/j.scitotenv.2022.154388.

Abbadie, L., Hatté, C. and Heulin, T. (2017). 4 per 1000, an effective climate change adaptation strategy. *In*: Euzen A., Laville B., Thiébault S. (Eds.), *Adaptation to Climatic Change*, 283–90.. Sous la direction scientifique de CNRS Editions, ISBN: 978-2-271-09482-7.

Achouak, W., Christen, R., Barakat, M., Martel, M.H. and Heulin, T. (1999). *Burkholderia caribensis* sp. nov., an exopolysaccharide-producing bacterium isolated from vertisol microaggregates in Martinique. *Int. J. Syst. Bacteriol.*, 49: 787–94. https://doi.org/10.1099/00207713-49-2-787.

Alami, Y., Achouak, W., Marol, C. and Heulin, T. (2000). Rhizosphere soil aggregation and plant growth promotion of sunflowers by an exopolysaccharide-producing *Rhizobium sp.* strain isolated from sunflower roots. *Appl. Environ. Microbiol.*, *66*(8): 3393–3398. https://doi.org/10.1128/AEM.66.8.3393-3398.2000.

Amellal, N., Burtin, G., Bartoli, F. and Heulin, T. (1998) Colonization of wheat rhizosphere by an exopolysaccharide-producing *Pantoea agglomerans* strain and its effect on rhizosphere soil aggregation. *Appl. Environ. Microbiol.*, *64*(10): 3740–47. https://doi.org/10.1128/AEM.64.10.3740-3747.1998.

Balesdent, J. (1996). The significance of organic separates to carbon dynamics and its modelling in some cultivated soils. *Eur. J. Soil Sci.*, *47*(4): 485–93. https://doi.org/10.1111/j.1365-2389.1996.tb01848.x.

Bezzate, S., Aymerich, S., Chambert, R., Czarnes, S., Berge, O. and Heulin, T. (2000). Disruption of the *Paenibacillus polymyxa* levansucrase gene impairs its ability to aggregate soil in the wheat rhizosphere. *Environ. Microbiol.*, *2*(3): 333–42. https://doi.org/10.1046/j.1462-2920.2000.00114.x.

Callendar, G.S. (1938). The artificial production of carbon dioxide and its influence on temperature. *Q. J. R. Meteorol. 64*(275): 223–40. https://doi.org/10.1002/qj.49706427503.

de la Fuente Cantó, C., Diouf, N.M., Ndour, P.M.S., Debieu, M., Grondin, A., Passot, S., Champion, A., Barrachina, C., Pratlong, M., Gantet, P., Assigbetse, K., Kane, N., Cubry, P., Diedhiou, A.G., Heulin, T., Achouak, W., Vigouroux, Y., Cournac, L. and Laplaze, L. (2022). Genetic control of rhizosheath formation in pearl millet. *Sci. Rep.*, *12*: 9205. https://doi.org/10.1038/s41598-022-13234-w.

Foote E. (1856). Circumstances Affecting the Heat of the Sun's Rays. *The American Journal of Science and Arts*, *22*(66): 383–84.

George, T.S., Brown, L.K., Ramsay, L., White, P.J., Newton, A.C., Bengough, A.G., Russell, J. and Thomas, W.T.B. (2014). Understanding the genetic control and physiological traits associated with rhizosheath production by barley (*Hordeum vulgare*). *New Phytol.*, *203*: 195–205. https://doi.org/10.1111/nph.12786.

Gouzou, L., Burtin, G., Philippy, R., Bartoli, F. and Heulin, T. (1993). Effect of inoculation with *Bacillus polymyxa* on soil aggregation in the wheat rhizosphere: Preliminary examination. *Geoderma*, *56*: 479–91. https://doi.org/10.1016/B978-0-444-81490-6.50038-8.

Graves, R.A., Pearson, S.M. and Turner, M.G. (2016). Landscape patterns of bioenergy in a changing climate: implications for crop allocation and land-use competition. *Ecol. Appl.*, *26*(2): 515–29. https://doi.org/10.1890/15-0545.

IPCC. (2022). *Climate Change 2022: Impacts, Adaptation, and Vulnerability*. Contribution of Working Group II to the Sixth Assessment Report of the Intergovernmental Panel on Climate Change ((Eds. Pörtner, H.-O., Roberts, D.C., Tignor, M., Poloczanska, E.S., Mintenbeck, K., Alegría, A., Craig, M., Langsdorf, S., Löschke, S., Möller, V., Okem, A., Rama, B.), 3056 pp.. Cambridge University Press, Cambridge, UK, and New York, NY, USA. https://doi:10.1017/9781009325844.

Le Quéré, C., et al. (2014). Global carbon budget 2014. *Earth Syst. Sci. Data Discuss.*,*7*: 521–610. https://doi.org/10.5194/essdd-7-521-2014.

Minx, J.C. et al. (2018). Negative emissions—Part 1: Research landscape and synthesis. *Environ. Res. Lett.*, *13*: 063001. https://doi.org/10.1088/1748-9326/aabf9b.

Mo, X., Wang, M., Zeng, H., Wang, J. (2023). Rhizosheath: Distinct features and environmental functions. *Geoderma*, *435* : 16500. https://doi.org/10.1016/j.geoderma.2023.116500.

Ndour, P.M.S., Gueye, M., Barakat, M., Ortet, P., Bertrand-Huleux, M., Pablo, A.L., Dezette, D., Chapuis-Lardy, L., Assigbetsé, K., Kane, N.A., Vigouroux, Y., Achouak, W., Ndoye, I., Heulin, T. and Cournac, L. (2017). Pearl millet genetic traits shape rhizobacterial diversity and modulate rhizosphere aggregation. *Front. Plant Sci.*, *8*: 1288. https://doi.org/10.3389/fpls.2017.01288.

Ndour, P.M.S., Heulin, T., Achouak, W., Laplaze, L. and Cournac, L. (2020). The rhizosheath: From desert plants adaptation to crop breeding. *Plant Soil*, *456*: 1–13. https://doi.org/10.1007/s11104-020-04700-3.

Ndour, P.M.S., Barry, C.M., Tine, D., Gueye, M., Barakat, M., Ortet, P., Achouak, W., Ndoye, I., Bassirou, S., Laplaze, L., Heulin, T. and Cournac, L. (2021). Pearl millet genotype impacts microbial diversity and enzymatic activities in the rhizosphere, in relation to soil aggregation. *Plant Soil*, *4164*(1–2). https://doi.org/10.1007/s11104-021-04917-w.

Ndour, P.M.S., Hatté, C., Achouak, W., Heulin, T. and Cournac, L. (2022). Rhizodeposition efficiency of pearl millet genotypes assessed on short growing period by carbon isotopes (δ13C and F14C). *Soil*, 8: 49–57. https://doi.org/10.5194/soil-8-49-2022.

Nguyen, C. (2003). Rhizodeposition of organic C by plants: Mechanisms and controls. *Agronomie*, *23*(5–6) : 375–396. https://doi.org/10.1007/978-90-481-2666-8_9.

Smith, P., Cotrufo, M.F., Rumpel, C., Paustian, K., Kuikman, P.J., Elliott, J.A., et al. (2015). Biogeochemical cycles and biodiversity as key drivers of ecosystem services provided by soils. *Soil Discussions*, *2*: 537–86.

Smith, P., Arneth, A., Barnes, D.K.A., Ichii, K., Marquet, P.A., Popp, A., Pörtner, H.O., Rogers, A.D., Scholes, R.J., Strassburg, B., Wu, J. and Ngo, H. (2022). How do we best synergize climate mitigation actions to co-benefit biodiversity? *Glob. Chang. Biol.*, *28*(8): 2555–77. https://doi.org/10.1111/gcb.16056.

Chapter 6

Carbon Footprint of Rice Production in Rice Production Systems

Md. Khairul Alam[1*], *Richard W. Bell*[2], *Wahidul K. Biswas*[3] and *S. Akhter*[4]

Introduction

Bangladesh government has targeted to secure food security, particularly by increasing rice production, in response to the global climate change, increasing food demand, and political outlook. To achieve this, they are using all-strategies that include improved crop management practices, cultivated area expansion, and genetic improvements. To talk about on rice alone, as of July 18, 2022, the production of rice was over 38.4 million tons (Mt), but the government aims to increase it to 47 Mts by 2030 and then to 54 Mts by 2040. They also target to produce more than 60 Mt by 2050, while in 2021, rice was grown on 78% of Bangladesh's total arable land. In the context of Bangladesh's intensive agricultural production system, agriculture remains the predominant sector of employment, engaging a substantial 40.6% of the labor force in 2022 (BBS, 2022).

The conventional practice of puddling soils for rice, in addition to intensive tillage for dryland non-rice crops, the common practice of crop production in the Gangetic Plains, has led to several challenges, including soil degradation, excessive water consumption, and the depletion of soil organic carbon (SOC) and essential nutrients like nitrogen (N). Globally, the adoption of conservation agriculture (CA) has been steadily growing since 2008/2009, with cropland used for CA expanding at a rate of over 10 M ha yr^{-1}. By 2018/2019, CA encompassed 14.7% of the world's cropland area, indicating its increasing adoption in Asia, Australia, Africa, and Europe (Kassam et al., 2022). Catching up to the annual tillage dependent agricultural system

[1] Commonwealth Scientific and Industrial Research Organization, Wembley, WA, Australia.
[2] Agriculture Discipline, Murdoch University, Perth, WA, Australia.
[3] Sustainable Engineering Group, Curtin University, WA, Australia.
[4] Bangladesh Agricultural Research Institute, Gazipur, Bangladesh.
* Corresponding author: Khairul.Alam@csiro.au

with CA in rice-based intensive cropping and fitting CA into the intensive cropping system has been a long-term endeavor as many challenges have to be overcome (Alam et al., 2014, 2016a).

While CA associated with the upland crops, the wet tillage-based operations (puddling) are employed for rice transplanting and establishment. Consequently, whatever benefits may accrue by integrating CA in upland crops get disappeared after puddling of soil for growing rice (Alam et al., 2016a, 2016b; Alam, 2018). A novel crop establishment practice employing strip tillage for upland crops and non-puddled transplanting of rice has been developed for CA to be integrated in rice-based intensive cropping systems in the Gangetic Plains (Bell et al., 2018). The set of CA practices has performed well in yield, economic return, saving fuel consumption, and improving soil health improvement (Bell et al., 2018; Alam et al., 2020a, 2020b; Salahin et al., 2021a, 2021b). In addition, the non-puddled (NP) rice production has been developed that ensure timeliness of planting, labor shortage, and savings of water (Alam et al., 2018; Bell et al., 2018; Alam et al., 2019a, b).

The significance of wetland rice in global GHG emissions cannot be overlooked, and agricultural practices have been refined to mitigate their contribution to GHGs. However, a comprehensive assessment of the carbon (C) footprint of component crops, including irrigated rice, monsoon rice, and upland crops, is yet to be explored. Studies on the global warming potential mitigation of the CA practices have been conducted in the Eastern Gangetic Plain (Bangladesh) that can be replicated across geographic locations and cropping systems (Alam, 2018). Such evaluations are vital in identifying potential GHG emission mitigation strategies and minimizing the trade-off between methane (CH_4) and nitrous oxide (N_2O) emissions (Miyata et al., 2000; Chakraborty et al., 2017; Salahin et al., 2017).

Despite being a highly climate vulnerable country, whilst contributing less than 0.35% of global GHG emissions, Bangladesh should actively play its part in the global collective action to reduce future GHG, especially CH_4 and N_2O emissions. The updated Nationally Determined Contributions (NDCs) of Bangladesh put total emissions at 169.1 Mt of GHGs, which is estimated to be increased by 2030 to around 409.4 Mt. Agriculture, livestock, and forestry contribute 27.3%, amounting to 46.2 Mt. By using default values, it was estimated that agriculture, forestry, and other land uses in Bangladesh were responsible for around 44% of total CH_4 emission during 2007–2016. The support to Global Methane Pledge would require 13.2% of the 44% CH_4 from agriculture. Sapkota et al. (2021) has demonstrated that about 17% of the CH_4 reduction can be achieved from implementing existing measures in agriculture. Despite of being among the lowest emitters in the world, Bangladesh proposes to cut CO_2 emissions by 89.5 Mts, equivalent to 21.8% of CO_2 eq by 2030 as part of global efforts to reduce the emissions that are responsible for climate change.

Studies estimating C footprints (also N footprints) of crops in a rice-based cropping system using the LCA approach is particularly important for the Government of Bangladesh as wetland rice is the dominant crop in the country and a major contributor to national C accounts. Estimating C footprints ((also N footprints) through adopting LCA of the cropping systems could help identify hotspots and stages of crop production which could help policy makers and farmers to identify the most promising options for GHGs mitigation and global warming potential (GWP). Bangladesh has been a signatory of the Copenhagen Accord since 2010 (MOEF, 2010). Though the country has no legal binding actions for mitigating GHG emission, with the increasing growth and development in economy and industrial progress, the country must take mitigation actions for showing their responsibility to the world for the financial, technical, and capacity building support. Actions to mitigate GHG emissions from agriculture must be considered for its role in the economy. Agriculture was attributed to poverty reduction by more than 90% during 2005–2010 in Bangladesh. Around 50% of all workers in the country and 75% in rural Bangladesh are employed in agriculture, while almost 87% of rural households depend on agriculture for at least part of their income (WB, 2016).

SOC sequestration has emerged as the key mitigation strategy, offering opportunities to improve agricultural soils through maximizing C inputs and minimizing outputs in terms of GHG emission. In Bangladesh, existing practices and climatic conditions have proven inadequate in maintaining high SOC levels, leading to detrimental effects on microbial diversity, soil structure, water holding capacity, and nutrient availability associated with soil health. The practices of non-puddled rice and strip planting for upland crops, coupled with residue retention and integrated nutrient management, show potential for improving soil health, warranting investigation into their efficacy in mitigating global warming potential (Alam et al., 2016, 2018; Alam, 2018).

To develop proactive mitigation strategies under the current climate scenario, an in-depth understanding of SOC and N dynamics under intensive rice-based cropping systems, as well as GHG implications, is essential. However, the ultimate potential for SOC sequestration remains elusive due to a lack of comprehensive databases and a fundamental understanding of SOC dynamics at various scales. Hence, this book chapter aims to present the C and N footprints of CA versus conventional agriculture in upland rice-based cropping systems in Bangladesh, shedding light on the best management practices to enhance C and N storage in soil while mitigating global warming potential (Sommer and Bossio, 2014; Zomer et al., 2017).

To align well with the context and scope of the book chapter, the objectives are listed as:

1. To understand the potential of strip planting for upland crops and non-puddled transplanting of rice to minimize GHG emissions from soil in monsoon and irrigated seasons,
2. To review the effect of CA implemented in rice-based intensive cropping systems on global warming mitigation potential, and
3. To review the existing C footprints estimated for rice and other cropping systems and the relative contribution of the individual GHG emissions to the net GWP from growing crops in intensive cropping systems estimated by using LCA approach.

Brief Methodology of Carbon Footprint Estimation

The carbon footprint (CF) could be quantified by taking into account the GHG emissions directly from the soils and from agricultural inputs used for crop production and protection and farm machinery in a single whole cycle of crop production (Adler et al., 2007; Alam et al., 2016, 2019a, b). Contrary to non-biological production processes, agriculture involves emissions from inputs and their interactions from the soil-water-plant ecosystems via biogeochemical cycles. Identifying hotspots and stages of crop production by characterizing the CFs through LCA are crucial for finding the solutions to mitigate GHG emissions for improving crop productivity. Therefore, CF may be defined as, "the quantity of GHGs (CO_2eq), emitted into the atmosphere by an individual, organization, process, product, or event from within a specified boundary". The CO_2eq mass has been expressed as unit of CF due to convenient calculations for wider acceptance (Lynas, 2007). However, estimating crop and regional-specific CFs would be important for understanding the crop and regional variation and identifying crop and regional-specific mitigation options (ISO 14044, 2006 and 2012; Pandey et al., 2010).

Beginning with the use of fossil fuels for mining, processing, and transportation of all inputs to the paddock, agricultural production processes emit CO_2 and other GHGs into the atmosphere. The physiological process of photosynthesis allows the plant to fix and convert gaseous carbon into C-enriched biomass. Therefore, an estimate of the net C footprint may be obtained if crop-specific calculations are considered on a unit of land by adding up the C released from inputs as well as the C emitted from the bio-geological interface by the practice adopted. As shown in the adapted Fig. 1, various LCA approaches can be used to calculate the C footprint of agricultural output (Biswas et al., 2008; Brentrup and Palliere, 2008).

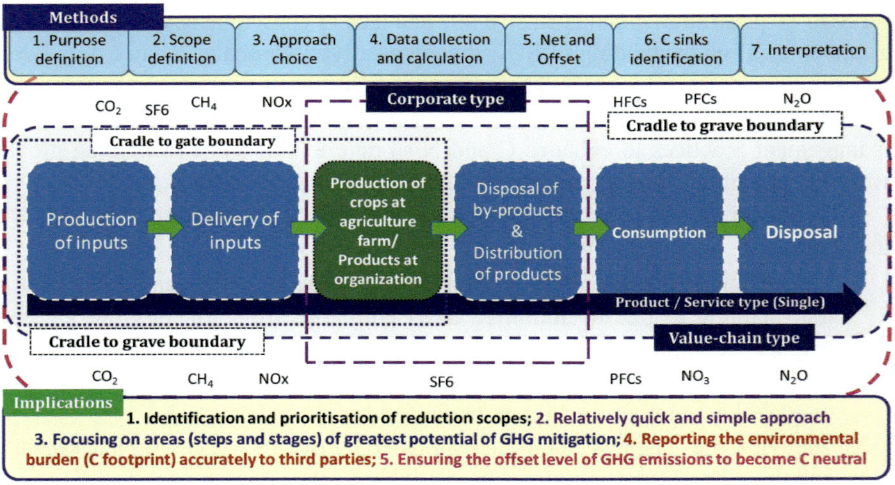

Fig. 1 This is a conceptual layout of the workflow, different types and boundaries of the carbon footprint, estimating techniques, and the allocation of the footprint to climate change, modified from ISO 14044, 2006 and ISO, 2012.

While ignoring the variability of the rice production systems in terms of crop establishment practices, residue, water and fertilizer management, all research on LCA of rice production or crop production in rice-upland cropping systems has placed an emphasis on GWP. Most studies did not evaluate the entire rice and upland

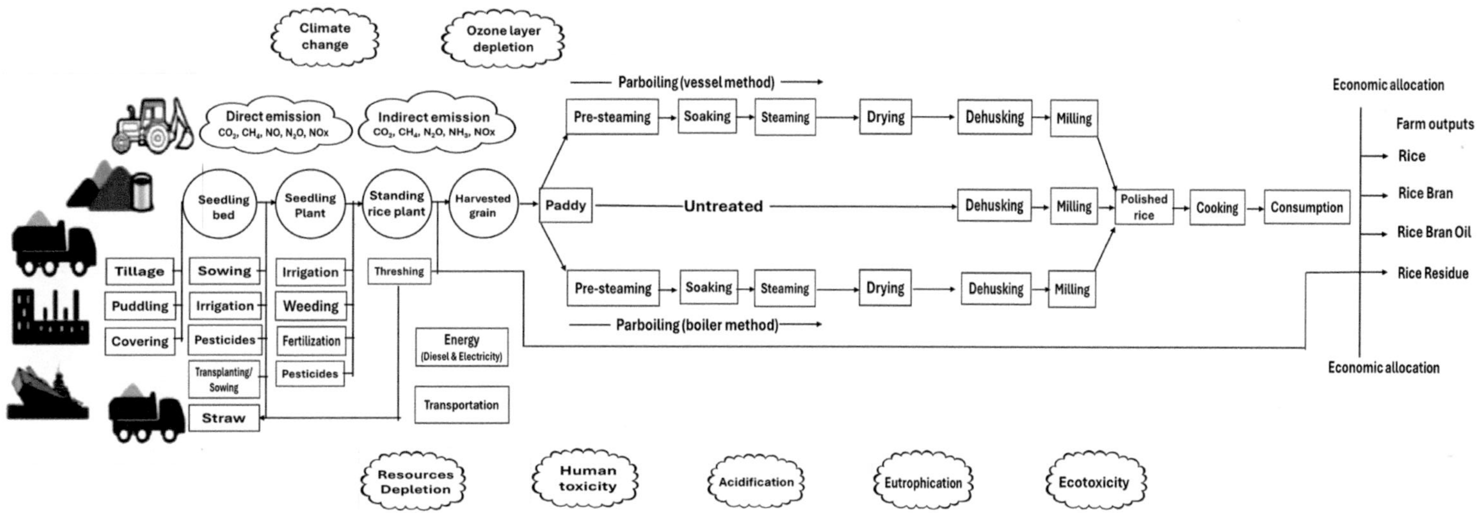

Fig. 2 Conceptual diagram of processes and stages of life cycle assessment of rice crop production from cradle to grave border and its various allocations. (Adapted from Roy et al., 2007; Alam 2018, and Alam et al., 2020).

crop production chain (cradle to grave border). All research took the GWP into account, and some estimated eutrophication and acidification, but most researchers still neglected to consider factors like water use, biodiversity, toxicity, and energy use. To comprehend the distribution of the LCA study, a conceptual picture is provided (Figs. 1, 2). Studies can be conducted using a single allocation or a compound allocation (Figs. 1, 2; Table 1).

Greenhouse Gas Emission Impacts of Cropping Systems

The long-term effects of SP/NP along with two levels of residue retention on C footprint estimation of a rice-based cropping system was conducted in north-western region of Bangladesh (FRG, 2018). The study sites are located at about 24°28′ N lat. and 88°46′ E long (Table 1). The details of the locations, methods, soil properties and other necessary information were collected from published studies (Alam et al., 2016, 2018, 2019a, b).

Table 1 Study sites at Alipur village, Durgapur Upazilla, Rajshahi division in the Agro-Ecological Zone level at Barind Tract, Bangladesh.

Crops and Location	*Soil Management*	*Residue Management*
Mustard	• Conventional tillage (CT) • Strip tillage (SP)	• Low residue retention (LR) • Increased residue retention (HR)
Rice (Irrigated/ Monsoon)	• Conventional puddling (CT) • Non-puddling (NP) followed by SP	• LR • HR

In the study, the GHG emissions from a rice-based cropping system were calculated (in CO_2eq) for four farming practices in the EGP (Durgapur, Rajshahi, northwest Bangladesh). The streamlined LCA approach was adopted to quantify LCA GHG of: (1) conventional practices with low residue retention (CTLR); (2) high residue retention (CTHR); (3) novel crop establishment practices (strip planting for upland crop and NP transplanting following strip tillage for wetland crops) with low residue retention (SP/NPLR); and (4) with high residue retention (SP/NPHR). The LCA, covering cradle-to-farm gate GHG emissions, comprises four steps: goal setting and scope definition, life cycle inventory, impact assessment, and interpretation (ISO 14040-44, 2006). The details of crop varieties, their intercultural operations, input applied, soil properties. and climate of the experimental site were described (Alam et al., 2016, 2019a, b).

The goal was accomplished with a functional unit which is the production of one ton (t) of paddy rice grain. Though, the system boundary can be from cradle to grave (pre-farm, on-farm and post-farm up to consumption) depending on the objectives of intended research, the system boundary of the study consisted of pre-farm and on-farm life cycle stages. The input and output data of these life cycle stages for producing one ton of rice are then quantified to form life cycle inventories (LCI) for CT, and NP with LR and HR retention. The LCI of monsoon rice, irrigated rice, and monsoon rice-irrigated rice-mustard cropping system can be found in the studies published in Alam et al. (2016a, 2019a, b). The GHG emissions from pre-farm stage involves the

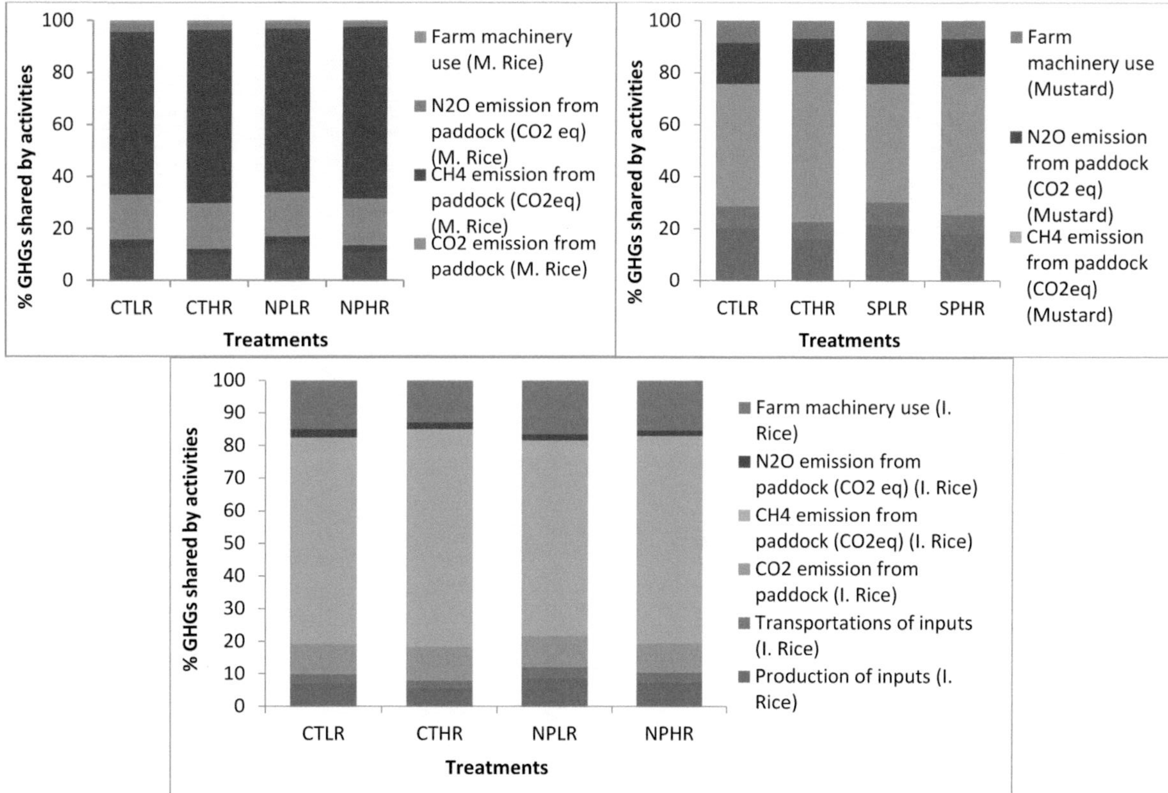

Fig. 5 Contribution of different activities in terms of agricultural GHG emissions due to crop establishment and residue retention practices in the monsoon rice-mustard-irrigated rice cropping system ($p \leq 0.05$). The emissions are converted to CO_2eq with the use of the global warming potentials of CO_2, CH_4, and N_2O across 100-yr time horizons. [*Legend:* CT–Conventional puddled transplanting of rice; NP–Non-puddled transplanting of rice; LR–Low residue retention level; HR–Increased residue retention level]. (Adapted from Alam, 2018).

remains unclear (Adviento-Borbe et al., 2013). However, no comprehensive studies with LCA have been carried out to evaluate the impacts of DSR and drill-seeded rice systems on GHG emission.

Shifting from flooded rice systems to drier crop production systems might lead to change in GHG emission from CH_4 under wet soil to abundant N_2O emissions under drier soil conditions. Weller et al. (2014) reported that though CH_4 emissions from the upland rice were lower than the paddy rice system, yield was below average, and also a shift from the paddy system to the upland system was only amount to pollution swapping as N_2O emissions were higher in the upland system. The GWP of the non-flooded crops was lower compared to flooded rice, whereas high CH_4 emissions under flooded conditions override increased N_2O emissions in the upland systems. Weller et al. (2014) indicated that the yield-scaled GWP favored maize over aerobic rice, due to lower yields of aerobic rice. However, the lower GHG emissions of upland systems are only beneficial if they are not overwhelmed by accelerated loss of SOC. To date there is a lack of LCA studies on the performance of aerobic vs wetland rice in terms of yield-scaled GWP on the EGP.

Accounting Carbon Sequestration in Life Cycle Analysis of Greenhouse Gas Estimation

There was an over-estimation by 6–40% of life cycle GHG by not including C sequestration in the C footprint equation. Production of 1 ton of REY of the rice–based system was caused for 0.73, 0.74, 0.98, and 1.12 ton of CO_2–eq actual LCA GHG emission with taking sequestered C into account. Production of 1 ton of irrigated rice in the EGP after accounting for C sequestered in soils accounted for 0.91-, 0.95-, 1.25- and 1.41-ton CO_2–eq for NPLR, NPHR, CTLR, and CTHR, respectively, whereas the LCA GHGs for the production of 1 ton of monsoon rice were 1.1, 1.21, 1.4, and 1.65 ton, respectively. For RE mustard, NPLR, NPHR, CTLR, and CTHR are responsible for 0.09, 0.18, 0.31, and 0.29 ton, respectively (Fig. 6).

The relative contribution of crops changed with C sequestration accounting. The relative contribution of monsoon rice ranged from 47–55% of the actual CS LCA GHG of which the lowest portion was contributed by SPLR and the highest portion was by CTHR. The relative contribution of the irrigated rice was 41–44% of the actual CS LCA GHG, while REY of mustard contributed 4–11% of the actual CS LCA GHG, when grown under SPHR practice (Fig. 7). The majority of LCAs of agricultural products have not included changes in SOC sequestration due to diverse cropping or soil management practices. While agricultural ecosystems can emit C as CO_2 and CH_4, they can also sequester C (Zhang et al., 2017).

While monsoon rice is a high CH_4 emitter which can be offset in part by high C sequestration. The actual LCA GHG emissions of the current practice of rice production was similar to that of total LCA GHG of the CA practice, non-puddled transplanting of rice with increased residue retention (NPHR) (Fig. 7). Goglio et al. (2015) and Petersen et al. (2013) found that accounting for SOC using long-term cropping system is important for finding actual/net LCA GHGs for any crop production practices. Consequently, after accounting for SOC sequestration, the LCA GHG of

NPHR was significantly lower than the actual GHG of CTLR. The NPHR had 15.5% lower actual life cycle GHG, while NPLR had even 32% lower emissions due to the reduced contribution of CH_4 emission and the increased SOC sequestration (Fig. 7).

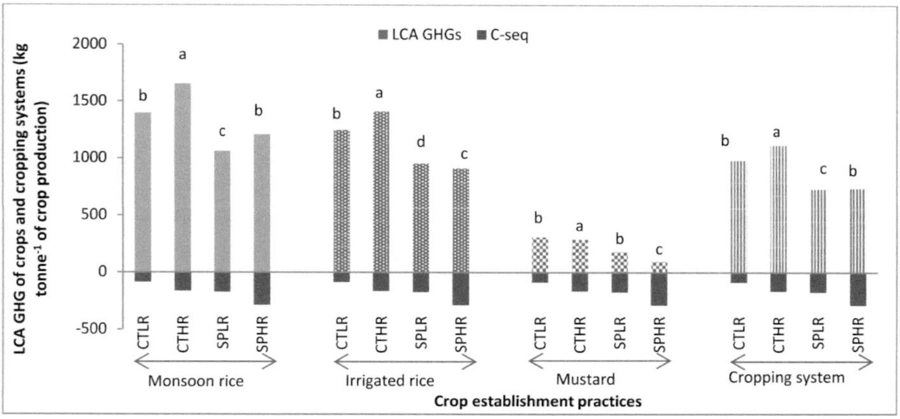

Fig. 6 Actual life cycle assessment based estimated GHG emissions (CS LCA GHG) with C sequestration t^{-1} production of component crops (irrigated rice, monsoon rice and mustard) and cropping system as influenced by crop establishment techniques and residue retention. Bars with the same letter above them are not significantly different at $p \leq 0.05$. Before comparisons, all emissions were converted to CO_2eq according to GWP of CO_2, CH_4 and N_2O over 100-year time horizons. [*Legend:* CT–Conventional puddled transplanting of rice; NP–Non-puddled transplanting of rice; LR–Low residue retention level; HR–Increased residue retention level] (Adapted from Alam, 2019).

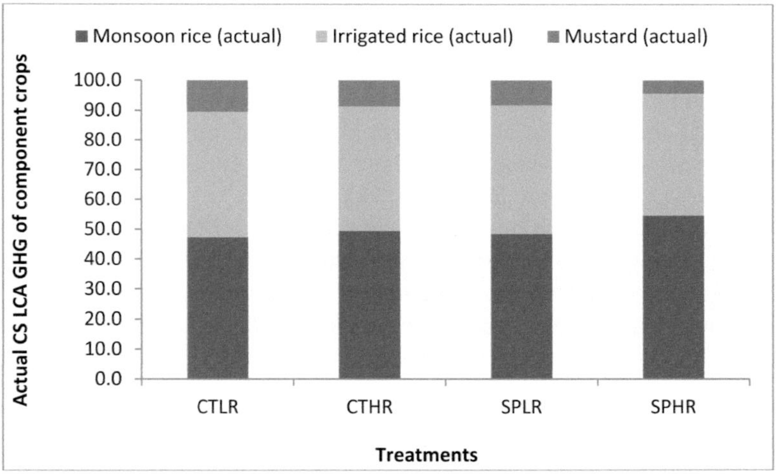

Fig. 7 Percent contributions of component crops on actual cropping system life cycle greenhouse gas emission (CS LCA GHG). [*Legend:* CT–Conventional puddled transplanting of rice; NP–Non-puddled transplanting of rice; LR–Low residue retention level; HR–Increased residue retention level] [Adapted from Alam et al., 2019].

Life Cycle Analysis of Rice Production: State-of-the-Art

The rice-based cropping systems in the EGP, notably the double cropping rice-wheat systems, are now diversified with crops receiving low to high inputs. In addition, most of the rice growing regions around the world grow more than one crop in a yearly rotation. Comparisons on hotspots, relative contribution of crops to cropping system LCA GHG between single rice crop LCA and LCA on rice-anchored multi-crop LCA are absent in the literature.

Carbon footprints have been evaluated for major crops (Hillier et al., 2009) around the world, under different crop establishment practices (West and Marland, 2002) and farming systems (Dubey and Lal, 2009) According to Hillier et al. (2009), the average C footprint of the main crops grown in the UK was 1.290-ton CO_2eq ha^{-1} $year^{-1}$, with the usage of N fertilizer accounting for 75% of this value. The study by West and Marland (2002) on agricultural practice revealed a lower C footprint in crop yield with no-tillage than under CT in the USA.

Harada et al. (2007) used LCA approach to compare GHG between CT and no-till rice farming in Japan, while Breiling et al. (2005) used the top-down LCA approach based on economic input output database to estimate GHG (CH_4 and N_2O) emissions of rice production in Japan. While CO_2 and N_2O emissions did not differ between the crop planting/establishment techniques, the cumulative CH_4 emissions from no-till planting were 43% lower than those from CT practice. According to Brodt et al. (2008), LCA analysis of GHG emissions estimated that 2.82 kg CO_2eq kg^{-1} milled rice were generated from rice production in California, USA. With potentials for eutrophication and global warming as indicators in Japan and with the data of over two growth seasons collected from over 55 ha area of a large farm over two growth seasons (2007 and 2008) for LCI, Hokazono et al. (2009) employed LCA approach to assess the production of conventional, organic, and sustainable (environmentally friendly) rice. The findings showed that the functional unit (FU)—in that study, the FUs were area, mass, and monetary value—determines the environmental performances.

The GHG emissions (kg of CO_2eq) per kilogram of brown rice in conventional rice systems were 1.52, which was greater than the emissions from sustainable farming (1.62 kg) and organic farming (1.34 kg).

Based on the implications of delivering white milled rice from cradle to supermarket, Blengini and Busto (2009) performed LCA with the functional unit kg^{-1} to the Italian rice manufacturing chain. The $unit^{-1}$ of farmed area could be less affected by organic and upland farming techniques. For the post-farm stage up till consumption, Roy et al. (2007) calculated the energy consumption and the environmental burden (CO_2 emission) of several parboiled rice processes in Bangladesh (vessel medium-boiler and untreated methods).

Using data on 400 farms, 24 milling facilities, and various national and international databases, Yossapol and Nadsataporn (2008) applied LCA approach to rice production in Thailand using un-milled grain as an FU and evaluated the environmental impact categories of GWP, Acidification, Eutrophication, Energy Consumption, and Abiotic Depletion. Water was the input to the system as a resource, but the output (impact compartment) was ignored. Kasmaprapruet et al. (2009) employed LCA approach to milled rice production so that environmental load can be

Table 2 Summarizing previous studies on life cycle analysis of rice production system.

Studies	*1*	*2*	*4*	*6*	*7*	*8*	*9*	*10*	*11*
References	Breiling et al., 1999	Roy et al., 2007	Brodt et al., 2008	Blengini & Busto, 2009	Kasmaprapruet et al., 2009	Hokazono & Hayashi, 2012	Wang et al., 2012	Xue et al., 2015	Alam et al., 2016
Functional unit	1 million yen of rice	1 tonne of milled & head rice	1 kg of milled rice	1 kg of delivery rice packed	1 kg of milled rice	1 kg of brown rice of organic, environment friendly & CT farming	1 ton of paddy rice	1 ton of paddy rice $year^{-1}$	1 ton of paddy rice
System boundary	Cradle-to farm-gate	Post-harvest to cooking	Cradle-to-mill gate	Cradle-to supermarket	Cradle-to-mill gate	Cradle-to-farm gate	Cradle-to-farm gate	Cradle-to-farm gate	Cradle-to-farm gate
Allocation rule	Economical Allocation	Mass Allocation	Mass allocation	Economical allocation	Economical allocation	Mass and Economic allocations	Not mentioned	Mass allocation	Mass allocation
Technical farm data	Secondary Data	Secondary Data	Interviewing industry representative & cooperative staff, secondary data	Site-wise records, interviewing farmers, and secondary data	Average practices observed	Farmers' interviews and secondary data	Local experts and farmer interviews, Secondary data	Average practices observed	Primary and secondary data
Direct field emission	Own model, National References	Secondary Data	Specific models, secondary data, & Ecoinvent	Primary& Secondary data, Ecoinvent	Asian and Thai references	Secondary data	Secondary data	Direct field measurements	Field measurements and secondary data

Contd.

Table 2 *Contd.*

Studies	*1*	*2*	*4*	*6*	*7*	*8*	*9*	*10*	*11*
Climate change	1.7-3.2 ton CO_2 emission	920-1320 kg CO_2eq	1.93-2.82 Kg CO_2eq	2.76-2.88 Kg CO_2eq	2.93 kg CO_2eq	1.46-2.0 kg CO_2-eq	1,570 kg CO_2eq	For both rice growing seasons 1.27, 1.85, and 1.40 kg CO_2eq year^{-1} for NT, RT, and CT, respectively	1.11-NP; 1.57 t CO_2eq/t rice-CT
Eutrophication	–	–	–	328.3-334.7 g CO_2eq	12.90 g NO_3-q	0.0076-0.0104 kgPO_4^{-3}eq	13.2 kg PO_4-eq	–	–
Acidification	–	–	–	0.25-0.28 mol H+	3.19 g SO_2eq	0.001-0.0023 kg SO_2eq	25.17 kg SO_2eq	–	–
Energy consumption	–	1028 – 1144 kWh of electricity	–	15.72 – 17.81 MJ	–	4.6 – 6.1 MJ of Non-renewable energy	3526 MJ	–	–

Legend: NT-No-tillage; RT-Reduced tillage; CT-Conventional tillage.

Table 3 Summary of life cycle greenhouse gas emission data of studies on rice production in the rice growing areas around the world.

Study (Ref.)	*Cultivation Practices*	*Emission ($t\ CO_2eq\ t^{-1}$ rice)*	*Yield ($t\ ha^{-1}$)*	*Growing Environment*
Alam et al. (2016), Bangladesh	Conventional puddling Non-puddling	Total net life cycle GHG emissions to farm gate (1.11-non-puddling; 1.57-puddling)	6.36 (puddling) 6.68 (non-puddling)	Irrigated (dry season)
Alam et al. (2019b), Bangladesh	Conventional puddling Non-puddling	NPLR-0.72 NPHR-0.72 CTLR-0.97 & CTHR-1.09	REY: CTLR-18.16 CTHR-19.33 NPLR-18.78 NPHR-20.38 $t\ ha^{-1}$	Irrigated and monsoon rice and upland crop (mustard)
Brodt et al. (2014), USA (California)	Direct water-seeding practices	100-year GWP: 1.47 kg $CO_2eq\ t^{-1}$ of milled rice (to farm-gate 1.01); IPCC Tier 1 estimates: 3.60 (to farm-gate 1.09).	9.3 (dried paddy rice)	Continuously flooded (rain-fed)
Hokazono and Hayashi (2012), Japan	Conventional, environment-friendly, and organic rice farming	Total net life cycle GHG of milled rice Conventional-1.46 Environmentally friendly-1.58 Organic-2.0	Organic (3.38), environment friendly (4.44), and conventional rice (4.36), respectively	Rain-fed
Ecoinvent Centre (2008)	Existing/ traditional	Total net life cycle GHG to farm gate (0.47)	7.5	Rain-fed
Blengini and Busto (2009), Italy	Traditional rice establishment	Total net life cycle GHG to milling 2.52–2.66	6.1	Rice cultivated without flooding and grown under a reduced water regime.
Thanawong et al. (2014), NE Thailand	Sowing by dry seeded and wet seeded/ transplanting (nursery)	Total net life cycle GHG to farm-gate 2.97–5.55	2.36–3.02	Both rain-fed and irrigated systems
Wang et al. (2010), China	Traditional rice establishment	Total net life cycle GHG to farm-gate (1.50)	8.8	Rice–wheat system where rice grown in monsoon season

Contd.

Table 3 *Contd.*

Study (Ref.)	*Cultivation Practices*	*Emission ($t\ CO_2eq\ t^{-1}$ rice)*	*Yield ($t\ ha^{-1}$)*	*Growing Environment*
Bautista and Saito (2015), Philippines	Traditional rice establishment	Total net life cycle GHG to farm gate (0.93) Total net life cycle GHG to farm gate (0.47)	4.21 (Irrigated) 2.93 (rain-fed)	Irrigated and rain-fed
Harada et al. (2007)	Puddling No-tilling, Non-puddling	Net life cycle GHG to milling (Brown rice) Puddling-0.94 No-tilling-0.44 Non-puddling-0.76	Puddling-4.43 No-tilling-5.49 Non-puddling-5.63 t ha^{-1}	Irrigated

Legend: REY-Rice Equivalent Yield

ɸ Life cycle GHG-Life cycle greenhouse gas emission

[a]Puddled transplanting with low residue retention (CTLR); [b]Puddled transplanting with high residue retention (CTHR); [c]non-puddled transplanting with low residue retention (NPLR); and [d]non-puddled transplanting with high residue retention (NPHR).

determined. The GWP of ton^{-1} milled rice production was 2.93-ton CO_2eq, followed by 3.19 kg SO_2eq for acidification and 12.9 kg NO_3eq for eutrophication and around 95% of the GWP inputs to the system are associated with the cultivation process, 2% with the harvesting process, and 2% with the seeding and milling processes.

The LCA of Jimmy et al. (2017) has highlighted the magnitude of impact kg^{-1} of paddy produced, including a CO_2eq emission of 3.15 kg as GWP, fossil resource depletion of 0.68 kg oileq, 14-DCB-kg oileq emission of 1.15 kg as human toxicity and a non-methane volatile organic compounds emission of 0.016 kg. According to contribution analysis, the irrigation and paddy field emissions are the two primary effects that are categorized as the most detrimental stages to the environment. As summarized in Tables 2 and 3, most of the LCA studies were conducted in rain-fed conditions in other rice growing areas except for Harada et al. (2007) in Japan. By contrast, Bautista and Saito (2015) in Philippines and Thanawong et al. (2014) in Northeast Thailand conducted studies in both rain-fed and irrigated conditions and showed that LCA GHGs up to farm-gate stage were lower under rain-fed conditions (Tables 2, 3). Thanawong et al. (2014) reported that total net LCA GHG up to farm-gate system boundary was ranged from 2.97–5.55-ton ha^{-1}. Most of the LCA studies conducted in rice/rice-based systems have examined the effects of rice crop establishment and production systems like direct water seeding, organic rice, environment-friendly rice, dry and wet direct seeding of rice, while Harada et al. (2007) contrasted no-tilling and non-puddling practices for irrigated rice production with puddling practices (Tables 2, 3). The net LCA GHG up to milling (brown rice) for puddling, no-tilling and non-puddling were 0.94-, 0.44-, and 0.76-ton CO_2eq ton^{-1} brown rice, respectively. The rice production practices adopted in the study of Harada et al. (2007) were conventional tillage and planting without puddling. The elimination of puddling, therefore, saved 0.18-ton CO_2eq ton^{-1} brown rice. The emerging non-puddled transplanting of rice following minimal disturbance of soil (strip tillage) in a rice-based triple cropping system (where other upland crops are growing following strip planting) has performed well in both biogenic GHGs and yield scale LCA GHG reduction (Tables 1, 2) under flooded, irrigated conditions (Alam et al., 2016, 2019b).

Life Cycle Analysis of Greenhouse Gas (C footprint) in Rice-Upland Cropping Systems: Sate-of-the-Art

Due to challenges of yield stagnation, water and labor shortages, soil, water and air pollution, farmers need alternatives to help conserve energy and water resources, reduce GHG emissions, and improve the quality of life (Singh et al., 2009). However, alternative practices to conventional systems in the rice-based cropping systems on the EGP need to be assessed (Chakraborty et al., 2017). The adoption of the CA practices depends very much on the performances of the practices in terms of mechanical suitability, agronomic outputs and environmental impact. The intensive cropping systems practiced on the EGP offer many dynamics / alterations in soils and biogeochemical processes when CA are introduced for crop production replacing conventional practice (Haque et al., 2016; Islam, 2017; Bell et al., 2018; Alam et al., 2019b). The emerging practices of CA (SP for upland crops, NP for rice and residue retention) being adopted in the EGP have been studied for their mechanization potential and agronomic performance (Haque et al., 2016; Islam, 2017; Salahin et al., 2017; Bell et al., 2018). To strengthen the dissemination and adoption of the CA practices, the performances of the novel practices in terms of C and N turnover and storage potentials are critical. The GHG implications of the emerging and conventional rice-based cropping systems are scarce for the IGP, let alone on the EGP. While studies have reported higher SOC sequestration with the adaptation of CA practices (Sharma et al., 2008), it is also required to estimate the relative contribution of each crop of rice-based cropping systems to the CS LCA GHG, so that the crop effects in the intensive cropping systems on the actual CS LCA GHG can be determined.

The soil disturbances, crop types and growth stages, chemical inputs, irrigation, seasonal wetting-drying events may contribute up to 70% of the total annual GHG flux (Barton et al., 2008; Wolf et al., 2010; Barton et al., 2013). When rice and upland crops are cultivated in rotation, variations in temperature and water regimes, cropping duration, biomass output, energy usage efficiency, inputs, amount and quality of residue, and other inputs all affect crop productivity. The wetland rice-upland cropping system has diverse residue return practices developed for fitting CA or intensifying crop cultivation and simultaneously reducing resource uses. While upland crops are typically grown on well-drained soils, rice is typically transplanted in soils that have been repeatedly puddled. According to Duxbury et al. (1993), chemicals used for pest control and as fertilizers, residue management, and tillage practices all have a significant impact on GHG emissions globally and ultimately on SOC sequestration.

While CH_4 was the main GHG during wetland rice cropping and CO_2eq emissions linked with irrigation contribute significantly, N_2O is the primary contributor to the total GWP during upland crop cultivation (Smith et al., 2007). Growing crops in rotation with minimal soil disturbance and increased residue retention enhance the regenerative processes, while C inputs help to increase SOC sequestration (Freibauer et al., 2014).

Nemecek et al. (2015) indicated that diverse cropping system with reduced chemicals use are a promising way to reduce the environmental impacts of intensive agricultural production systems. Lemke et al. (2007) suggested that inclusion of pulses in cereal-based cropping systems influenced the net GHG balance of those systems with variable pesticide and fertilizer requirements and return variable quality and quantity of residues to soil compared with the mono-cultured cereal crops; however, crops with high N requirements increase N_2O emissions and that, N_2O production increases non-linearly with increased N fertilization (McSwiney and Robertson, 2005; Burton et al., 2008; Zebarth et al., 2008). Badey et al. (2013) found N_2O accounted for 44–54% of total GHG emitted by oilseed crops. In contrast, N_2O is the most dominant GHG, being responsible for 80% of wheat GHG emissions under arable agriculture (Woods et al., 2008). Eshun et al. (2013) in an LCA showed that N_2O contributed the highest proportion (about 70%) of GWP for rice production, followed by CO_2. The CH_4 emissions from wetland rice contribute from 40–67% of the total LCA GHG up to farm-gate boundary of LCA (Fusi et al., 2014). The higher emissions under both irrigated and flooded rice cultivation can be attributed to CH_4 emissions which shared the major part of GHG in rice cultivation (Takai and Kamura, 1966; Yu and Chen, 2004). Farag et al. (2013) in their LCA showed that CH_4 emission from the flooded rice fields was the main source of GHG emissions, contributing about 53%, while N fertilization added about 10%, and mechanical activities about 1% of the total emissions.

In irrigated rice systems with adequate water management, the anaerobic condition developed in soils within few hr. or few days of flooding (Bodelier, 2003), which started building constantly very low soil redox potential (Eh) for having restricted gas diffusion in flooded soils (Adhya et al., 2000). Eventually, the Eh in flooded soils stabilizes in a range of +200 to −300 mV depending on the soil, especially on the quantity and quality of organic matter and texture, and the presence and contents of reducible nutrient elements (Sahrawat, 2015). The reduced Eh condition in soil maintains an overall stable levels of GHG emissions apart from changes in GHG emission patterns during chemical fertilization.

Large amount of N fertilizer applications could significantly increase N_2O emissions due to volatilization of NH_3 and NO_3 via leaching and denitrification (Gan et al., 2014; Law et al., 2012). However, several studies have suggested that N_2 loss via NH_3 volatilization, nitrate leaching, and N_2O emission should also be accounted for in the LCA. Currently most studies do not measure N_2 loss directly (Chen et al., 2014a; Zhang et al., 2017), the loss and its contribution to GHG emissions are estimated based on models (Chen et al., 2014b). To improve the accuracy of C footprint estimation, there remains scope to measure N_2 loss and incorporate the related CF in the LCA to estimate more accurate GWP.

Additional Climate-Smart Practices to Reduce Greenhouse Gas Emissions

Improved agronomic practices can minimize GHG emissions and increase SOC, while reducing the environmental impacts. Changes in irrigation management, including alternate wetting-drying, drainage, drip irrigation, land levelling, and alternate furrow irrigation for upland crops, and shallow flooding for wetland rice, can save water and lower GHG emissions (Wassmann et al., 2000). Modified irrigation techniques are efficient in reducing GHG emissions while conserving water because antecedent soil moisture plays an important role in managing GHG emissions. One midseason drainage, for instance, can reduce the CH_4 emission from the irrigated rice fields by almost 50%. Alternate wetting-drying has been successful in reducing CH_4 emissions by 70% from rice fields (Wassmann et al., 2000; Pathak et al. 2003). However, because of the processes involved in denitrification, there can be an increase in N_2O emissions. The increase in N_2O emissions resulted from alternate wetting-drying or midseason drainage is non-significant when compared to the decline in CH_4 emissions. Several studies (Bhatia et al., 2012; Pathak et al., 2014, Sarker et al., 2020) have reported that irrigation management is a viable option to reduce CH_4 emissions and conserve water in rice production. Increased N_2O emissions partially balance the effect of midseason drainage and intermittent irrigation, which can reduce CH_4 flux by 40% and lower net GWP (Zou et al., 2005). Changes in irrigation scheduling from constant flooding of land to draining twice each season, resulting in a 60% reduction in CH_4 emissions (Wang et al., 2012). The direct N_2O flux was reduced by 50% in response to drip irrigation and optimal N fertilization (Sanz-Cobena et al., 2017).

The quantity of N provided and the effectiveness with which it is absorbed by plants are two important variables that regulate the flux of N_2O. Therefore, optimization of N fertilization is essential to reduce N_2O emissions (Vijayakumar et al., 2021). If N application corresponds with crop needs, mid-season drainage and intermittent irrigation can reduce the net GWP of paddy fields (Wassmann et al., 2009). According to Zou et al. (2005), rice fields receiving a higher amount of N fertilizer (277 kg N ha^{-1}) had an increase in N_2O flux with midseason drainage in China. It is reported that despite employing N fertilizer at lower rates (between 30 and 90 kg ha^{-1}), there was a modest rise in N_2O emissions as compared to the traditional water-management techniques. This implies that by optimizing environmental and agronomic parameters, the effect of water management on N_2O emissions may vary.

Nitrogen Footprints in Rice-based Cropping Systems and Their Implications

Chemical N fertilization is common in conventional agriculture to sustain crop productivity. According to Cameron et al. (2013), more than 50% of the N applied to the field (global average) in intensive agricultural systems moves beyond the root zone of crops. In Bangladesh, the N-use efficiency is even poorer than it is elsewhere (up to 65% of the N applied to the field is wasted). The excess N might be lost to the aquatic and atmospheric environments, where it could cause major pollution and pose a threat to conservation. In contrast to ammonium (NH_4^+), which binds to soil negative charges, nitrate (NO_3^-), and urea are highly soluble substances that can be susceptible to edge-of-field loss especially leaching into groundwater or runoff into surface water systems. Nitrogen loss from the soil to the atmosphere can be caused by NH_3 volatilization, nitrification to N_2O, and denitrification to N_2 (Cameron et al., 2013). A decline in biodiversity both inside and outside of agricultural systems, as well as eutrophication of freshwaters, estuaries, coastal waters, and ecosystems on nutrient-poor land, have all been linked to the application of N fertilization (Fowler et al., 2013).

To provide information on how individual and collective behavior leads in the loss of N to the environment, a tool called N-Calculator has recently been developed utilizing the Leach et al. (2012) proposed N 'footprint' model. The N-Calculator populate average capita^{-1} data for a nation to concentrate on food and energy consumption. There is a potential for minimizing environmental pollution and enhancing environmental sustainability in the development of N footprints for rice-based cropping systems. We can learn a lot about how effectively these systems use N as a whole and what effects that has on the environment by analyzing the N inputs, outputs, and related environmental implications within these systems. An N footprint offers a thorough understanding of N dynamics at different stages of crop production, from fertilization to harvest. This all-encompassing technique enables us to pinpoint key locations where N is lost, such as N_2O emissions, a potent GHG, and $NO3^-$ leaching into water systems. Based on this information, customized actions can be created to improve N management, reduce waste, and protect the environment.

The advantages of developing an N footprint for rice-based cropping systems are numerous. First, it encourages the adoption of sustainable N management practices by enabling farmers and policymakers to make well-informed decisions. Second, a smaller N footprint results in lesser emissions of N compounds that pollute ecosystems and human health by contributing to air and water pollution. Thirdly, improved N use efficiency decreases the need rate of chemical fertilizers. Furthermore, as N management and agricultural GHG emissions are closely related, a smaller N footprint also results in a smaller overall C footprint. Significant progress can be made towards reaching environmental sustainability goals, such as cleaner air and water, healthier ecosystems, and a more resilient agriculture system, by tackling N-related environmental challenges. To put it briefly, the development of an N footprint for rice-based cropping systems presents a users' friendly tool for reducing pollution, increasing N use efficiency, and advancing environmental sustainability. The foundation of an economically viable and environmentally compatible approach to support rice growing can be formed by understanding how N management interacts with its wider effects (Leach et al., 2012).

When compared to bed planting (BP) and conventional practice (CT) with increased HR, soil total N was increased by 9 and 32% at Rajshahi, Bangladesh (Alipur) by SP which included NP of rice. The employed CA practices (SP, and NP of rice, together with HR) have influence the N cycling by decreasing the concentration of available N to plants in the initial crop establishment with low crop N demand, and by increasing soil total N content and plant N uptake, which enhanced the synchrony between crop demand and available N supply (Alam et al., 2020a). So, the development of N footprint for intensive rice-based cropping systems would help identify the key issues of N management and loss where the greatest N loss reduction can be achieved. The adoption of CA practices would reduce the N footprint for crop production, while ensuring the savings of money and promoting the healthy environment. However, adopting N management practices includes, but not limited to, apply N fertilizer at optimum rates, applying N by maintaining crop demand synchrony, growing cover crops to utilize the residual mineral N, splitting N fertilization, using urease or nitrification inhibitors, using mixed cropping with legumes or intercropping/legume based cropping system, crop residue management, site-specific N fertilizer management, integrated nutrient management, and cropping diversity.

Conclusions

Incorporating Conservation Agriculture into rice-based systems is a strategic focus for scientists aiming to achieve economic viability and environmental compatibility. Practices like increased residue retention, strip planting of upland crops, and non-puddled transplanting of rice under CA have demonstrated positive effects on carbon and nitrogen dynamics. This includes heightened SOC sequestration and reduced GHG emissions. Additional benefits encompass balanced soil pH, improved soil fertility, and enhanced soil structural stability. The adoption of these practices in rice-based cropping systems has resulted in significant reductions in GHG emissions throughout both pre-farm and on-farm stages of crop production, contributing to global warming mitigation. Specifically, strip planting and non-puddled transplanting with increased residue retention exhibit the highest increase in SOC sequestration, leading to improved soil health. The incorporation of mustard into the cropping system further enhances the sustainability profile by reducing overall life cycle assessment GHG emissions. This reduction is attributed to practices like residue retention and minimal soil disturbance, leading to increased SOC sequestration and decreased CH_4 emissions. Despite variations in tillage systems or residue retention, CH_4 remains the predominant GHG emitted per ton of production for irrigated and monsoon rice, while for mustard, CO_2eq emissions during the on-farm stage take precedence. In summary, the pursuit of CA practices in rice-based systems not only aims for economic and environmental benefits but also results in reduced GHG emissions, improved soil health, and increased sustainability. The findings highlight the potential of these practices to contribute significantly to a more environmentally friendly and resilient approach to rice cultivation.

References

Adhya, T.K., Bharati, K., Mohanty, S.R., Ramakrishnan, B., Rao, V.R., Sethunathan, N. and Wassmann, R. (2000). Methane emission from rice fields at Cuttack, India. *Nutr. Cycling Agroecosys*, *58*: 95–105.

Adler, R.A., Del Grosso, S.J. and Parton, W.J. 2007. Life-cycle assessment of net greenhouse-gas flux for bioenergy cropping systems. *Ecol. Appl.*, *17*: 675–91.

Adviento-Borbe, M.A., Pittelkow, C.M., Anders, M., van Kessel, C., Hill, J.E., McClung, A.M., Six, J. and Linquist, B.A. (2013). Optimal fertilizer nitrogen rates and yield-scaled global warming potential in drill seeded rice. *J. Environ. Qual.*, *42*: 1623–34.

AIJ (Architectural Institute of Japan). (2003). LCA database based on inter–industry relations table in 1995. *In*: *Guiding Principle of LCA for Architectures*. Tokyo: Architectural Institute of Japan.

Alam, M.K., Bell, R.W., Haque, M.E. and Kader, M.A. (2018). Minimal soil disturbance and increased residue retention increase soil carbon in rice-based cropping systems on the Eastern Gangetic Plain. *Soil Till Res.*, *183*: 28–41.

Alam, M.K., Salahin, N., Islam, S., Begum, R.A., Hasanuzzaman, M., Islam, M.S. and Rahman, M.M. (2016a). Patterns of change in soil organic matter, physical properties and crop productivity under tillage practices and cropping systems in Bangladesh. *The Journal of Agricultural Science*, *155*: 216–38.

Alam, M.K., Biswas, W.K. and Bell, R.W. (2016b). Greenhouse gas implications of novel and conventional rice production technologies in the Eastern-Gangetic plains. *J. Clean. Prod., 112*: 3977–3987.

Alam, M.K., Biswas, W.K. and Bell, R.W. (2019a). Decreasing the carbon footprint of an intensive rice-based cropping system using conservation agriculture on the Eastern Gangetic Plains. *J. Clean. Prod., 218*: 259–72.

Alam, M.K., Biswas, W.K. and Bell, R.W. (2019b). Increases in soil sequestered carbon under conservation agriculture cropping decrease the estimated greenhouse gas emissions of wetland rice using life cycle assessment. *J. Clean. Prod.* (under review).

Alam, M.K., Islam, M.M., Salahin, N. and Hasanuzzaman, M. (2014). Effect of tillage practices on soil properties and crop productivity in wheat–mung bean–rice cropping system under subtropical climatic conditions. *The Scientific World Journal, 2014*: 1–10.

Alam, M.K. and Salahin, N. (2013). Changes in soil physical properties and crop productivity as influenced by different tillage depths and cropping patterns. *BJAR, 38*: 289–99.

Alam, M.K., Bell, R.W., Haque, M.E., Islam, M.A. and Kader, M.A. (2020a). Soil nitrogen storage and availability to crops are increased by conservation agriculture practices in rice–based cropping systems in the Eastern Gangetic Plains. *Field Crops Res., 250*: 107764.

Alam, M.K., Bell, R.W., Hasanuzzaman, M., Salahin, N., Rashid, M.H., Akter, N., Akhter, S., Islam, M.S., Islam, S., Naznin, S., Anik, M.F.A., Apu, M., Bin Saif, H., Alam, M.J. and Khatun, M.F. (). Rice (*Oryza sativa* L.) establishment techniques and their implications for soil properties, global warming potential mitigation and crop yields. *Agronomy-Basel, 10*(6): 888.

Andrews, M. and Lea, P.J. (2013). Our Nitrogen 'Footprint': The Need for Increased Crop Nitrogen Use Efficiency. *Ann. Appl. Biol., 163*: 165–69.

Arunrat, N. and Pumijumnong, N. (2017). Practices for reducing greenhouse gas emissions from rice production in Northeast Thailand. *Agriculture, 7*(1): 4.

Badey, L., Lahitte, N., Flenet, F. and Bosque, F. (2013). French environmental communication on sunflower and rapeseed oils based on life cycle assessment. *OCL, 20*(4): A401.

Barton, L., Kiese, R., Gatter, D., Butterbach-Bahl, K., Buck, R., Hinz, C. and Murphy, D.V. (2008). Nitrous oxide emissions from a cropped soil in a semiarid climate. *Glob. Change Biol., 14*: 177–92.

Barton, L., Murphy, D. and Butterbach–Bahl, K. (2013). Influence of crop rotation and liming on greenhouse gas emissions from a semi-arid soil. *Agric. Ecosyst. Environ., 167*: 23–32.

Barton, L., Wolf, B., Rowlings, D., Scheer, C., Kiese, R., Grace, P., Stefanova, K. and Butterbach-Bahl, K. (2015). Sampling frequency affects estimates of annual nitrous oxide fluxes. *Sci. Rep., 5*: 15912.

Bautista, E.G. and Saito, M. (2015). Greenhouse gas emissions from rice production in the Philippines based on life-cycle inventory analysis. *J. Food Agr. Environ., 13*: 139–44.

BBS. (2022). *Yearbook of Agricultural Statistics of Bangladesh*. Ministry of Planning, Government of the Peoples' Republic of Bangladesh, Dhaka.

Bhatia, A., Aggarwal, P.K., Jain, N. and Pathak, H. (2012). Greenhouse gas emission from rice-and wheat-growing areas in India: Spatial analysis and upscaling. *Greenhouse Gases: Science and Technology, 2*: 115–125.

Biswas, W.K., Barton, L. and Carter, D. (2008). Global warming potential of wheat production in Western Australia: A life cycle assessment. *Water Environ. J., 22*: 206–216.

Blengini, G.A. and Busto, M. (2009). The life cycle of rice: LCA of alternative agri–food chain management systems in Vercelli (Italy). *J. Environ. Manag., 90*: 1512–1522.

Bodelier, P.L.E. (2003). Interactions between oxygen releasing roots and microbial processes in flooded soils and sediments, *In*: de Kroon, H., Visser, E.J.W. (Eds.), *Root Ecology*, 331–62. (Run on)Springer, Berlin.

Breiling, M., Hashimoto, S., Sato, Y. and Ahamer, G. (2005). Rice-related greenhouse gases in Japan, variations in scale and time and significance for the Kyoto protocol. *Paddy Water Environ., 3*: 39–46.

Brentrup, F. and Pallière, C. (2008). GHG emissions and energy efficiency in European nitrogen fertilizer production and use. *International Fertiliser Society, Proceedings* 639pp. December 11, York, UK.

Brodt, S., Kendall, A., Mohammadi, Y., Arslan, A., Yuan, J., Lee, I. and Linquist, B. (2014). Life cycle greenhouse gas emissions in California rice production. *Field Crops Res., 169*: 89–98.

Burton, D.L., Zebarth, B.J., Gillam, K.M. and MacLeod, J.A. (2008). Effect of split application of fertilizer nitrogen on N2O emissions from potatoes. *Can. J. Soil Sci.*, *88*: 229–39.

Cai, Z., Xing, G., Yan, X., Xu, H., Tsuruta, H., Yagi, K. and Minami, K. (1997). Methane and nitrous oxide emissions from rice paddy fields as affected by nitrogen fertilizers and water management. *Plant Soil*, *196*: 7–14.

Cameron, K.C., Di, H.J. and Moir, J.L. (2013). Nitrogen losses from the soil/plant system: A review. *Ann. Appl. Biol.*, *162*: 145–73.

Carbon Trust. (2008). Carbon trust. Carbon trust, viewed 11th of September 2008. Retrieved from http://www.carbon-label.co.uk/.

Chakraborty, D., Ladha, J.K., Rana, D.S., Jat, M.L., Gathala, M.K., Yadav, S., Rao, A.N., Ramesha, M.S. and Raman, A. (2017). A global analysis of alternative tillage and crop establishment practices for economically and environmentally efficient rice production. *Sci. Rep.*, *7*: 9342.

Chen, B., Liu, E., Tian, Q., Yan, C. and Zhang, Y. (2014a). Soil nitrogen dynamics and crop residues: A review. *Agron. Sustain. Dev.*, *34*: 429–42.

Chen, X., Cui, Z. and Fan, M., et al. (2014b.) Producing more grain with lower environmental costs. *Nature*, *514*: 486–89.

Devakumar, A.S., Pardis, R. and Manjunath, V. (2018). Carbon Footprint of Crop Cultivation Process under Semiarid Conditions. *Agric. Res.*, *7*: 167.

Dubey, A. and Lal, R. (2009). Carbon footprint and sustainability of agricultural production systems in Punjab, India, and Ohio, USA. *J. Crop Improvement*, *23*: 332–50.

Duxbury, J.M., Harper, L.A. and Mosier, A.R. (1993). Contributions of agroecosystems to global climate change. *In*: Harper, L., Mosier, A.R., Duxbury, J.M. and Rolston, D.E. (Eds.), *Agroecosystem Effects on Radiatively Important Trace Gases and Global Climate Change*, 1–18. ASA Special Publication No: 55. Amer. Soc. Agron. Madison. WI.

Ecoinvent Centre. (2008). *Ecoinvent Data v2.0*. Swiss Centre for Life Cycle Assessment, Dubendorf, Switzerland.

Engelbrecht, D., Biswas, W., Pritchard, D. and Ahmad, W. (2015). Integrated spatial technology to mitigate greenhouse gas emissions in grain production. *RSASE*, *2*: 44–55.

Eshun, J.F., Apori, S.O. and Wereko, E. (2013). Greenhouse gaseous emission and energy analysis in rice production systems in Ghana. *African Crop Science J.*, *21*: 119–25.

Farag, A.A., Radwan, H.A., Abdrabbo, M.A.A., Heggi, M.A.M. and McCarl, B.A. (2013). Carbon footprint for paddy rice production in Egypt. *Nat. Sci.*,*11*: 36–45.

Flessa, H. and Beese, F. (1995). Effects of sugar beet residue on soil redox potential and nitrous oxide emission. *Soil Sci. Soc. Am. J.*, *59*: 1044–51.

Fluck, R.C. (1992). Energy in farm production. *In*: Fluck, R.C. (Ed.), *Energy in World Agriculture,* (Run on) 6th Edn., 218–67. New York: Elsevier.

Fowler D., Coyle M., Skiba U., Sutton M.A., Cape J.N., Reis S., Sheppard L.J., Jenkins A., Grizzetti B., Galloway J.N., Vitousek P., Leach A., Bouwman A.F., Butterbach- Bahl K., Dentener F., Stevenson D., Amann M. and Voss M. (2013). The global nitrogen cycle in the twenty-first century. *Philosophical Transactions of the Royal Society B*, *368*: 20130164.

Freibauer, A., Rounsevell, M.D.A., Smith, P. and Verhagen, J. (2004). Carbon sequestration in agricultural soils of Europe. *Geoderma*, *122*: 1–23.

FRG (2018). *Fertilizer Recommendation Guide:* (*2012*). Bangladesh Agricultural Research Council (BARC): Dhaka, Bangladesh, 274 pp.

Fusi, A., Bacenetti, J., González–García, S., Vercesi, A., Bocchi, S. and Fiala, M. (2014). Environmental profile of paddy rice cultivation with different straw management. *Sci. Total Environ.*, 494–95, 119–128.

Gan, Y., Liang, C., Chai, Q., Lemke, R.L., Campbell, C.A. and Zentner, R.P. (2014). Improving farming practices reduces the carbon footprint of spring wheat production. *Nat. Commun.*, *5*: 5012.

Goglio, P., Smith, W.N., Grant, B.B., Desjardins, R.L., McConkey, B.G., Campbell, C.A. and Nemecek, T. (2015). Accounting for soil carbon changes in agricultural life cycle assessment (LCA): A review. *J. Clean. Prod.*, *104*: 23–39.

Gupta, P.K., Gupta, V., Sharma, C., Das, S.N., Purkait, N., Adhya, T.K., Pathak, H., Ramesh, R., Baruah, K.K., Venkataraman, L., Singh, G. and Iyer, C.S.P. (2009). Development of methane emission factors for Indian paddy fields and estimation of national methane budget. *Chemosphere*, *74*: 590–98.

Gustafson, D.I. and Yildiz, F. (2017). Greenhouse gas emissions and irrigation water use in the production of pulse crops in the United States. *Cogent Food and Agriculture*, *3*: 1.

Haque, M.M., Biswas, J.C., Salahin, M., Alam, K., Akhter, S., Akhtar, S., Maniruzzaman, M. and Hossain, M.S. (2022). Tillage systems influence on greenhouse gas emission factor and global warming potential under rice-mustard-rice cropping system. *Arch. Agron. Soil Sci.*, *68*: 1–16.

Haque, M.E., Bell, R.W., Islam, M.A. and Rahman, M.A. (2016). Minimum tillage unpuddled transplanting: An alternative crop establishment strategy for rice in conservation agriculture cropping systems. *Field Crops Res.*, *185*: 31–39.

Harada, H., Kobayashi, H. and Shindo, H. (2007). Reduction in greenhouse gas emissions by no-tilling rice cultivation in Hachirogata polder, northern Japan: Life cycle inventory analysis. *Soil Sci. Plant Nutr.*, *53*: 668–77.

HBEFA (2014). *Handbook on Emission Factors of Road Transport (HBEFA). Version HBEFA 3.2.* INFRAS, Graz.

Hillier, J., Hawes, C., Squire, G., Hilton, A., Wale, S. and Smith, P. (2009). The carbon footprints of food crop production. *International Journal of Agricultural Sustainability*, *7*: 107–18.

Hokazono, S. and Hayashi, K. (2012). Variability in environmental impacts during conversion from conventional to organic farming: A comparison among three rice production systems in Japan. *J. Clean. Prod.*, *28*: 101–12.

Hokazono, S., Hayashi, K. and Sato, M. (2009). Potentialities of organic and sustainable rice production in Japan from a life cycle perspective. *Agron. Rese.*, *7*(Sp. issue I): 257–62.

INFRAS. (2010). *The Handbook Emission Factors for Road Transport (HBEFA), v. 3.1.* INFRAS in cooperation with further editors: FOEN/Switzerland; Umweltbundesamt Dessau/Germany; Umweltbundesamt Wien/Austria; Swedish Road Administration, ADEME/France; SFT/Norway. Bern.

Intergovernmental Panel on Climate Change (IPCC). (2007). Agriculture. *In*: B. Metz, O.R. Davidson, P.R. Bosch (Eds.), *Climate Change 2007: Mitigation.* Contribution of Working Group III to the Fourth Assessment Report of the Intergovernmental Panel on Climate Change. Cambridge, United Kingdom/ New York, NY, USA: Cambridge University Press.

Islam, A.K.M.S., Hossain, M.M., Saleque, M.A., Rabbani, M.A. and Sarker, R.I. (2013). Energy consumption in unpuddled transplanting of wet season rice cultivation in Northwest region of Bangladesh. *Progress. Agric.*, *24*: 229–37.

Islam, M.A. (2017). *Conservation Agriculture: Its Effects on Crop and Soil in Rice-based Cropping Systems in Bangladesh*, PhD thesis. School of Veterinary and Life Sciences, Murdoch University, Australia.

ISO. (2012). Environmental management-life cycle assessment-principles and framework. Geneva, Switzerland. <https://www.astandis.at/

ISO14040. (2006). *Environmental Management-Life Cycle Assessment: Principles and Framework.* ISO Geneva: Switzerland, 2006.

Jimmy, A.N., Khan, N.A., Hossain, M.N. and Sujauddin, M. (2017). Evaluation of the environmental impacts of rice paddy production using life cycle assessment: Case study in Bangladesh. *Model. Earth Syst. Environ.*, *3*: 1691.

Kasmaprapruet, S., Paengjuntuek, W., Saikhwan, P. and Phungrassami, H. (2009). Life cycle assessment of milled rice production: Case study in Thailand. *Eur. J. Sci. Res.*, *3*: 195–03.

Kassam, A., Friedrich, T. and Derpsch, R. (2022). Successful Experiences and Lessons from Conservation Agriculture Worldwide. *Agronomy, 12*: *769*.

Khan, S., Khan, M.A., Hanjra, M.A. and Mu, J. (2009). Pathways to reduce the environmental footprints of water and energy inputs in food production. *Food Policy*, *34*: 141–49.

Kitzes, J. (2013). An introduction to environmentally–extended input–output analysis. *Resources*, *2*: 489–03.

Lal, R. (2004). Carbon emissions from farm operations. *Environ. Int.*, *30*: 981–90.

Law, Y., Ye, L., Pan, Y. and Yuan, Z. (2012). Nitrous oxide emissions from wastewater treatment processes. *Philos. Trans. R. Soc. Lond.*, *367*: 1265–77.

Leach A.M., Galloway J.N., Bleeker A., Erisman J.W., Kohn R. and Kitzes J. (2012). A nitrogen footprint model to help consumers understand their role in nitrogen losses to the environment. *Environmental Development*, *1*: 40–66.

Lemke, R.L., Zhong, Z., Campbell, C.A. and Zentner, R. (2007). Can pulse crops play a role in mitigating greenhouse gases from North American agriculture? *Agron. J.*, *99*: 1719–25.

Liu, C., Cutforth H., Chai Q. and Gan Y. (2016). Farming tactics to reduce the carbon footprint of crop cultivation in semiarid areas. A review. *Agron. Sust. Dev.*, *36*: 69–85.

Lynas, M. (2007). *Carbon Counter*. Glasgow: Harper Collins Publishers.

McSwiney, C.P. and Robertson, G.H. 2005. Nonlinear response of N_2O flux to incremental fertilizer addition in a continuous maize (*Zea mays* L.) cropping system. *Global Change Biol.*, *11*: 1712–19.

Miyata, A., Leuning, R., Denmead, O.T., Kirn, J. and Harazono, Y. (2000). Carbon dioxide and methane fluxes from an intermittently flooded paddy field. *Agric. Forest Meteorol.*, *102*: 287–303.

MOEF (Ministry of Environment and Forest, Peoples Republic of Bangladesh). (2010). *Bangladesh Association with the Copenhagen Accord.*

Naser, H.M. (2005). *Evaluating the Status of Greenhouse Gas Budgets of Paddy Fields in Central Hokkaido, Japan*, PhD. thesis. Hokkaido University, Sapporo, Japan, 87 pp.

Nemecek, T., Hayer, F., Bonnin, E., Carrouée, B., Schneider, A. and Vivier, C. (2015). Designing eco-efficient crop rotations using life cycle assessment of crop combinations. *Euro. J. Agron.*, *65*: 40–51.

Nemecek, T., von Richthofen, J.S., Dubois, G., Casta, P., Charles, R. and Pahl, H. (2008). Environmental impacts of introducing grain legumes into European crop rotations. *Eur. J. Agron.*, *28*: 380–93.

O'Dea, J.K., Miller, P.R. and Jones, C.A. (2013). Greening summer fallow with legume green manures: On-farm assessment in North-Central Montana. *J. Soil Water Conserve.*, *68*: 270–82.

Pandey, D., Agrawal, M. and Pandey, J.S. (2010). Carbon footprint: Current methods of estimation. *Environ Monit. Assess.*, *178*(1–4): 135–160.

Pathak, H., Bhatia, A. and Jain, N. (2014). *Greenhouse Gas Emission from Indian Agriculture: Trends, Mitigation, and Policy Needs*, 39pp. Indian Agric. Res. Institute, New Delhi.

Pathak, H., Prasad, S., Bhatia, A., Singh, S., Kumar, S., Singh, J. and Jain, M.C. (2003). Methane emission from rice-wheat cropping system of India in relation to irrigation, farmyard manure, and dicyandiamide application. *Agric. Ecosys. Environ.*, *97*: 309–16.

Pathak, H., Sankhyan, S., Dubey, S.S., Bhatia, A. and Jain, N. (2013). Dry direct-seeding of rice for mitigating greenhouse gas emission: Field experimentation and simulation. *Paddy Water Environ.*, *11*: 593–601.

Petersen, B.M., Knudsen, M.T., Hermansen, J.E. and Halberg, N. (2013). An approach to include soil carbon changes in life cycle assessments. *J. Clean. Prod.*, *52*: 217–24.

Roy, P., Shimizu, N., Okadome, H., Shiina, T. and Kimura, T. (2007). Life cycle of rice: Challenges and choices for Bangladesh. *J. Food Eng.*, *79*(4): 1250–55.

Ryu, J.H., Lee, J.S., Kim, K.H., Kim, G.Y. and Choi, E.J. (2013). A case study to estimate the greenhouse–gas mitigation potential on conventional rice production system. *Korean J. Soil Sci. Fert.*, *46*: 502–09.

Sahrawat, K.L. (2015). Redox Potential and pH as Major Drivers of Fertility in Submerged Rice Soils: A Conceptual Framework for Management. *Commun. Soil Sci. Plant Anal.*, *46*: *13*: 1597–1606.

Salahin, N., Alam, M.K., Mondol, A.T.M.A.I., Islam, M.S., Rashid, M.H. and Hoque, M.A. (2017). Effect of tillage and residue retention on soil properties and crop yields in Wheat-Mungbean-Rice Crop rotation under subtropical humid climate. *OJSS*, *7*: 1–17.

Salahin, N., Jahiruddin, M., Islam, M. R., Alam, M. K., Haque, M. E., Sharif, A., Baazeem, A., Hadifa, A., Sabagh, A.E.L. and Bell, R.W. (2021a). Establishment of Crops under Minimal Soil Disturbance and Crop Residue Retention in Rice-Based Cropping System: Yield Advantage, Soil Health Improvement, and Economic Benefit. *Land*, *10*(6): 581.

Salahin, N., Jahiruddin, M., Islam, M., Alam, K., Haque, M., Ahmed, S., Baazeem, A., Hadifa, A., EL Sabagh, A. and Bell, R. (2021b). Establishment of Crops under Minimal Soil Disturbance and Crop Residue Retention in Rice-Based Cropping System: Yield Advantage, Soil Health Improvement, and Economic Benefit. *Land*, *10*: 581.

Sanz-Cobena, A., Lassaletta, L., Aguilera, E., Del Prado, A., Garnier, J., Billen, G., et al. (2017). Strategies for greenhouse gas emissions mitigation in Mediterranean agriculture: A review. *Agriculture, Ecosystems, and Environment, 238*: 5–24.

Sapkota, T.B., Khanam, F., Mathivanan, G.P., Vetter, S., Hussain, S.G., Pilat, A., Shahrin, S., Hossain, M.K., Sarker, N.R. and Krupnik, T.J. (2021). Quantifying opportunities for greenhouse gas emissions mitigation using big data from smallholder crop and livestock farmers across Bangladesh. *Science of The Total Environment, 786*: 147344.

Sarker, K.K., Hossain, A., Timsina, J., Biswas, S.K., Malone, S.L., Alam, M.K., Loescher, H.W. and Bazzaz, M. (2020). Alternate furrow irrigation can maintain grain yield and nutrient content and increase crop water productivity in dry season maize in sub-tropical climate of South Asia. *Agric Water Manag., 238*: 106229.

Sharma, R.K., Chhokar, R.S., Jat, M.L, Singh, S., Mishra, B. and Gupta, R.K. (2008). Direct drilling of wheat into rice residues: Experiences in Haryana and western Uttar Pradesh. *In*: Humphreys, E. and Roth, C.H. (Eds.), *Permanent Beds and Rice–Residue Management for Rice–Wheat System of the Indo–Gangetic Plain. In*: Australian Centre Int. Agric. Res. (ACIAR) Proc. *127*: 147–58.

Smith, P., Martino, D., Cai, Z., Gwary, D., Janzen, H., Kumar, P., Mccarl, B., et al. (2007). Agriculture. *In*: *Climate Change 2007: Mitigation of Climate Change.* Contribution of Working Group III to the Fourth Assessment Report of the Intergovernmental Panel on Climate Change (Metz, B., Davidson, O.R., Bosch, P.R., Dave, R., Meyer, L.A. (Eds.), pp. 497–540. Cambridge, UK: Cambridge University Press.

Sommer, R. and Bossio, D. (2014). Dynamics and climate change mitigation potential of soil organic carbon sequestration. *J. Environ. Manage., 144*: 83–87.

Stagnari, F., Maggio, A, Galieni, A. and Pisante, M. (2017). Multiple benefits of legumes for agriculture sustainability: An overview. *Chem. Biol. Technol. Agric., 4*(2): 1–13.

Suh, S. (2004). *Material and Energy Flows in Industry and Ecosystem Network.* Centre for Environmental Science, University of Leiden, The Netherlands.

Takai, Y. and Kamura, T. (1966). The mechanism of reduction in waterlogged paddy soil. *Folia Microbiol., 11*: 304–313.

Thanawong, K., Perret, S.R. and Basset-Mens, C. (2014). Eco-efficiency of paddy rice production in Northeastern Thailand: a comparison of rain-fed and irrigated cropping systems. J. *Cleaner Prod., 73*: 204–217.

Wang, M., Xia, X., Zhang, Q. and Liu, J. (2010). Life cycle assessment of a rice production system in Taihu region, China. *Int. J. Sust. Dev. World, 17*: 157–61.

Wang, X.Y., Sun, B., Mao, J.D., Sui, Y.Y. and Cao, X.Y. (2012). Structural convergence of maize and wheat straw during two-year decomposition under different climate conditions. *Environ. Sci. Technol., 46*: 7159–65.

Wang, Y. and Hsieh, Y.P. (2002). Uncertainties and novel prospects in the study of the soil carbon dynamics. *Chemosphere, 49*: 791–804.

Wassmann, R. and Dobermann, A. (2009). Greenhouse gas emissions from rice fields: What do we know and where should we head for? Paper presented at: *The 2nd Joint International Conference on "Sustainable Energy and Environment"*. Bangkok, Thailand. 21–23 Nov. Paper D–030 (O).

Wassmann, R., Neue, H.U., Lantin, R.S., Makarim, K., Chareonsilp, N., Buendia, L.V. and Rennenberg, H. (2000). Characterization of methane emissions from rice fields in Asia: II. Differences among irrigation, rain-fed, and deep-water rice. *Nutrient Cycling in Agroecosystems, 58*: 13–22.

WB. (2016). *Bangladesh: Growing the Economy through Advances in Agri*culture. From http://www.worldbank.org/en/results/2016/10/07/bangladesh-growing-economy-through-advances-in-agriculture. Accessed on 31 August 2018.

Weller, S., Kraus, D., Ayag, K.R.P., Wassmann, R., Alberto, C.R., Butterbach-Bahl, K. and Kiese, R. (2014). Methane and nitrous oxide emissions from rice and maize production in diversified rice cropping systems. *Nutr. Cycl. Agroecosyst., 101*: 37–53.

West, T.O. and Post, W.M. (2002). Soil organic carbon sequestration rates by tillage and crop rotation. *Soil Sci. Soc. Am. J., 66*: 1930–46.

Wolf, B., Zheng, X.H., Brueggemann, N., Chen, W.W., Dannenmann, M., Han, X.G., Sutton, M.A., Wu, H.S., Yao, Z.S. and Butterbach-Bahl, K. (2010). Grazing-induced reduction of natural nitrous oxide release from continental steppe. *Nature*, *464*: 881–84.

Woods, J., Brown, G., Gathorne–Hardy, A., Sylvester–Bradley, R., Kindred, D. and Mortimer, N. (2008). *Facilitating Carbon (GHG) Accreditation Schemes for Biofuels and Feedstock Production*. Report 435, Part 1, Home Grown Cereals Authority, London.

WRI (World Resource Institute), WBCSD. (2013). *Technical Guidance for Calculating Scope 3 Emissions*. Supplement to the Corporate Value Chain (Scope 3) Accounting & Reporting Standard. (Version 1.0), 2013. World Resources Institute & World Business Council for Sustainable Development, 2013. Accessed on 12.10.14.

Xu, X., Zhang, B., Liu, Y., Xue, Y. and Di, Y. (2013). Carbon footprints of rice production in five typical rice districts in China. *Acta Ecol. Sin.*, *33*: 227–32.

Xue, J.F., Pu, C., Liu, S.L., Chen, Z.D., Chen, F., et al. (2015). Effects of tillage systems on soil organic carbon and total nitrogen in a double paddy cropping system in Southern China. *Soil Till Res.*, *153*: 161–68.

Yagi, K., Tsuruta, H. Kanda, K. and Miami, K. (1991). Manual of CH_4 and N_2O flux measuring. *Collected Papers on Environmental Planning*. *7*: 143–58.

Yossapol, C. and Nadsataporn, H. (2008). Life Cycle Assessment of, Rice Production in Thailand, Paper presented at: *The 6th International Conference on Life Cycle Assessment in the Agri-Food Sector, Towards a Sustainable Management of the Food Chain*. Zurich, Switzerland, November 12–14.

Yu, K. and Chen, G. (2004). Reduction of global warming potential contribution from a rice field by irrigation, organic matter, and fertilizer management. *Global Biogeochem. Cycles*, *18*: GB3018.

Zebarth, B.J., Rochette, P. and Burton, D.L. (2008). N_2O emissions from spring barley production as influenced by fertilizer nitrogen rate. *Can. J. Soil Sci.*, *88*: 197–205.

Zhang, D., Shen, J.B., Zhang, F.S., Li, Y.E. and Zhang, W.F. (2017). Carbon footprint of grain production in China. *Sci. Rep.*, *7*: 4126.

Zomer, R.J., Bossio, D.A., Sommer, R. and Verchot, L.V. (2017). Global Sequestration Potential of Increased Organic Carbon in Cropland Soils. *Sci. Rep.*, *7*: 1–8.

Zou, J, Huang, Y., Jiang, J., Zheng, X. and Sass, R.L. (2005). A 3-year field measurement of methane and nitrous oxide emissions from rice paddies in China: Effects of water regime, crop residue, and fertilizer application. *Glob. Biogeochem. Cy.*, *19*: GB2021.

Chapter 7

Impacts of Climate Change on Epidemic Outbreak of Wheat Blast Disease in Asia and Its Threat to Global Food and Nutritional Security

Tofazzal Islam[1*] and *Dipali Rani Gupta*[1]

Introduction

Wheat, a crucial crop that sustains billions of people worldwide, plays a vital role in ensuring global food security and economic stability. Its multifaceted applications, ranging from fundamental sustenance to industrial usage, highlight its significance in both developing and developed nations. However, the consistent production of wheat is currently threatened by the insidious wheat blast disease. This destructive ailment, caused by the fungus *Magnaporthe oryzae Triticum* (MoT), has emerged as a formidable adversary capable of rapidly annihilating entire wheat fields (Islam et al., 2023). For instance, during the initial outbreak of wheat blast in Brazil in 1985, affected areas recorded devastating yield losses of up to 100% (Valent et al., 2021; Singh et al., 2021).

In Bangladesh, where the disease first appeared in 2016, various regions reported yield losses ranging from 20–100% (Islam et al., 2016). More recently, wheat blast was also detected in an experimental plot in Zambia between 2017–2018, with disease incidence ranging from 50–100% (Tembo et al., 2020). These losses can significantly impact food availability, as well as the incomes and livelihoods of farmers, exerting far-reaching consequences. Genomics surveillance revealed that wheat blast outbreaks in both Bangladesh and Zambia were caused by a South American clonal lineage of

[1] Institute of Biotechnology and Genetic Engineering, Bangabandhu Sheikh Mujibur Rahman Agricultural University, Bangladesh

* Corresponding author: tofazzalislam@bsmrau.edu.bd

a pandemic fungus MoT introduced into these countries in two separate occasions through grain trading (Latorre et al., 2023). This devastating wheat disease manifests as a sudden, rapid collapse of the vascular system and the destruction of wheat plants. Initial symptoms start with water-soaked lesions on the leaves, which later develop into characteristic eye-shaped spots. The fungus blocks the vascular system when it attacks the neck, resulting in bleached heads and shriveled grains. It produces numerous asexual conidia at the site of infection, which are later dispersed through wind and contaminate adjacent fields.

Plant pathogen infection risk tracks global crop yields under climate change (Chaloner et al., 2021). Climate change plays a pivotal role in shaping the dynamics of wheat blast disease spread. The fungus favors high temperatures (18–25°C) and prolonged moisture (25–40 hours) for successful infection and disease outbreak. Higher temperatures create favorable conditions for fungal growth and development, while shifts in humidity levels contribute to increased spore production and dissemination. These changing climatic factors can extend the geographic range of wheat blast, potentially allowing it to invade new regions where it was previously limited. Furthermore, climate-induced stress weakens the natural defenses of wheat plants, making them more susceptible to infections.

While some diagnostic protocols have been established for wheat blast disease, their convenience and reliability are not without limitations (Pieck et al., 2017; Gupta et al., 2019; Kang et al., 2021). Consequently, there is a need for the development of a more user-friendly, rapid, and cost-effective diagnostic tool. Validation of this tool in the field is essential for effective surveillance and monitoring of this wheat pathogen. Attempts to control wheat blast through fungicide sprays after the manifestation of disease symptoms have proven unpredictable due to the development of resistance in the pathogen (McGrath, 2001; Fernández-Ortuño et al., 2006; Castroagudin et al., 2015; Poloni et al., 2020; Samara et al., 2022). Although classical breeding techniques offer an effective avenue for developing wheat varieties resistant to the disease, the limited availability of resistance genes and the eventual breakdown of resistance over time pose significant challenges (Anh et al., 2015, 2018; Ceresini et al., 2018; Wang et al., 2018). In addition to stacking resistance genes, target deletion of susceptible (*S*-) genes is another smart approach to developing a durable resistant variety. To date, only a few *S*-genes homologous to rice *S*-genes have been identified in wheat plants. Furthermore, a comprehensive approach such as the use of biocontrol agents and their metabolites could be a facilitating way to mitigate the disease. Exploring the local environment for the identification of novel plant probiotic bacteria with antagonistic properties could present an alternative to the use of less-effective fungicides in combating wheat blast. Recent research highlights the potential of these beneficial bacteria and their metabolites as promising solutions for disease management (Surovy et al., 2017; Dutta et al., 2018; Chakraborty et al., 2020a, b; Rabby et al., 2022; Paul et al., 2022).

Despite wheat blast posing a threat to South American wheat production since 1985, the pathogen's biology remains inadequately understood (Islam et al., 2023). As such, innovative molecular and plant breeding approaches are imperative to counteract this emerging menace to global food security. Numerous recent reviews have shed light on different facets of wheat blast disease (Cruz and Valent, 2017;

Ceresini et al., 2018, 2019; Vales et al., 2018; Islam et al., 2019; Islam et al., 2023). This comprehensive assessment serves to enhance our understanding of the ongoing advancements in mitigating wheat blast through innovative approaches. This chapter provides an update on current knowledge and critically discusses strategies for the mitigation of this climate-driven, potentially catastrophic wheat killer fungal disease.

Understanding the Biology of the Wheat Blast Mungus *M. oryzae Triticum* (*MoT*)

The haploid filamentous blast fungus reproduces both sexually and asexually. It produces enormous pyriform asexual conidia in appropriate medium or in host tissue under favorable environmental conditions. Conidia are hyaline to pale gray-colored arises in single or cluster on coniodiophore. Conidia are mostly three-celled and two septate, pyriform, round at the base with a tapering tip, and 20–22 × 10^{-12} µm in size (Cruz and Valent, 2017; Gupta et al., 2020). In the presence of moisture, conidia of the wheat blast fungus undergo a series of transformations. They initially produce small germ tubes that gradually swell into distinctive hook-shaped structures. These structures eventually differentiate into specialized infection sites known as appressoria, crucial for penetrating the host plant. The fungus engages in sexual reproduction through mating between compatible and fertile individuals. Fully fertile strains are self-sterile hermaphrodites, capable of mating with hermaphroditic strains of the opposite mating type. Mating types are determined by two alternate *MAT* alleles, *MAT1-1* and *MAT1-2*, present in opposing strains. These alleles typically contain several genes. The blast fungus requires specific temperature ranges (around 20–22°C) and prolonged light exposure for successful mating. Genes like *MAT1-1-1*, *MAT1-1-3*, and *MAT1-2-1* play essential roles in forming the reproductive structure called perithecium, while *MAT1-1-2* influences asci and ascospore production. The *MAT1-2-2*, on the other hand, is not crucial for sexual development (Wang et al., 2021). Sexually fertile strains of *M. oryzae* also generate a Phialophora-like anamorph. This stage produces small, crescent-shaped microconidia, emerging from structures called phialides (Chuma et al., 2009). Although these microconidia can infect plants through wounds, their specific role in nature remains uncertain (Zhang et al., 2014). This intricate interplay of reproductive structures and strategies contribute to the complexity of the wheat blast fungus's life cycle and overall biology. Evidence has demonstrated the ability of sexual spores from blast pathogens to successfully infect a range of plants, including rice, wheat, turf grass, and finger millet (Consolo et al., 2005; Le et al., 2010; Takan et al., 2012; Saleh et al., 2012; Marciel et al., 2014). This phenomenon underscores the significance of sexual recombination, which contributes to the substantial variability of the pathogen and the emergence of novel pathotypes capable of causing diseases in previously unaffected hosts through host shifts.

Disease Symptoms in the Field

The symptoms of wheat blast by the fungus *MoT* can manifest in various parts of the wheat plant and at different growth stages. These symptoms can be quite distinctive and are crucial for accurate disease identification. The initial symptoms often appear as small, oval, or diamond-shaped lesions on the leaves. The lesions gradually turn tan to brown in color and may have a brownish halo surrounding them. As the disease progresses, the lesions can expand and coalesce, causing widespread necrosis of leaf tissue (Igarashi et al., 1986; Igarashi, 1990). One of the most characteristic and devastating symptoms of wheat blast is the infection of the spike or head of the wheat plant. When the fungus infects rachis, it blocks the translocation of photosynthates in the upper part of the spike/head resulting in head blast symptoms (Fig. 1a). Infected spikelet may turn brown, become shriveled, and fail to develop normally, resulting in reduced grain filling and yield loss (Goulart et al., 2007; Islam et al., 2016). In a recent study, Surovy and colleagues demonstrated that the pathogen colonizes in the infected seeds and serves as the inoculum for long distance dispersal of the pathogen (Surovy et al., 2023a). On the infected spikelet or other plant parts, the fungus may produce fluffy, white spore masses (conidial sporulation) under high humidity conditions (Fig. 1c).

Fig. 1 Blast disease symptoms and pathogen morphology. (a) Head blast symptom in a wheat field in Zambia, Africa (this image was generously provided by Dr. Batiseba Tembo of Zambia Agricultural Research Institute, Mt. Makulu Central Research Station, Lusaka, Zambia). (b) Dark-gray sporulation (arrow) of the *MoT* fungus on the rachis. (c) Microscopic view of sporulation of *MoT* conidia on rachis. (d) Arrangement of asexual conidia on conidiophore. (e) Pyriform three-celled conidia under the microscope.

These spore masses contain numerous fungal spores and contribute to the spread of the disease to nearby plants. In severe cases, the entire panicle or inflorescence of the wheat plant can become infected and blighted, leading to severe yield losses (Islam et al., 2016). Although the appearance and severity of symptoms can vary based on factors such as environmental conditions, wheat cultivar susceptibility, and the specific strain of the fungus. Proper disease diagnosis requires a careful examination of symptoms and, if needed, laboratory tests to confirm the presence of the wheat blast pathogen. Early detection and prompt management strategies are crucial to minimize the impact of the disease on wheat production.

Lifecycle and Interaction of *MoT* with Host Plant

The life cycle of the wheat blast pathogen is closely tied to environmental conditions, host susceptibility, and cultural practices. The cycle begins when *MoT* spores encounter susceptible parts of the wheat plant, such as leaves or spikes. These spores are usually produced on infected seeds, crop residues from previous disease outbreaks (Goulart and Paiva, 1990). When favorable conditions of high humidity and moisture are present, the spores germinate, sending out thin threads called germ tubes. These germ tubes then develop specialized infection structures called appressoria. Appressoria attach to the plant surface and generate mechanical pressure to penetrate the plant's outer protective layers, enabling the fungus to enter the plant tissues (Talbot, 2003). A turgor-sensing histidine–aspartate kinase, Sln1, allows the appressorium to sense when a critical turgor threshold has been reached and thereby facilitates host penetration (Ryder et. al., 2019). Once inside the plant, the fungus grows and spreads within the plant's cells, forming a network of thread-like structures called hyphae. As the hyphae spread, they cause damage to plant tissues, leading to the formation of visible lesions, which are characteristic symptoms of the disease (Kankanala et al., 2007; Eseola et al., 2021; Cruz-Mireles et al., 2021).

Intriguingly, a very recent study reported that a set of 863 genes encoding secreted proteins differentially express at specific stages of infection among which 546 were predicted to be effectors and named MEP (*Magnaporthe* effector protein) genes (Yan et al., 2023). As the disease progresses, the fungus starts to produce asexual spores (conidia) on the surface of the lesions (Fig. 1c). These spores are easily dislodged and can be spread by wind, rain, or human activities, contributing to the rapid spread of the disease to other plants. The released spores can infect nearby healthy plants or even travel longer distances to initiate new infections. This process contributes to the further spread of the disease within a field or to other areas. In some cases, the fungus can survive on infected crop debris or other plant hosts during the off-season. These overwintering structures serve as sources of infection for the next growing season, contributing to the persistence of the disease (Urashima et al., 1999; Talbot, 2003; Wilson and Talbot, 2009; Islam 2018). Interestingly, in a study, Pizolotto et al. (2019) demonstrated that the *MoT* fungus recovered from Brazil did not survive more than five months in plant tissue. Therefore, it is difficult to presume that the crop residue serve as a source for wheat blast in the subsequent spring wheat season in Brazil where overwinter period is more than that. Following the demise of the host, *MoT* typically survives through the winter on alternate hosts, often weed grasses Subsequently, it spreads to new host plants through mechanisms like dewdrop splash, wind, infected seeds, and crop residues (Maciel et al., 2014; Cruz and Valent, 2017; Gupta et al., 2021).

Distribution of the *MoT* Fungus and Epidemiology of Disease

Historically, wheat blast was primarily considered a disease of rice (known as rice blast) until a severe outbreak occurred on wheat in Brazil in the mid-1980s. Since then, wheat blast has become a significant threat to wheat production, particularly in regions with warm and humid climates. The disease has been reported in South American countries, including Brazil, Paraguay, Bolivia, and Argentina.

Having originated in the Paraná province of Brazil in 1985, the *MoT* fungus gradually spread to other neighboring countries including Bolivia (1996), Paraguay (2002), and Argentina (2007). Presently, *MoT* poses a significant potential threat to wheat production across approximately 3 million hectares of wheat-growing areas in South America. Particularly impacted are central and southern regions of Brazil, the lowland Santa Cruz area in Bolivia, as well as southern and southeastern parts of Paraguay and northeastern Argentina (Kohli et al., 2011). The consequences of this infection have led to substantial yield losses, with estimates ranging from 10 100% (Goulart et al., 2007; Urashima et al., 2009; Duveiller et al., 2016).

While the exact origin of Brazilian wheat blast remains a subject of debate, research by Inoue et al. (2017) suggested a probable transition to wheat from a native grass through a phenomenon known as "host jump". In 2011, a singular spike of wheat displaying symptoms similar to wheat blast was identified in an experimental field located in Kentucky, USA (Farman et al., 2017). Interestingly, the wheat blast occurrence in the United States stemmed from the 'host jump' of the *M. oryzae* Lolium pathotype originating from a native Lolium species, rather than being introduced from South America (Maekawa and Schulze-Lefert, 2017; Inoue et al., 2017). Notably, in February 2016, the presence of wheat blast was initially detected in eight southwestern districts in Bangladesh, believed to have been introduced from South America via grain trading (Fig. 2) (Islam et al., 2016; Islam et al., 2019). The filed pathogenomics study revealed the genetic identity and origin of the isolates which was a South American lineage of wheat blast fungus (Islam et al., 2016; Malaker et al., 2016). Over the past seven years, wheat blast has spread to 15 new districts in Bangladesh as fresh infections (Singh et al., 2021). While there's no official report on its occurrence, the Indian government implemented a "wheat holiday" near the border (about 2,200 km) and restricted wheat grain movement from the infected area to prevent further spread (source: https://bit.ly/2JcE9nK, Last accessed: May 16, 2020). This strategy aims to reduce the fungus's ability to rapidly move to new wheat-growing areas, even though it can affect other plants too.

In February 2018, the presence of wheat blast was initially documented in Zambia, specifically in the Mpika district of the Muchinga province during the rain-fed season. The disease gained significant traction during the 2017–2018 season, with high incidence and severity attributed to favorable weather conditions conducive to disease proliferation. However, contrasting conditions prevailed during the subsequent 2018–2019 season, marked by hot and dry weather, leading to diminished disease severity in both experimental and farmer fields within the Mpika district. In the crop season spanning from 2019–2020, wheat blast experienced a resurgence, displaying heightened occurrence and severity in experimental locations (Mt. Makulu, Mpika, and Mpongwe) as well as within fields cultivated by farmers in Mpika. Interestingly, regions characterized by hot and arid climatic conditions exhibited a diminished impact of the disease. This trend persisted into the following crop season of 2020–2021, with the disease persisting in experimental fields (Mt. Makulu, Mpika, and Mpongwe) and a significant occurrence observed in a farmer's field situated in the Kafue district. Remarkably, the susceptible Coucal variety of wheat was being cultivated in this affected field. However, the transmission dynamics of wheat blast in Zambia appear to be linked to seed-borne mechanisms, rather than air-borne dispersal. This suggests that the disease's propagation is more closely associated with the movement of infected seeds than airborne spores. Such insights into the mode of spread are critical for devising effective strategies to manage and control the further

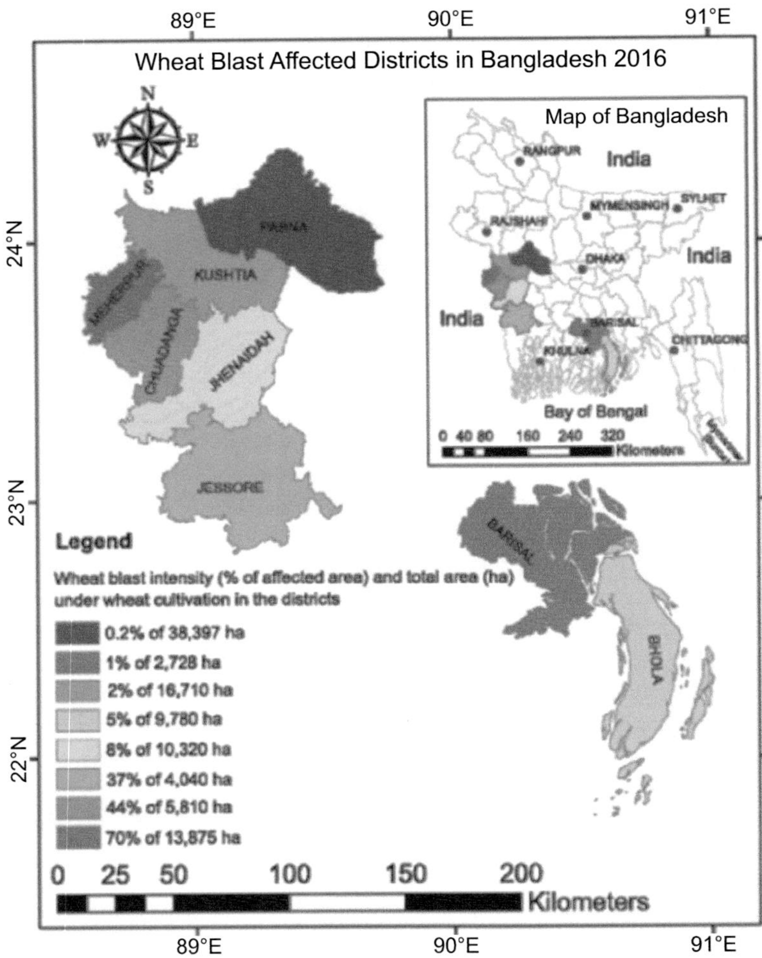

Fig. 2 Wheat blast disease outbreak in eight south-western districts of Bangladesh in the year 2016 (Islam et al., 2016). The map describes the intensity of wheat blast outbreak across Bangladesh. The percentage of affected area and the total area (hectares) under cultivation are shown for each district based on the color chart.

dissemination of wheat blast in Zambia (Tembo et al., 2020; Latorre et al., 2023). Distribution of wheat blast disease in different regions of the world is shown in the Fig. 3.

Conducive Weather for Wheat Blast Epidemic

The interplay of weather elements holds significant sway over the wheat blast's emergence. Notably, global warming and ill-timed rains during the plant's heading stage have been pinpointed as catalysts for the onset and amplification of this affliction. The degree of head blast severity hinges on factors such as prevailing environmental conditions, the susceptibility of specific cultivars, and the plant organs affected (Goulart et al., 2007; Urashima et al., 2009).

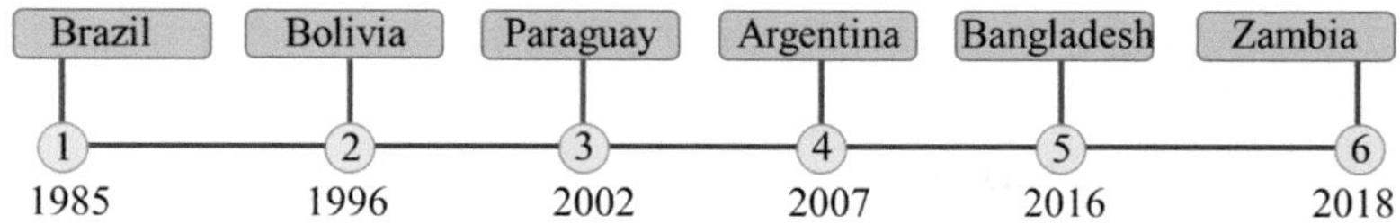

Fig. 3 Timeline of spread of wheat blast to different regions in the world.

Factors such as temperature, humidity, rainfall, and crop management practices play a pivotal role in influencing the disease favoring index (DFI). Higher values of the DFI indicate conditions that are more conducive to the proliferation of the wheat blast pathogen (Islam et al., 2019). The most profound instances of field infections manifest during seasons characterized by continuous rainfall and an average temperature ranging from 18–25°C around the anthesis period, succeeded by a stretch of sunny, sweltering, and moisture-laden conditions (Kohli et al., 2011). Cardoso et al. (2008) underscored that the optimal temperature window spans from 25–30°C, coupled with spike wetness spanning 25–40 hr., which serves as a catalyst for severe disease outbreaks. However, temperatures above 30°C can limit disease development, while temperature below 17°C is less favorable for the fungus (Valent et al., 2021).

In regions of South America, the disease has flourished in humid and warmer areas, such as Bolivia, Paraguay, and northeastern Argentina (Kohli et al., 2011). Geographically, the weather condition of Bangladesh is hot and humid in summer but not in winter. However, anticipated temperature increases in Bangladesh, particularly during the winter season, as predicted by various reports (Hossain and da Silva, 2012), could potentially elevate the susceptibility to wheat blast in 2016. Another environmental factor conducive to disease propagation is the substantial dewfall observed during the winter months in Bangladesh. This extended period of moisture retention, lasting around 16–17 hours on wheat plants, provides a favorable setting for fungal sporulation. Examination of weather data sourced from the Bangladesh Meteorological Department reveals a notable temperature rise across all regions in 2016, primarily attributed to a 1.8–6.5°C increase in minimum temperatures compared to the period from 2011–2015 (Fig. 4) (Islam et al., 2019). This warming trend, coupled with precipitation during the flowering stage, contributed to the 2016 epidemic of wheat blast in the affected districts of Bangladesh.

In an ex-ante impact assessment, Mottaleb et al. (2018) delved a potential wheat-blast scenario across Bangladesh, India, and Pakistan. By considering the agro-climatic conditions at the epicenter where the disease was initially identified in Bangladesh in 2016, susceptible regions were pinpointed in India, Pakistan, and Bangladesh, collectively covering around 7 million hectares. Employing a conservative estimate of 5–10% for potential wheat production loss attributed to wheat blast, the study projected an annual wheat yield reduction ranging from 0.89–1.77 million tons across these countries. This renders a substantial economic impact of USD 132264 million. These projected losses compound the existing challenges in national food security, exacerbating the reliance on wheat imports and elevating wheat prices in South Asian region. Wheat blast is a complex disease influenced by interactions between the pathogen, host plant, and environment. As climate patterns change, there is a concern that the geographic range of wheat blast may expand, impacting wheat production in new regions.

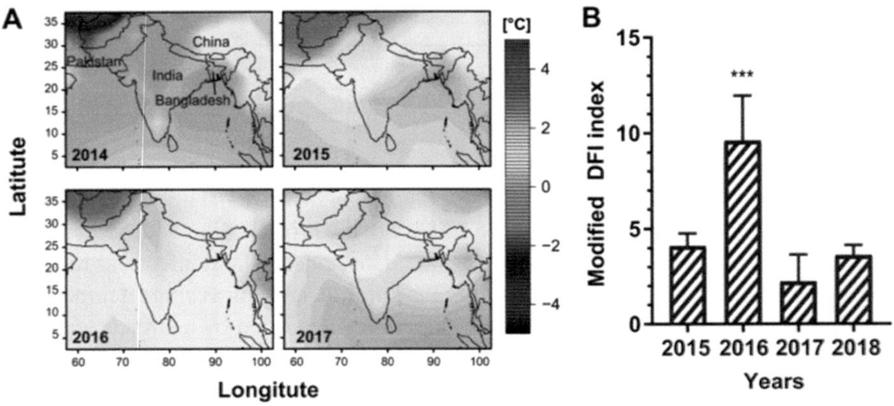

Fig. 4 Effects of climate change on the wheat blast epidemic in Bangladesh in 2016 (Islam et al., 2019). (a) Differences of temperature in the month of February from 2014–2017 in the India-Bangladesh region. (b) The modified DFI (Disease favoring index) for the wheat blast-infected regions in 2016.

Diagnosis and Detection of Wheat Blast Fungus

Ensuring the accurate detection of pathogens within field samples stands as a crucial strategy for proactively managing diseases before they reach catastrophic levels. Identifying wheat blast disease during the heading stage presents challenges due to its resemblance to Fusarium head blight (FHB) symptoms (Pieck et al., 2017). Molecular diagnosis, distinguished by its remarkable sensitivity and specificity, has demonstrated significant promise in boosting plant disease control and prevention efforts. While the Pot 2 transposable elements have proven valuable for detecting *M. oryzae* pathogens associated with blast, they do not differentiate between *M. oryzae* pathotypes (Farman, 2002). Later, Pieck et al. (2017) adopted a genome-based approach to develop specific molecular markers for *MoT* isolate detection.

Among these markers, MoT3 emerged and was derived from a retinol dehydrogenase gene, discerned through analysis of 258 *M. oryzae* isolates from diverse hosts. This primer successfully amplified a 368bp PCR product exclusive to *MoT* isolates, setting them apart from other *M. oryzae* pathotypes. Based on the Pot2 and MoT3 markers, Yasuhara-Bell et al. (2018) created loop-mediated isothermal amplification (LAMP) assays tailored for the specific detection of *MoT*, both within laboratory and field samples. However, a contradictory finding emerged from Gupta et al. (2019), which challenged the earlier outcomes presented by Pieck et al. (2017). Gupta et al. (2019) observed that, under certain conditions, the MoT3 primer also amplified sequences from both rice and wheat blast isolates. Their investigation led them to conclude that this assay relies on primers that match the WB12 sequence, corresponding to a fragment of the *M. oryzae* MGG_02337 gene annotated as a short-chain dehydrogenase, a trait found in *M. oryzae* isolates affecting both rice and wheat. Intriguingly, all eight rice blast isolates subjected to their study generated the WB12 amplicon. Furthermore, a comparative genomic analysis of the WB12 nucleotide sequence unveiled a multifaceted underlying genetic arrangement with interconnected sequences spanning various *M. oryzae* strains, encompassing both rice and wheat blast isolates.

In a separate study, Thierry et al. (2019) identified distinctive genetic variations within the Triticum lineage by analyzing 81 *Pyricularia* genomes. They subsequently devised C17, a real-time PCR-based detection method boasting high sensitivity. Later, they adopted a comparative genomics approach and identified new loci specific to the *MoT* isolates and designed a set of new markers. The newly developed markers were sensitive for the detection of *MoT* isolates from infected seeds (Thierry et al., 2020). However, the utility of this method for identifying the pathogen in field samples remains uncertain, given the absence of validation using actual samples. As such, there persists a need for the development of an efficient, cost-effective, and swift detection technique for early-stage wheat blast infections, encompassing alternate hosts and seed lots. Recently, a rapid, specific and convenient diagnostic method was developed for wheat blast pathogen, *M. oryzae Triticum* pathotype using genome-specific primers and Cas12a-mediated technology (Kang et al., 2021).

To develop an accurate and sensitive method to detect MoT at the seedling and vegetative stages for disease control, Kang et al. (2021) sequenced the genomes of two *MoT* isolates from Brazil and identified two DNA fragments, *MoT*-6098 and *MoT*-6099, that are present in the *MoT* genome but not in the genome of the rice-infecting *M. oryzae Oryzae* (*MoO*) pathotype. Using polymerase chain reaction (PCR), they confirmed the specificity of the two markers in 53 *MoT* and *MoO* isolates from South America and Bangladesh.

To test the efficiency of the two markers, they first established a loop-mediated isothermal amplification (LAMP) method to detect *MoT* at isothermal conditions, without the use of a PCR machine. To make their method more convenient, they used the Cas12a protein and guide RNAs (gRNAs) to target the *MoT*-6098 and *MoT*-6099 sequences. The activated Cas12a showed indiscriminate single-stranded deoxyribonuclease (ssDNase) activity. They combined target-dependent Cas12a ssDNase activation with recombinase polymerase amplification (RPA) and nucleic acid lateral flow immunoassay (NALFIA) to develop a method that accurately, sensitively, and cost-effectively detects *MoT*-specific DNA sequences in infected wheat plants. This novel technique can be easily adapted for the rapid detection of wheat blast and other important plant diseases in the field. The simple method developed by Kang et al. (2021) can easily be adapted by plant pathologists, quarantine specialists, and agriculture extension workers to rapidly detect *MoT* in infected plants, seed lots, and alternate hosts. However, a further improvement of this rapid test is needed for cost-effectiveness and convenience that are important for its worldwide application by researchers, quarantine offices, and farmers. Genome-specific primers and Cas12-mediated technology is becoming popular method for point-of-care specific and rapid disease diagnosis of various diseases and pathogens including phytopathogens (Lu et al., 2020; Matthew and Yang, 2021; Yao et al., 2022).

The approach of field pathogenomics presents a potent approach in the detection and comprehension of wheat blast, a pernicious affliction afflicting wheat crops. Unlike conventional methods reliant solely on visible symptoms, field pathogenomics delves into the genetic makeup of the fungus responsible for wheat blast, facilitating a deeper understanding of the disease's mechanisms. Islam et al. (2016) successfully applied field pathogenomics, open data sharing, and open science approaches for precisely determining the genetic identity and origin of the first epidemic outbreak of wheat blast in Bangladesh (Islam et al., 2016; Islam and Kamoun, 2018; Kamoun et al., 2019). Unlike traditional isolation and culture-based time-consuming method, field pathogenomics is based on RNA-seq data generated directly formed pathogen-infected plant leaf samples collected in the field, providing a unique opportunity to characterize the pathogen population and its host directly in their natural environment (Bueno-Sancho et al., 2017). The field pathogenomics provides a high-resolution representation of pathogen population dynamics directly in the field, providing new insights into pathogen biology, population structure, and pathogenesis. Therefore, this approach is suggested to be applied for determining the genetic identity and origin of any epidemic outbreak of blast disease in any cereal crop.

Using genomic surveillance can help quickly detect and track wheat blast, which is important due to its potential harm. The World Health Organization has a strategy for tracking dangerous pathogens globally (Carter et al., 2022). Wheat blast is a good candidate for this kind of surveillance, even though there are challenges. Latorre et al. (2023) emphasize the importance of using genomic surveillance to monitor and control the spread of Wheat Blast beyond South America. They successfully show how genomic data can help track the disease's movement (Latorre et al., 2023). By looking at specific markers in the genome (84 SNPs), they confirm that a particular type of Wheat Blast, called B71, has spread from different South American populations to Zambia and Bangladesh on two separate occasions. This shows that it has the potential to become a widespread problem.

Control Strategies of Wheat Blast Disease

Managing wheat blast presents a complex challenge, and a single approach alone is insufficient to attain effective control. In blast-free zones, stringent quarantine measures hold utmost importance, acting as a vital barrier against the introduction of the disease, which, once established, becomes extremely challenging to eliminate. Adjacent areas to endemic regions could consider implementing a "wheat holiday" concept, temporarily suspending wheat cultivation to impede disease spread. For regions already affected by endemic wheat blast, a comprehensive and integrated management strategy is advised. This multifaceted approach encompasses various techniques, including the utilization of resistant crop varieties, judicious fungicide application, non-fungicidal chemical treatments, and the application of biocontrol methods. The ensuing sections elaborate on these strategies, highlighting their roles in effectively tackling the complex issue of wheat blast and contributing to enhanced disease management.

Chemical Fungicides to Control Wheat Blast.

In the current scenario, fungicides have emerged as a critical component in the management of wheat blast (WB), particularly given the limited impact of varietal resistance. However, the efficacy of fungicides can decrease under conditions of

Blast community (http://openwheatblast.net). The urgency of this situation cannot be emphasized enough, as neighboring countries like India and China, the top two wheat producers in the world, are in close proximity to the wheat blast-contaminated Bangladesh. A large portion of their wheat-growing regions have conditions favorable for the outbreak of wheat blast. Therefore, swift action is required to prevent the spread of this disease to these countries before it becomes truly catastrophic (Islam and Kamoun, 2018).

Major Challenges and Strategies for Combating Wheat Blast

Controlling wheat blast presents a complex challenge, requiring comprehensive strategies for effective management (Islam et al., 2019, 2020). One of the main difficulties is the rapid spread of the disease, as the fungus has the ability to infect various hosts and adapt to different environments (Fig. 5). The issue is further compounded by the looming threat of climate change, which could potentially expand the reach of wheat blast (Callaway, 2023).

Limitations

- Lack of resistance sources.
- Poor monitoring in the evolution of the pathogen and forecasting system.
- Rudimentary early disease detection system.
- Low efficacy fungicides.
- Insufficient research on biological agents and their metabolites for controlling wheat blast.
- Limited success in developing transgenic cultivars.
- Lack of collaborative efforts.

Future directions

- Exploring wheat biodiversity and identification and incorporation of more QTL for resistance.
- Improving disease surveillance and early warning systems.
- Developing sensitive and reliable early disease detection methods.
- Identification of biocontrol agents and its metabolites with new mode of actions.
- Complete understanding of the pathogen and host mechanisms.
- Identifications of *R*- or *S*- genes for developing transgenic cultivars.
- Collaborative efforts among researchers, policymakers and farmers.

Fig. 5 Limitations and future directions in managing wheat blast disease.

While developing resistant wheat varieties is crucial, achieving long-term resistance is complicated due to the pathogen's genetic diversity. Moreover, the timely and accurate detection of the disease remains problematic, particularly in regions with limited resources. Overcoming these obstacles necessitates a comprehensive approach that combines resilient cultivars, use of smart and advanced diagnostics (Islam and Kasfy, 2023), improved farming techniques, and precise disease monitoring. Establishing sustainable and robust wheat production systems to effectively combat the evolving threat posed by wheat blast will be facilitated by collaborative research and global cooperation (Islam et al., 2020).

Conclusions

In conclusion, the management of wheat blast disease remains a challenge due to the complex nature of the pathogen. While progress has been made in understanding its epidemiology, there are still gaps in our knowledge regarding the pathogen's biology, mechanism of genetic resistance, and its rapid evolution. It is imperative to further enhance our understanding of the pathogen and host mechanisms of resistance. Additionally, adopting climate-resilient wheat varieties, implementing integrated pest management strategies, improving disease surveillance, and promoting sustainable farming practices are necessary measures to combat this disease. Collaboration between researchers, policymakers, and farmers is vital in mitigating the potential consequences of climate change on wheat blast and ensuring global food security. Moreover, emphasizing open science methods and open data sharing will prove instrumental in tackling wheat blast disease and achieving food security in the face of a changing climate.

References

Anh, V.L., Anh, N.T., Tagle, A.G., Vy, T.T., Inoue, Y., Takumi, S., et al. (2015). Rmg8, a new gene for resistance to Triticum isolates of *Pyricularia oryzae* in hexaploid wheat. *Phytopathology*, *105*(12): 1568–72.

Anh, V.L., Inoue, Y., Asuke, S., Vy, T.T., Anh, N.T. and Wang. S., et al. (2018). Rmg8 and Rmg7, wheat genes for resistance to the wheat blast fungus, recognize the same avirulence gene AVR-Rmg8. *Molecular Plant Pathology*, *19*(5): 1252–56.

Bhowmik, P., Ellison, E., Polley, B., Bollina, V., Kulkarni, M., Ghanbarnia, K., et al. (2018). Targeted mutagenesis in wheat microspores using CRISPR/Cas9. *Sci. Reports*, *8*: 6502.

Bueno-Sancho, V., Persoons, A., Hubbard, A., Cabrera-Quio, L.E., Lewis, C.M., Corredor-Moreno, P., Bunting, D.C.E., Ali, S., Chng, S., Hodson, D.P., Burrows, R.M., Bryson, R., Thomas, J., Holdgate, S. and Saunders, D.G.O. (2017). Pathogenomic analysis of wheat yellow rust lineages detects seasonal variation and host specificity. *Genome Biology and Evolution*, *9*(12): 3282–96.

BWMRI. (2020). *Annual Research Report 2019–2020*. Dinajpur: Bangladesh Wheat and Maize Research Institute.

Callaway, E. (2023). Wheat disease's global spread concerns researchers. *Nature*, April 11.

Cardoso, C.A.A., Reis, E.M. and Moreira, E.N. (2008). Development of a warning system for wheat blast caused by Pyricularia grisea. *Summa Phytopathology*, *34*: 216–21.

Carter, L., Yu, M.A., Sacks, J., Barnadas, C., Pereyaslov, D., Cognat, S., et al. (2022). Global genomic surveillance strategy for pathogens with pandemic and epidemic potential 2022–32. *Bulletin of World Health Organization*, *100*(04): 239–239A.

Castroagudin, V.L., Ceresini, P.C., de Oliveira, S.C., Reges, J.T.A., Maciel, J.L.N., Bonato, A.L.V., et al. (2015). Resistance to QoI fungicides is widespread in Brazilian populations of the wheat blast pathogen *Magnaporthe oryzae*. *Phytopathology*, *105*: 284–94.

Ceresini, P.C., Castroagudín, V.L., Rodrigues, F.A., Rios, J.A., Aucique-Pérez, C.E., Moreira, S.I., et al. (2018). Wheat blast: Past, present, and future. *Ann. Rev. Phytopathol.*, *56*: 427–56.

Ceresini, P.C., Castroagudín, V.L., Rodrigues, F.A., Rios, J.A., Aucique-Pérez, C.E., Moreira, S.I., et al. (2019). Wheat blast: From its origins in South America to its emergence as a global threat. *Molecular Plant Pathology*, *20*(2): 155–72.

Chakraborty, M., Mahmud, N.U., Muzahid, A.N.M., Rabby, S.M.F. and Islam, T. (2020b). Oligomycins inhibit *Magnaporthe oryzae Triticum* and suppress wheat blast disease. *PLOS One*, *15*(8): e0233665.

Chakraborty, M., Mahmud, N.U., Gupta, D.R., Tareq, F.S., Shin, H.J. and Islam, T. (2020a). Inhibitory effects of linear lipopeptides from a marine Bacillus subtilis on the wheat blast fungus *Magnaporthe oryzae Triticum*. *Frontiers in Microbiology*, *11*: 665.

Chaloner, T.M., Gurr, S.J. and Bebber, D.P. (2021). Plant pathogen infection risk tracks global crop yields under climate change. *Nat. Clim. Chang.*, *11*: 710–15.

Chuma, I., Shinogi. T., Hosogi. N., Ikeda. K-I., Nakayashiki. H., Park. P. and Tosa, Y. (2009). Cytological characteristics of microconidia of *Magnaporthe oryzae*. *Journal of General Plant Pathology*, *75*: 353–58.

Consolo, V.F., Cordo, C.A. and Salerno, G.L. (2005). Mating-type distribution and fertility status in *Magnaporthe grisea* populations from Argentina. *Mycopathologia*, *160*: 285–90.

Cruz, C.D., Kiyuna, J., Bockus, W.W., Todd, T.C., Stack, J.P. and Valent, B. (2015). *Magnaporthe oryzae* conidia on basal wheat leaves as a potential source of wheat blast inoculum. *Plant Pathology*, *64*: 1491–98.

Cruz, C.D., Santana, F.M., Todd, T.C., Maciel, J.L.N., Kiyuna, J., Cruz, A.P., et al. (2019). Multi-environment assessment of fungicide performance for managing wheat head blast (WHB) in Brazil and Bolivia. *Tropical Plant Pathology*, *44*: 183–91.

Cruz, C.D. and Valent, B. (2017). Wheat blast disease: Danger on the move. *Tropical Plant Pathology*, *42*: 210–22.

Cruz, C.D., Peterson, G.L., Bockus, W.W., Kankanala, P., Dubcovsky, J., Jordan, K.W., et al. (2016). The 2NS translocation from Aegilops ventricosa confers resistance to the *Triticum* pathotype of *Magnaporthe oryzae*. *Crop Science*, *56*: 990–1000.

Cruz-Mireles, N., Eseola, A.B., Osés–Ruiz, M., Ryder, L.S. and Talbot, N.J. (2021). From appressorium to transpressorium-defining the morphogenetic basis of host cell invasion by the rice blast fungus. *PLoS Pathogen*, *17*: e1009779.

Dutta, S., Surovy, M.Z., Gupta, D.R., Mahmud, N.U., Chanclud, E., Win, J., et al. (2018). Genomic analyses reveal that biocontrol of wheat blast by Bacillus spp. may be linked with production of antimicrobial compounds and induced systemic resistance in host plants. *Figshare*, *17*: 48.

Duveiller, E., He, X. and Singh, P.K. (2016). Wheat blast: An emerging disease in South America potentially threatening wheat production. *In*: Bonjean A, van Ginkel M (Eds.), *World Wheat Book: A History of Wheat*, Vol. 3, 1107–22. Lavoisier, Paris,

Eseola, A.B., Ryder, L.S., Osés-Ruiz M., Findlay, K., Yan, X., Cruz-Mireles, N., Molinari, C., Garduño-Rosales, M. and Talbot, N.J. (2021). Investigating the cell and developmental biology of plant infection by the rice blast fungus *Magnaporthe oryzae*. *Fungal Genetics and Biology*, *18*: 103562.

Farman, M., Peterson, G., Chen, L., Starnes, J., Valent, B., Bachi, P., Murdock, L., Hershman, D., Pedley. K., Fernandes, J.M. and Bavaresco, J. (2017). The Lolium pathotype of *Magnaporthe oryzae* recovered from a single blasted wheat plant in the United States. *Plant Disease*, *101*: 684–92.

Fernandes, J.M.C., Nicolau, M., Pavan, W., Hölbig, C.A., Karrei, M., Karrei de Vargas, F., et al. (2017). A weather-based model for predicting early season inoculum build-up and spike infection by the wheat blast pathogen. *Tropical Plant Pathology*, *42*: 230–37.

Fernández-Ortuño, D., Pérez-García, A., López-Ruiz, F., Romero, D., de Vicente, A. and Torés, J.A. (2006). Occurrence and distribution of resistance to QoI fungicides in populations of Podosphaera fusca in south-central Spain. *European J. Plant Pathology*, *115*: 215–22.

Goddard, R., Steed, A., Chinoy, C., Ferreira, J.R., Scheeren, P.L., Maciel, J.N.L., et al. (2020). Dissecting the genetic basis of wheat blast resistance in the Brazilian wheat cultivar BR 18–Terena. *BMC Plant Biology*, *20*: 398.

Goulart, A.C.P. and de Paiva, F.A. (1990). Transmission of *Pyricularia oryzae* by wheat (*Triticum* aestivum) seeds. *Fitopatologia Brasileira*, *15*: 359–62.

Goulart, A.C.P., Sousa, P.G. and Urashima, A.S. (2007). Damages in wheat caused by infection of Pyricularia grisea. *Summa Phytopathology*, *33*: 358–63.

Gupta, D.R., Avila, C.S.R., Win, J., Soanes, D.M., Ryder, L.S., Croll, D., Bhattacharjee, P., Hossain, M.S., Mahmud, N.U., Mehbub, M.S., Surovy, M.Z., Rahman, M.M., Talbot, N.J., Kamoun, S. and Islam, M.T. (2019). Cautionary notes on use of the MoT3 diagnostic assay for *Magnaporthe oryzae* wheat and rice blast isolates. *Phytopathology*, *109*: 504–508.

Gupta, D.R., Khanom, S., Rohman, M.M., Hasanuzzaman, M., Surovy, M.Z., Mahmud, N.U., et al. (2021). Hydrogen peroxide detoxifying enzymes show different activity patterns in host and non-host plant interactions with *Magnaporthe oryzae* Triticum pathotype. *Physiology and Molecular Biology of Plants*, *27*: 2127–39.

Gupta, D.R., Surovy, M.Z., Mahmud, N.U., Chakraborty, M., Paul, S.K., Hossain, M.S., Bhattacharjee, P., Mehebub, M.S., Rani, K., Yeasmin, R., Rahman, M. and Islam, M.T. (2020). Suitable methods for isolation, culture, storage and identification of wheat blast fungus *Magnaporthe oryzae* Triticum pathotype. *Phytopathology Research*, *2*: 30.

Dutta, S., Surovy, M.Z., Gupta, D.R., Mahmud, N.U., Chanclud, E., Win, J., Kamoun, S. and Islam, T. (2018) Genomic analyses reveal that biocontrol of wheat blast by Bacillus spp. may be linked with production of antimicrobial compounds and induced systemic resistance in host plants. Figshare. doi: 10.6084/m9.figshare.5852661.v1

Haque, E., Taniguchi, H., Hassan, M. M., Bhowmik, P., Karim, M.R., Śmiech, M., et al. (2018). Application of CRISPR/Cas9 genome editing technology for the improvement of crops cultivated in tropical climates: Recent progress, prospects, and challenges. *Frontiers in Plant Science*, *9*: 617.

He, M., Su, J., Xu, Y., et al. (2020). Discovery of broad–spectrum fungicides that block septin–dependent infection processes of pathogenic fungi. *Nature Microbiology*, *5*: 1565–75.

He, X., Juliana, P., Kabir, M.R., Roy, K.K., Islam, R., Marza, F., et al. (2021). Screening and mapping for head blast resistance in a panel of CIMMYT and South Asian bread wheat germplasm. *Frontiers in Genetics*, *12*: 679162.

Hossain. A. and da Silva, J.A.T. (2012). Wheat production in Bangladesh: Its future in the light of global warming. *AoB Plants*, *5*: pls042.

Hu, Z., Zuo-hua, R., Xue, Z., Xi-yue, Y., Hua-jun, Z., Xiao-juan, L., Jie, Z. and Er-ming, L. (2021). Efficacy of plant growth–promoting bacteria Bacillus cereus YN917 for biocontrol of rice blast. *Frontiers in Microbiology*, *12*.

Igarashi, S. (1990). Update on wheat blast (*Pyricularia oryzae*) in Brazil. *In*: Saunders, D.A. (Ed.), *Wheat for the Nontraditional Warm Areas,* 480–83. *In*: *Proceedings of the International Conference*. Foz do Iguaçu, Brazil.

Igarashi, S., Utiamada, C.M., Igarashi, L.C., Kazuma, A.H. and Lopes, R.S. (1986). Pyricularia em trigo. 1. Ocorrência de Pyricularia sp. no estado do Paraná. *Fitopatologia Brasileira*, *11*: 351–52.

Inoue, Y., Vy, T.T.P., Yoshida, K., Asano, H., Mitsuoka, C., Asuke, S., Anh, V.L., Cumagun, C.J.R., Chuma, I., Terauchi, R., Kato, K., Mitchell. T., Valent, B., Farman, M. and Tosa, Y. (2017). Evolution of the wheat blast fungus through functional losses in a host specificity determinant. *Science*, *357*: 80–83.

Islam T. and Kasfy, S.H. (2023). CRISPR-based point-of-care plant disease diagnostics. *Trends in Biotechnology*, *41*: 144–46.

Islam, M.T. and Kamoun, S. (2018). Open science and international collaboration to tackle the fearsome wheat blast in Asia and beyond. *Phytopathology*, *108*(S1): 276.

Islam, M.T., Croll, D., Gladieux, P., Soanes, D.M., Persoons, A., Bhattacharjee, P., Hossain, M.S., Gupta, D.R., Rahman, M.M., Mahboob, M.G., Cook, N., Salam, M.U., Surovy, M. Z., Sancho, V.B., Maciel, J.L.N., Nhani, A., Castroagudin, V.L., Reges, J.T.D., Ceresini, P.C., Ravel, S., Kellner, R., Fournier, E., Tharreau, D., Lebrun, M.H., Mcdonald, B.A., Stitt, T., Swan, D., Talbot, N.J., Saunders, D.G.O., Win, J. and Kamoun, S. (2016). Emergence of wheat blast in Bangladesh was caused by a South American lineage of *Magnaporthe oryzae*. *BMC Biology*, *14*: 11.

Islam, M.T., Gupta, D.R., Hossain, A., Roy, K.K., He, X., Kabir, M.R., Singh, P.K., Khan, M.A.R., Rahman, M. and Wang, G-L. (2020). Wheat blast: A new threat to food security. *Phytopathology Research*, *2*: 28.

Islam, M.T., Kim, K.H. and Choi, J. (2019). Wheat blast in Bangladesh: The current situation and future impacts. *Plant Pathology Journal*, *35*(1): 1–10.

Islam, M.T. (2018). A new pathotype of *Magnaporthe oryzae* causing devastating wheat blast disease in multiple continents. *Phytopathology*, *108*(S2): 16.

Islam, T., Ansary, M.W.R. and Rahman, M.M. (2023). *Magnaporthe oryzae* and its pathotypes: A potential plant pandemic threat to global food security. *In*: Scott, B., Mesarich, C. (Eds.), *Plant Relationships. The Mycota*, Vol. 5. Springer, Cham. https://doi.org/10.1007/978-3-031-16503-0_18.

Islam, T. (2019). CRISPR-Cas technology in modifying food crops. *CAB Rev.*, *14*: 50.

Juliana, P., Poland, J., Huerta–Espino, J., Shrestha, S., Crossa, J., Crespo–Herrera, L., et al. (2019). Improving grain yield, stress resilience, and quality of bread wheat using large–scale genomics. *Nature Genetics*, *51*: 1530–39.

Kamoun, S., Talbot, N.J. and Islam, M.T. (2019). Plant health emergencies demand open science: Tackling a cereal killer on the run. *PLoS Biology*, *17*(6): e3000302.

Kang, H., Peng, Y., Hua, K., Deng, Y., Bellizzi, M., Gupta, D.R., Mahmud, N.U., Urashima, A.S., Paul, S.K., Peterson, G. and Zhou, Y. (2021). Rapid detection of wheat blast pathogen *Magnaporthe Oryzae Triticum* pathotype using genome–specific primers and cas 12a–mediated technology. *Engineering*, *7*(9): 1326–35.

Kankanala, P., Czymmek, K. and Valent, B. (2007). Roles for rice membrane dynamics and plasmodesmata during biotrophic invasion by the blast fungus. *Plant Cell*, *19*: 706–24.

Kim, Y.A., Moon, H. and Park, C.J. (2019). CRISPR/Cas9–targeted mutagenesis of Os8N3 in rice to confer resistance to *Xanthomonas oryzae* pv. Oryzae. *Rice*, *12*: 10–13.

Kohli, M.M., Cazal, C. and Chavez, A. (2020). Integrated management of wheat blast disease. *In*: S. Kumar, P.L. Kashyap, G.P. Singh (Eds.), *Wheat Blast*, 175–190. Boca Raton, FL: CRC Press.

Kohli, M.M., Mehta, Y.R., Guzman, E., Viedma, L. and Cubilla, L.E. (2011). Pyricularia blast: A threat to wheat cultivation. *Czech Journal of Genetics and Plant Breeding*, *47*: S130–S134.

Latorre, S.M., Were, V.M., Foster, A.J., Langner, T., Malmgren, A., Harant, A., et al. (2023). Genomic surveillance uncovers a pandemic clonal lineage of the wheat blast fungus. *PLoS Biology*, *21*: e3002052.

Le, M. T., Arie, T. and Teraoka, T. (2010). Population dynamics and pathogenic races of rice blast fungus, *Magnaporthe oryzae* in the Mekong Delta in Vietnam. *Journal of General Plant Pathology*, *76*: 177–82.

Lu, Y., Xue, J., Deng, T., et al. (2020). Safety and feasibility of CRISPR-edited T cells in patients with refractory non-small-cell lung cancer. *Nature Medicine*, *26*: 732–40.

Maciel, J.L.N., Ceresini, P.C., Castroagudin, V.L., Zala, M., Kema, G.H.J. and McDonald, B.A. (2014). Population structure and pathotype diversity of the wheat blast pathogen *Magnaporthe oryzae* 25 years after its emergence in Brazil. *Phytopathology*, *104*: 95–107.

Maekawa, T. and Schulze-Lefert, P. (2017). Caught in the jump. *Science*, *357*(6346): 31–32.

Malaker, P.K., Barma, N.C., Tewari, T.P., Collis, W.J., Duveiller, E., Singh, P.K., Joshi, A.K., Singh, R.P., Braun, H.J., Peterson, G.L. and Pedley, K.F. (2016). First report of wheat blast caused by *Magnaporthe oryzae* pathotype Triticum in Bangladesh. *Plant Disease*, *100*: 2330.

Matthew, S.W. and Yang, Y. (2021). Versatile Applications of the CRISPR/Cas Toolkit in Plant Pathology and Disease Management. *Phytopathology*, *111*(7): 1080–90.

McGrath, M.T. (2001). Fungicide resistance in cucurbit powdery mildew: Experience and challenges. *Plant Disease*, *85*: 236–45.

Mottaleb, K.A., Singh, P.K., Sonder, K., Kruseman, G., Tiwari, T.P., Barma, N.C.D., et al. (2018). Threat of wheat blast to South Asia's food security: An ex-ante analysis. *PLOS One*, *13*: e0197555.

Nga, N., Hau, V.T. and Tosa, Y. (2009). Identification of genes for resistance to a Digitaria isolate of *Magnaporthe grisea* in common wheat cultivars. *Genome*, *52*: 801–809.

Park, J-H., Choi, G.J., Jang, K.S., Lim, H.K., Kim, H.T., Cho, K.Y. and Kim, J-C. (2005). Antifungal activity against plant pathogenic fungi of chaetoviridins isolated from Chaetomium globosum. *FEMS Microbiology Letters*, *252*(2): 309–13.

Patel, A., Sahu, K.P., Mehta, S., Javed, M., Balamurugan, A., Ashajyothi, M., Sheoran, N., Ganesan, P., Kundu, A., Gopalakrishnan, S., et al. (2023). New insights on endophytic Microbacterium-assisted blast disease suppression and growth promotion in rice: Revelation by polyphasic functional characterization and transcriptomics. *Microorganisms*, *11*: 362.

Paul, S.K., Chakraborty, M., Rahman, M., Gupta, D.R., Mahmud, N.U., Rahat, A.A.M., Sarker, A., Hannan, M.A., Rahman, M.M., Akanda, A.M., Ahmed, J.U. and Islam, T. (2022). Marine natural product antimycin A suppresses wheat blast disease caused by *Magnaporthe oryzae Triticum*. *Journal of Fungi*, *8*: 618.

Pieck, M.L., Ruck, A., Farman, M.L., Peterson, G.L., Stack, J.P., Valent, B. and Pedley, K.F. (2017). Genomics–based marker discovery and diagnostic assay development for wheat blast. *Plant Disease*, *101*(1): 103–109.

Pizolotto, C.A., Maciel, J.L.N., Fernandes, J.M.C. and Boller, W. (2019). Saprotrophic survival of *Magnaporthe oryzae* in infested wheat residues. *Eur. J. Plant Pathology*, *153*: 327–39.

Poloni, N.M., Carvalho, G., Vicentini, S.N.C., Dorigan, A.F., Maciel, J.L.N., McDonald, B.A., Moreira, S.I., Hawkins, N., Fraaije, B.A., Kelly, D.E., Kelly, S.L. and Ceresini, P.C. (2020). Widespread distribution of resistance to triazole fungicides in Brazilian populations of the wheat blast pathogen. *Plant Pathology*, *70*(2): 436–48.

Portz, K., Casanova, F., Jordine, A., et al. (2021). Wheat blast caused by *Magnaporthe oryzae* pathovar Triticum is efficiently controlled by the plant defense inducer isotianil. *Journal of Plant Disease and Protection*, *128*: 249–59.

Rabby, S.M.F., Chakraborty, M., Gupta, D.R., Rahman, M., Paul, S.K., Mahmud, N.U., Rahat, A.A.M., Jankuloski, L. and Islam, T. (2022). Bonactin and feigrisolide C inhibit *Magnaporthe oryzae Triticum* fungus and control wheat blast disease. *Plants*, *11*(16): 2108.

Ryder, L.S., Dagdas, Y.F., Kershaw, M.J., et al. (2019). A sensor kinase controls turgor-driven plant infection by the rice blast fungus. *Nature*, *574*: 423–27.

Sadat, M.A. and Choi, J. (2017). Wheat blast: A new fungal inhabitant to Bangladesh threatening world wheat production. *Plant Pathology Journal*, *33*: 103–108.

Saleh, D., Xu, P., Shen, Y., Li, C., Adreit, H., Milazzo, J., Ravigné, V., Bazin, E., Nottéghem, J., Fournier, E. and Tharreau, D. (2012). Sex at the origin: An Asian population of the rice blast fungus *Magnaporthe oryzae* reproduces sexually. *Molecular Ecology*, *21*: 1330–44.

Samara, N.C., Vicentini, S.N.C., Casado, P.S., de Carvalho, G., Moreira, S.I., Dorigan, A.F., Silva, T.C., Silva, A.G., Custódio, A.A.P., Gomes, A.C.S., Maciel, J.L.N., Hawkins, N., McDonald, B.A., Fraaije, B.A. and Ceresini, P.C. (2022). Monitoring of Brazilian wheat blast field populations reveals resistance to QoI, DMI, and SDHI fungicides. *Plant Pathology*, *71*(2): 304–21.

Shan, H., Zhao, M., Chen, D., Cheng, J., Li, J., Feng, Z., Ma, Z. and An, D. (2013). Biocontrol of rice blast by the phenaminomethylacetic acid producer of Bacillus methylotrophicus strain BC79. *Crop Protection*, *44*: 29–37.

Singh, P.K., Gahtyari, N.C., Roy, C., Roy, K.K., He, X., Tembo, B., et al. (2021). Wheat blast: A disease spreading by intercontinental jumps and its management strategies. *Frontiers in Plant Science*, *12*: 710707.

Surovy, M.Z., Dutta, S., Mahmud, N.U., Gupta, D.R., Farhana, T., Paul, S.K., Joe, W., Dunlap, C.A., Oliva, R., Rahman, M. and Islam, T. (2022). Probiotic Bacillus species: Promising biological control agents for managing worrisome wheat blast disease. *Preprints*, 2022110382. https://doi.org/10.20944/preprints202211.0382.v1.

Surovy, M.Z., Gupta, D.R., Chanclud, E., Win, J., Kamoun, S. and Islam, T. (2017). Plant probiotic bacteria suppress wheat blast fungus *Magnaporthe oryzae Triticum* pathotype. *Figshare*. doi:10.6084/m9.figshare.5549278.v1.

Surovy, M.Z., Islam, T. and von Tiedemann, A. (2023a). Role of seed infection for the near and far distance dissemination of wheat blast caused by *Magnaporthe oryzae pathotype Triticum. Frontiers in Microbiology, 14:1040605.*

Surovy, M.Z., Rahman, S., Rostás, M., Islam, T. and von Tiedemann, A. (2023b). Suppressive effects of volatile compounds from Bacillus spp. on *Magnaporthe oryzae Triticum* (MoT) pathotype, causal agent of wheat blast. *Microorganisms*, *11*(5): 1291.

Tagle, A.G., Chuma, I. and Tosa, Y. (2015). Rmg7, a new gene for resistance to Triticum isolates of *Pyricularia oryzae* identified in tetraploid wheat. *Phytopathology*, *105*: 495–99.

Takabayashi, N., Tosa, Y., Oh, H.S. and Mayama, S. (2002). A gene-for-gene relationship underlying the species-specific parasitism of Avena/Triticum isolates of Magnaporthe grisea on wheat cultivars. *Phytopathology*, *92*(11): 1182–88.

Takan, J.P., Chipili, J., Muthumeenakshi, S., Talbot, N.J., Manyasa, E.O., Bandyopadhyay, R., Sere, Y., Nutsugah, S.K., Talhinhas, P., Hossain, M., Brown, A.E. and Sreenivasaprasad, S. (2012). *Magnaporthe oryzae* populations adapted to finger millet and rice exhibit distinctive patterns of genetic diversity, sexuality, and host interaction. *Molecular Biotechnology*, *50*: 145–58.

Talbot, N.J. (2003). On the trail of a cereal killer: Exploring the biology of Magnaporthe grisea. *Annual Review of Microbiology*, *57*(1): 177–202.

Tembo, B., Mulenga, R.M., Sichilima, S., M'siska, K.K., Mwale, M., Chikoti, P.C., et al. (2020). Detection and characterization of fungus (*Magnaporthe oryzae* pathotype Triticum) causing wheat blast disease on rain-fed grown wheat (*Triticum aestivum* L.) in Zambia. *PLOS One*, *15*: e0238724.

Thierry, M., Chatet, A., Fournier, E. and Tharreau, D. (2020). A PCR, qPCR, and LAMP toolkit for the detection of the wheat blast pathogen in seeds. *Plants*, *9*: 277.

Thierry, M., Gladieux, P., Fournier, E., Tharreau, D. and Ioos, R. (2019). A genomic approach to develop a new qPCR test enabling detection of the *Pyricularia oryzae* lineage causing wheat blast. *Plant Disease*, *104*: 60–70.

Urashima, A.S., Hashimoto, Y., Don, Y., Kusaba, M., Tosa, Y., Nakayashiki, H. and Mayama, S. (1999). Molecular analysis of the wheat blast population in Brazil with a homolog of retrotransposon MGR583. *Annals of Phytopathological Society in Japan*, *65*: 429–36.

Urashima, A.S., Grosso, C., Stabili, A., Freitas, E., Silva, D., Netto, D., Franco, I. and Bottan, M. (2009). Effect of Magnaporthe grisea on seed germination, yield, and quality of wheat. *In*: Wang, G.L., Valent, B. (Eds.), *Advances in Genetic, Genomics, and Control of Rice Blast Disease*, 267–77. Springer, Netherlands,

Valent, B., Cruppe, G, Stack, J.P., Cruz, C.D., Farman, M.L., Paul, P.A., Peterson, G.L. and Pedley, K.F. (2021). Recovery plan for wheat blast caused by *Magnaporthe oryzae* Pathotype Triticum. *Plant Health Progress*, *22*(2): 182–12.

Vales, M., Anzoátegui, T., Huallpa, B. and Cazon, M.I. (2018). Review on resistance to wheat blast disease (*Magnaporthe oryzae* Triticum) from the breeder point-of-view: use of the experience on resistance to rice blast disease. *Euphytica*, *214*: 1.

Vy, T.T.P., Hyon, G.S., Nga, N.T.T., Inoue, Y., Chuma, I. and Tosa, Y. (2014). Genetic analysis of host–pathogen incompatibility between Lolium isolates of *Pyricularia oryzae* and wheat. *Journal of General Plant Pathology*, *80*: 59–65.

Wang, F., Wang, C., Liu, P., Lei, C., Hao, W., Gao, Y., et al. (2016). Enhanced rice blast resistance by CRISPR/Cas9-targeted mutagenesis of the ERF transcription factor gene OsERF922. *PLOS One*, *2016*: 11(4).

Wang, J.Y., Wang, S.Z., Zhang, Z., Hao, Z.N., Shi, X.X., Li, L., Zhu, X.M., Qiu, H.P., Chai, R.Y., Wang, Y.L., Li, L., Liu, X.H., Feng, X.X., Sun, G.C. and Lin, F.C. (2021). MAT loci play crucial roles in sexual development but are dispensable for asexual reproduction and pathogenicity in rice blast fungus *Magnaporthe oryzae*. *J. Fungi*, *7*: 858.

Wang, S., Asuke, S., Vy, T.T.P., Inoue, Y., Chuma, I., Win, J., et al. (2018). A new resistance gene in combination with Rmg8 confers strong resistance against Triticum isolates of *Pyricularia oryzae* in a common wheat landrace. *Phytopathology*, *108*: 1299–1306.

West, J. and Kimber, R.B.E. (2015). Innovations in air sampling to detect plant pathogens. *Annals of Applied Biology*, *166*(1): 4–17.

Wilson, R.A. and Talbot, N.J. (2009). Under pressure: Investigating the biology of plant infection by *Magnaporthe oryzae*. *Nature Reviews. Microbiology*, *7*(3): 185–95.

Yan, X., Tang, B., Ryder, L. S., MacLean, D., Were, V.M., Eseola, A.B., Cruz–Mireles, N., Ma, W., Foster, A.J., Osés–Ruiz, M. and Talbot N.J. (2023). The transcriptional landscape of plant infection by the rice blast fungus *Magnaporthe oryzae* reveals distinct families of temporally co-regulated and structurally conserved effectors. *Plant Cell*, *35*(5): 1360–85.

Yao, X., Yang, Q., Wang, Y., Bi, C., Du, H. and Wu, W. (2022). Dual-enzyme-based signal-amplified aptasensor for zearalenone detection by using CRISPR-Cas12a and Nt. AlwI. *Foods*, *11*: 487.

Yasuhara-Bell, J., Pedley, K.F., Farman, M., Valent, B. and Stack, J.P. (2018). Specific detection of the wheat blast pathogen (*Magnaporthe oryzae Triticum*) by loop-mediated isothermal amplification. *Plant Disease*, *102*: 2550–59.

Yesmin, N., Jenny, F., Abdullah, H.M., Hossain, M.M., Kader, M.A., Solomon, P.S., et al. (2020). A review on South Asian wheat blast: The present status and future perspective. *Plant Pathology*, *69*: 1618–29.

Zhan, S.W., Mayama, S. and Tosa, Y. (2008). Identification of two genes for resistance to Triticum isolates of *Magnaporthe oryzae* in wheat. *Genome*, *51*: 216–21.

Zhang, H.L., Wu, Z.S., Wang, C.F., Li, Y. and Xu, J.R. (2014). Germination and infectivity of microconidia in the rice blast fungus *Magnaporthe oryzae*. *Nature Communication*, *5*: 4518.

CHAPTER 8

Food and Nutrition Security in Relation to Climate Change

Mostafa Faruq Al Banna[1]*

Introduction

Government is responsible for ensuring sustainable food and nutrition security for all as stated in Article 15(A) and Article 18 (1) of the Bangladesh Constitution. Bangladesh has achieved a remarkable progress in food and nutrition security. The country has achieved self-sufficiency in food especially in rice, fish, meat, and vegetables. The production of other food has also increased. However, climate change is a great challenge to ensure sustainable food and nutrition security for all in Bangladesh. It is important to mitigate the negative impact of climate change on food and nutrition security in Bangladesh.

Definition of food and nutrition security: "All people at all times have physical, social and economic access to healthy and nutritious food, which is consumed in sufficient quantity and quality to meet their dietary needs and food preferences, and is supported by an environment of safe water, adequate sanitation, health services and care, allowing for a healthy and active life" (The Committee on World Food Security, 2012).

Definition of climate change: Climate change can be defined as any long-term significant change in the expected pattern of average weather of a specific region over an appropriately significant period of time (normally accepted as 30 years).

Impact of climate change on agriculture, food security, and health: Evidence shows that the factors of agricultural production (that is water, labour, energy, and equipment) might be affected by climate change. Climate change has adverse effects on crop production globally due to an increase in temperature, storm, salinity, and rainfall. Climate change will affect those countries adversely that are still more dependent on agriculture and have lower financial and technological strength to cope

[1] FPMU, Ministry of Food, Dhaka, Bangladesh.
* Corresponding author:

with the impact of climate change (Schmidhuber and Tubiello, 2007). The poor people will be more vulnerable as the adverse effect of climate change will increase the dependency of developing countries on imports and as a result, price of the essential products will be increased.

Recent study (Alpino et.al., 2022) has shown that there are impacts of climate change on food security (four dimensions of food security including availability, access, utilization, and stability), which are described as follows:

Table 1 Impact of climate change on food security.

Dimension of Food Security	*Consequences of Climate Changes*
Availability	• Climate changes directly affecting the availability of food, especially in what concerns the production of food. • Due to alterations on the agroecological conditions (increases in temperature, frequency and severity of extreme events, river levels, CO2 concentrations in the atmosphere, soil erosions, changes to the water cycle, alterations in the cycles of vectors, etc.) that reduce the productivity of plantations, such as corn, wheat, sorghum, and rice, and loss of animals (fish, cattle, pigs, and poultry). • Studies on populations' perception also discussed the relationship between the effects of climate changes and the production of food, as well as the population growth, higher competition for resources, and higher demand of food, besides a lower availability of food due to climate variabilities.
Access to Food	• Lack of income arising from the unemployment of farmers who lose their animals (cattle raising), their lands and/or suffer from the reduced productivity of these. • Increase in prices of basic foods resulting from productivity decrease and demand increase. • Difficulties in distributing and storing the production, resulting in lower access to food.
Utilization	• Climate changes may affect the consumption of food, the nutritional quality of foods, and their social value. • Nutritional quality may be affected by a reduced concentration of several micronutrients (minerals). • Lower ingestion of fruits, vegetables, nuts, seeds, and fish, which may cause deficiencies of zinc, iron, vitamin A, vitamin B12, vitamin D, protein, and omega 3.
Stability	• Relationship between climate changes affecting the reduction of agricultural production as well as the quantity of food available in the market, which consequently increases the food prices in the market, especially in rural areas that tend to be more sensitive to weather variations.

Source: Alpino et Al., 2022.

Climate change has multiple negative effects on global public health; reduced quality and quantity of crops result in increased food and financial insecurities leading to malnutrition (undernutrition and obesity) and diet-related non-communicable diseases, such as diabetes mellitus and cardiovascular diseases. The health effects of these disruptions include increased respiratory and cardiovascular disease, injuries and premature deaths related to extreme weather events, changes in the prevalence and geographical distribution of food- and water-borne illnesses, and other infectious diseases, and threats to mental health.

Materials and Methods

This report is based on the secondary literature on food security and climate change. Data from National and International Survey have been used to interpret the findings from the research report.

Searched Strategy

During the research work, PubMed, Google Scholar, Google, and Banglajol databases have been searched to identify relevant publications on Food Security and Climate Change in Bangladesh and other countries in the world. Following keywords have been searched in relevant articles using: "food security", "climate change", "resilience", "adaptation", "disaster", "nutritioPnal status", "prevalence", "health", "Bangladesh". In addition, the bibliography of the finally selected articles or reports (snowballing) has been manually searched to identify more potential articles and reports related to climate change and food security in Bangladesh.

Inclusion and Exclusion Criteria

All the articles were screened based on predefined inclusion and exclusion criteria.

Inclusion criteria for this review were: (a) study or report published between 1st January 1990 and 31st December 2022 to get the most recent view on the topic, (b) published in English language.

Exclusion criteria for this review were: (a) Unpublished report, and (b) absence of nationally representative data.

Results and Discussion

Findings from different research papers and survey reports have been analyzed carefully. Findings shows that there is negative impact of climate change on food security, agriculture, health, and nutritional status in Bangladesh and other countries.

Effect of Climate Change Globally

Climate Change and Food Security: Research shows that there is a significant relationship between climate change and food security. "Climate change affects the key dimensions of food security: availability, stability, access, and utilization of food. Availability of food is affected by climate change directly (through its impact on crop yield, crop pests and diseases, soil fertility) and indirectly (through its impact on economic growth, income distribution, and agricultural demand)." (Braun, 2008, p.14). Research shows that the food access and utilization will be affected indirectly via collateral effects on household and individual incomes, and food utilization could be impaired by loss of access to drinking water and damage to health (Tim Wheeler and Joachim von Braun, Climate Change, 2013).

A research review conducted by Alpino et al., (2022) shows that climate changes affect the dimensions of Food and Nutrition Security, especially in poorer populations in situations of social inequality. Evidence also shows that climate change may affect food production system including agriculture, fisheries, and livestock—as well as the socioeconomic forces that may influence equitable distribution (Myers, et al., 2017).

A research review shows that climate change and related phenomena will increasingly threaten the safety of the food supply and bring about a steady rise in the incidence of foodborne infection and intoxication in the coming years (Duchenne-moutien and Neetoo, 2021).

Findings from a review article shows that climate change will put demands on our current and future supply to meet the needs of a growing and increasingly affluent population (Leisner, 2020).

Climate Change and Agriculture

Findings from a review research suggest that agriculture is one of the most vulnerable sectors in terms of risks associated with climate change and distinct programmatic initiatives are essential to overcome the situation (Yi-Yuan et al., 2009).

Climate Change and Vulnerability

Findings from a review research conducted by Al-Jawaldeh et al. (2022) suggest that climate change will exert a significant adverse effect on water and food security and showed that the nutritional status of the population, which is already characterized by the triple burden of malnutrition, is likely to worsen via three main pathways mediated by climate change, namely, its impact on food security, care, and health. This study also finds that women, infants, children, those living in poor households, and those experiencing displacement will be among the most vulnerable to the nutritional impacts of climate change.

Climate Change and Health

Climate change can deteriorate the health situation in different regions of the world. Evidence shows that climate change can be associated with adverse health outcomes (Jonathan A. Patz et al. 2014). Climate change, as a global disturbance, can exacerbate many environmental health risks familiar to clinicians and public health professionals (Patz et al., 1996, 2000). The nature of risks and population vulnerability will vary by region; indirect consequences such as ecosystem collapse may overshadow more direct health effects, yet are more difficult to estimate.

Recent evidence shows that extreme weather events, meteorological factors, and air pollution have affected the health of children and women in South Asia including Bangladesh (Ishwar Tiwari et al., 2022). This review also found evidence of a significant association between exposure to air pollution and the risk of childhood stunting in Bangladesh (Goyal and Canning, 2018; Kurata, et al., 2020).

Effects of Climate Change in Bangladesh

Bangladesh is one of the most adversely affected countries to climate change in the world. According to the Global Climate Risk Index 2021 Report, Bangladesh is the 7th most vulnerable country to climate adversities. Agriculture of Bangladesh, especially in the coastal areas, has been affected by the adverse effect of climate change such as salinity, soil erosion, drought, and prolonged inundation. Cyclone Aila and Cyclone Sidr are examples of the negative impact of climate change. In the last 30 years, the country has experienced nearly 200 climate-related disasters including drought, extreme temperatures, floods, and storms. These events have killed thousands of people, destroyed homes, and livelihoods.

Evidence shows that the **factors of agricultural production** (that is water, labour, energy, and equipment) might be affected by climate change. Climate change has adverse effects on crop production globally due to an **increase in temperature, storm, salinity, and rainfall**.

Climate change will affect those countries adversely that are still more **dependent on agriculture** and have lower financial and technological strength to cope with the impact of climate change (Schmidhuber and Tubiello, 2007).

A recent field study conducted in coastal region of Bangladesh shows that the knowledge of the total health system is highly affected by global climate change. The study also found that the natural calamities like flood, drought, and salinity will occur frequently because of the changes in the climatic system. As consequence of sea level rising, coastal regions and their health will face great risk since cyclones may hit now and then (Ashrafuzzamana and Furini, 2019).

The poor people will be more vulnerable as the adverse effect of climate change will increase the dependency of developing countries on imports and as a result price of the essential products will be increased.

Effect on Agriculture: Agriculture is one of the most sensitive sectors to climate change, particularly changes in temperature, rainfall patterns, droughts, and floods. Coastal areas of Bangladesh are occupying more than 30.0% of net cultivable land which constitutes the 52.5% (0.828 of 2.85 million hectares) for net cultivable area (Paul and Rashid, 1993). But the expected 30 cm and 50 cm Sea Level Rise by 2030 and 2050 will lead to loss up to 3.0% and 6.0% of coastal agricultural lands in the respective years (Tanner et al., 2007) along with declining 8.0–17.0% of rice production (IPCC, 2007). Apart from extreme temperature, natural disasters such as cyclones and tidal surges, cause immense loss and suffering to livestock through destruction of forage crops as well as shelter.

Effect on Fish: Fish production in Bangladesh is mainly steered by the aquaculture sector, which is dependent on private hatchery-based fish seed production to a great extent. There are 260 species of fish in Bangladesh, all of which are sensitive to particular salt and freshwater conditions. The changes in tidal patterns, as well as increasing saline intrusion into the freshwater rivers, associated with climate change, will impact on fish populations. Recent research review shows that there are indisputable negative impacts of climate change on fish hatchery production in Bangladesh (Siddique et al., 2022). There are no cost-effective proven strategies that have been developed to minimize the effects of climate change on Bangladesh's fish hatchery production. It is important to do research on climate change on fish hatchery production in Bangladesh as the aquaculture industry is very much dependent on fish hatchery production (Siddique et al., 2022).

Effect on Livestock: Climate change has negative impact on humans, animals, plants, and environment. Livestock sector is one of the important and vulnerable sectors due to the negative effects from climate change. "Research has established that changes in global or regional climate patterns due to climate change are affecting livestock health directly and indirectly" (Ali et al., 2020).

Bangladesh has been considered as one of the most vulnerable countries for global climate change due to its distinctive geographical location. To fulfill the increased demand of nutritious food and animal protein, livestock sectors are growing rapidly in Bangladesh. Recent research has shown that the livestock sector in Bangladesh has been suffering with increased disease loads and often reporting emerging and reemerging infectious diseases, and climate changes are increasing heat stress by a lingering hot season and increased heat waves in the summer, resulting in an increased disease incidence for homoeothermic farm animals (Ali et al., 2020).

Health Affect: Climate change will also affect the distribution of climate sensitive diseases (Malaria prevalence increases in line with the warmer). Other diseases such as dysentery, diarrhea, dengue, hypertension associated with heat stress, asthma and skin diseases are also increasing in Bangladesh. Increased flooding as well as drought is resulting in a decline in the availability of clean water, causes water-borne diseases. Findings from a study show that the climate change has a potential adverse impact on human health in Bangladesh and the magnitude of malaria, dengue, childhood diarrhea, and pneumonia was high among the vulnerable communities (Kabir et al., 2016).

Effect on Coastal Areas: A recent study conducted in the climate-vulnerable coastal areas of Bangladesh shows the influence of climate change on household vulnerability and income diversity; however, the study also reveals that climate change, especially the increase in salinity, has affected crop production in Bangladesh, resulting in increased income vulnerability of small and marginal farmers who are highly reliant on farm income (Jalal et al., 2021).

Climate Change and Sustainable Development

A recent study conducted by Mondal (2018) shows the implications of population growth and climate change on sustainable development of Bangladesh and the study also depicts that sustainable development may be difficult to attain if climate change continues to jeopardize economic growth, environmental stability, and the social progress of Bangladesh.

Initiatives from Bangladesh to Mitigate Climate Change

Policy and strategy level: Government of Bangladesh has framed several policies, plan of action and strategies to mitigate the negative impact of climate change. These are as follows:

- **National Adaptation Programme of Action (NAPA) 2005 and Updated NAPA in 2009:** These are documents to identify immediate and urgent needs for adaptation to climate change.
- **Bangladesh Climate Change Strategy and Action Plan-2009:** This is the policy document of the government to address both adaptation and mitigation of climate change.
- **Delta Plan 2100:** The aim of the delta plan is to ensure water and food security, economic progress, and sustainable climate. Adaptation is a critical component of this plan.
- **Five Years Plans (6th FYP, 7th FYP, 8th FYP):** All the FYPs give emphasis on adaptation and mitigation of climate change.
- **National Sustainable Strategy 2010–2021:** This strategy has been prepared to meet the formidable environmental challenges that Bangladesh faces in the way to development.
- **Perspective Plan 2021 and Perspective Plan 2041:** Both the plans give emphasis on climate change.
- **Country Investment Plan for Environment, Forestry, and Climate Change:** This plan responds to the urgent need to address environmental degradation in Bangladesh and to improve the country's ability to meet the threats posed by climate change.
- **Third Country Investment for Sustainable, Nutrition-Sensitive, and Resilient Food Systems (Final Draft):** Consider climate change issues.
- **Mujib Climate Prosperity Plan:** Hon'ble Prime Minister of Bangladesh has announced the "Mujib Climate Prosperity Plan – Decade 2030" to build resilience for achieving prosperity, at the first Climate Finance Summit of the Climate Vulnerable Forum in July 2021. The Mujib Plan is specific to the climate change agenda and to securing Bangladesh's prosperity and independence path amid the major and rapidly increasing threats of global climate change. The Plan proposes a number of hallmarks, new initiatives that will help to anchor the Mujib Plan and accomplish its goals and objectives.

Programme Level: The Government has taken several programs and projects on the basis of following thematic areas: (i) Food security, social protection, and health; (ii) Comprehensive Disaster Management; (iii) Infrastructure; (iv) Research and knowledge management; (v) Mitigation and low carbon development; (vi) Capacity building and institutional strengthening.

Climate Change and International Forum

Climate change issues have been considered one of the major concerns in the recent International Forum like 1) UN Food Systems Summit 2021; 2) Nutrition for Growth Summit (N4G Summit-2021), and 3) The UNFCCC's Conference of the Parties (COP).

UN Food Systems Summit: The ambition of the 2021 UN Food Systems Summit is to launch a collective journey of transforming our food systems to give us the best possible chance of delivering on the 2030 Agenda. To address this, Action Track 5 (AT5) – building resilience to vulnerabilities, shocks and stresses – is proposing actions to ensure that food systems, which are affected by conflict, environmental, and economic shocks and stresses, can maintain functionality, recover from the effects of harmful events, and improve to a better-off state.

Nutrition for Growth Summit (N4G-2021): There are three thematic areas like Health, Food, and Resilience.

COP26 (The UNFCCC's 26th Conference of the Parties): Decisions were also taken to begin work on the next set of financial commitments by developed countries to support adaptation in climate-vulnerable developing countries, and on rules on common timeframes for updating Nationally Determined Contributions (NDCs). It is important to harmonize between the commitments of the international forums to mitigate climate change.

Conclusions

Adapting to climate change in Bangladesh is crucial. The Coastal Embankment Project (CEP) intended for disaster adaptation in the southwest has proven counterproductive for shrimp aquaculture (Kais and Islam, 2023). Sustainable strategies are needed, considering government policies; however, challenges lie in effective implementation and institutionalization. Collaboration involving government, NGOs, and locals is essential for resilience building (Rahman, 2017). Existing policies must be prioritized over creating new ones. Bangladesh faces heightened vulnerability to climate-induced disasters, necessitating ongoing innovation in adaptation, particularly in coastal regions. Implementation should empower vulnerable groups, requiring collaborative efforts. Research on climate-related hazards is vital. Commitments from the UN Food Systems Summit 2021 and other international forums should guide climate-resilient development pathways in Bangladesh.

References

Alpino, T.M.A., Mazoto, M.L., de Barros, D.C. and de Ferries, C.M. (2022). The impacts of climate change on Food and Nutritional Security: A literature review, *Ciência & Saúde Coletiva*, *27*: 273–86.

Ashrafuzzaman, M. and Furini, G.L. (2019). Climate change and human health linkages in the context of globalization: An overview from global to southwestern coastal region of Bangladesh. *Environment International*, *127*: 402–11.

Ayoub Al-Jawaldeh, Maya Nabhani, Mandy Taktouk and Lara Nasreddine. (2022). Climate Change and Nutrition: Implications for the Eastern and Mediterranean Region. *International Journal of Environmental Research and Public Health*, *19*: 17086.

Braun, J.V. (2008). The impact of rising food prices and climate change on the poor. *Rural, 21*: 5.

Global Climate Risk Index (2021). https://www.germanwatch.org/en/19777.

Goyal, N. and Canning, D. Exposure to ambient fine particulate air pollution in utero as a risk factor for child stunting in Bangladesh. *Int. J. Environ. Res. Publ. Health*, *15*(1).

Intergovernmental Panel on Climate Change. (2007). http://www.ipcc.ch/ipccreports/index.htm.

Ishwar Tiwari, McKenzie Tilstra, Sandra M. Campbell, Charlene C. Nielsen, Stephen Hodgins, Alvaro R. Osornio Vargas, Kyle Whitfield, Bhim Prasad Sapkota and Shelby S. Yamamoto (2022). Climate change impacts on the health of South Asian children and women subpopulations: A scoping review. *Heliyon, 8*: e10811.

Jahid, E.J.D., Akhtaruzzaman, M.K., Md. Emran Hossain, Sudhakar Yedla, G.M. and Monirul Alam. (2021). Does climate change stimulate household vulnerability and income diversity? Evidence from southern coastal region of Bangladesh. *Heliyon, 7*: e07990.

Jonathan A. Patz, Howard Frumkin, Tracey Holloway, Daniel J. Vimont and Andrew Haines, (2014). Climate Change: Challenges and Opportunities for Global Health. *JAMA, 312*(15): 1565–80.

Kais, S.M. and Md. Saidul Islam. (2023). Climate Change, Ecological Modernization, and Disaster Management: The Coastal Embankment Project in southwestern Bangladesh. *International Journal of Environmental Research and Public Health, 20*: 6086.

Kurata, M., Takahashi, K. and Hibiki, A. (2020). Gender differences in associations of household and ambient air pollution with child health: Evidence from household and satellite-based data in Bangladesh. *World Dev., 128*: 104779.

Leisner, P. Courtney. (2020). Review: Climate change impacts on food security: Focus on perennial cropping systems and nutritional value. *Plant Science, 293.*

Md. Iqbal Kabir, Md Bayzidur Rahman, Wayne Smith, Mirza Afreen Fatima Lusha and Abul Hasnat Milton. (2016). Climate change and health in Bangladesh: A baseline cross-sectional survey. *Global Health Action, 9*(1): 29609.

Md. Zulfekar Ali, Gemma Carlile and Mohammad Giasuddin. (2020). Impact of global climate change on livestock health: Bangladesh perspective. *Open Veterinary Journal, 10*: 178–88.

Md. Sanaul H. Mondal. (2018). The implications of population growth and climate change on sustainable development in Bangladesh. *Journal of Disaster Risk Studies, 11*: a535.

Mohammad Abu Baker Siddique, A.K. Shakur Ahammad, Abul Bashar, Neaz A. Hasan, Balaram Mahalder, Md. Mehedi Alam, Jatish Chnadra Biswas and Mohammad Mahfujul Haque. (2022). Impacts of climate change on fish hatchery productivity in Bangladesh: A critical review. *Heliyon, 8*: e11951.

Patz, J.A., Epstein, P.R., Burke, T.A. and Balbus, J.M. (1996). Global climate change and emerging infectious diseases. *JAMA, 275*: 217–23.

Patz, J.A., McGeehin, M.A., Bernard, S.M., et al. (2000). The potential health impacts of climate variability and change for the United States: Executive summary of the report of the health sector of the US National Assessment. *Environ Health Perspective, 108*: 367–76.

Paul, B.M. and Rashid, H. (1993). Climatic Hazards in Coastal Bangladesh, chapter 6. *In*: *Land Use Change and Coastal Management*, 183–207.

Rahman, A. (2017). *Climate Change Challenges and Its Preparation in Bangladesh*. Dhaka: Business Times.

Ramona, A.D.M. and Neetoo, H. (2021). Climate Change and Emerging Food Safety Issues: A Review. *Journal of Food Protection, 84*: 1884–97.

Samuel S. Myers, Matthew R. Smith,1 Sarah Guth, Christopher D. Golden,1,2 Bapu Vaitla, Nathaniel D. Mueller, Alan D. Dangour and Peter Huybers. (2017). Climate Change and Global Food Systems: Potential Impacts on Food Security and Undernutrition. *Annu. Rev. Public Health, 38*: 259–77.

Schmidhuber J. and Tubiello, F.N. (2007). Global food security under climate change. *Proceedings of the National Academy of Science* (*PNAS*), *104* (50): 19703–19708.

Tanner TM, Hassan A, Islam KMN, Conway, D, Mechler R, Ahmed AU, and Alam, M (2007) *ORCHID: Piloting Climate Risk Screening in DFID Bangladesh*. Summary Research Report. Institute of Development Studies, University of Sussex, UK.

Tim, W. and Joachim von Braun. (2013). Climate Change and Impacts on Global Food Security. *Science, 341*(6145): 508–13.

Yi-Yuan Su SJD, Yi-Hao Weng, M.D. and Ya-Wen Chiu. (2009). Climate change and food security in East Asia. *Asia Pac. J. Clin. Nutri., 18*: 674–78.

Chapter 9

Mainstreaming Adaptation into Development Agenda

M.V.K. Sivakumar[1]*

Introduction

A large human population, continuing high rate of population growth, and poverty characterize South Asia. The current population of South Asia is 2.005 billion people as of 2023, based on the latest United Nations estimates. South Asian population is equivalent to 24.9% of the total world population. Agriculture is a complex system, within which changes are driven by the joint effects of economic, environmental, political, and social forces (Olmstead 1970; Bryant and Johnston, 1992). Agriculture continues to play a critical role in terms of employment and livelihood security in the region, although its share has now considerably declined to 20–25% of gross domestic product (GDP). More than half of the population of the region is still engaged in conventional agriculture. Most farms in the region are small (less than one ha). In addition to the marginal and small farmers, there are a large number of landless households. The region houses more than 300 M hungry and malnourished people.

Earth's climate system is controlled by a complex set of interactions among the atmosphere, oceans, continents, and living systems. The United Nations Framework Convention on Climate Change (UNFCCC), in its Article 1, defines climate change as: "a change of climate which is attributed directly or indirectly to human activity that alters the composition of the global atmosphere and which is in addition to natural climate variability observed over comparable time periods". Evidence is pointing to the fact that human-driven changes in Earth's energy balance are driving a warmer and wetter atmosphere, with this trend superimposed on and magnifying natural variability. Over the last three decades, climate change has emerged as one of the most crucial issues for humankind, with serious implications for sustainable development. There is certainty from the surface temperature data, collected by World Meteorological Organization's Global Observing System (WMO-GBS), that the globally averaged surface temperatures are rising. According to records maintained by members of

1 Weather and Climate Extrêmes (Elsevier), Switzerland
* Corresponding author: mannavas@gmail.com

WMO, the global surface temperature has increased since the beginning of instrumental records in 1861. Over the 20th century that increase was about 0.6°C. The rate of change for the period since 1976 is roughly three times that for the past 100 years. According to the Sixth Assessment Report of the WMO/UNEP Intergovernmental Panel on Climate Change (IPCC), each of the last four decades has been successively warmer than any decade that preceded it since 1850 (IPCC, 2021). Global surface temperature in the first two decades of the 21st century (2001–2020) was 0.99 (0.84–1.10°C higher than 1850–1900.

Weather and climate are some of the biggest risk factors impacting on farming performance and management. Factors such as climate variability and change contribute to the vulnerability of individual farms, as well as on whole rural communities. More frequent and intense extreme weather events and increasing uncertainty in rainy season patterns are already having significant impacts on food production, food distribution infrastructure, livelihood assets, human health, and food emergencies, in both rural and urban areas (FAO, 2008). This also particularly impacts on regional and world food security.

South Asian population is predicted to increase by almost 700 M people in the next 40 years. This, and the rising income of people are expected to result in a large demand for food. The additional quantities will have to be produced from the same land resource, or less, due to the increase in competition for land and other resources by non-agricultural sectors. The situation will be further complicated by global climatic changes that directly affect agriculture and hence food supply.

Rapidly increasing population growth and diminishing arable lands, particularly in the developing countries, have increased the stress on the natural resource base. Agricultural production systems are expected to produce food for a global population that will amount to 9.1 billion people in 2050 and over 10 billion by end of the century (UNFPA, 2011). Most of the additional billions of people will live in developing countries. Lal (1991) calculates that the per capita arable land will progressively decline from about 0.3 ha in 1990 to 0.23 ha by the year 2000, 0.15 ha by 2050, and 0.14 ha by 2100. Combined with the growing concerns regarding the decline in the non-renewable sources of energy and the degradation of environment, it is certainly timely that the world is taking a hard look at the way agricultural adaptation to climate change and improved sustainability.

Climate Change: Current Trends

According to IPCC (2023), human activities, principally through emissions of GHGs, have unequivocally caused global warming, with global surface temperature reaching 1.1°C above 1850–1900 in 2011–2020. Global surface temperature has increased faster since 1970 than in any other 50-year period over at least the last 2,000 years. Temperatures during the most recent decade (2011–2020) exceed those of the most recent multi-century warm period, around 6,500 years ago (0.2°C–1°C relative to 1850–19000. Prior to that, the next most recent warm period was about 125,000 years ago, when the multi-century temperature (0.5°C to 1.5°C relative to 1850–1900) overlaps the observations of the most recent decade.

Worldwide GHGs have continued to increase, with unequal historical and ongoing contributions arising from unsustainable energy use, land use and land-use change, lifestyles and patterns of consumption and production across regions, between and within countries, and among individuals *(high confidence)*. The change in the global average temperature from 1850–2022 is shown in Fig. 1.

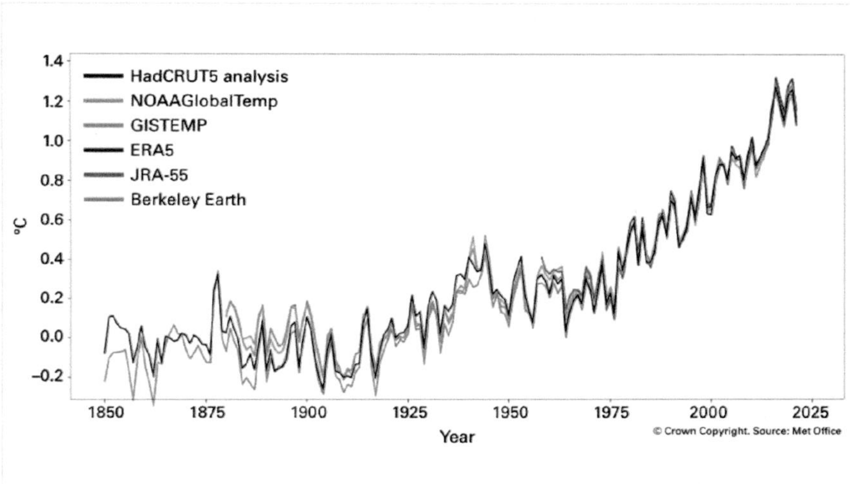

Fig. 1 The Change in Global Average Temperature 1850–2020 (*Source:* Met Office).

It is likely that well-mixed GHGs contributed a warming of 1.0°C–2.0°C, other human drivers (principally aerosols) contributed a cooling of 0.0°C–0.8°C, natural drivers changed global surface temperature by −0.1°C to +0.1°C, and internal variability changed it by −0.2°C to +0.2°C. Heating of the climate system has caused global mean sea level rise through ice loss on land and thermal expansion from ocean warming. Thermal expansion explained 50% of sea level rise during 1971–2018, while ice loss from glaciers contributed 22%, ice sheets 20%, and changes in land-water storage 8%. According to IPCC (2021), global mean sea level increased by 0.20 (0.15–0.25] m between 1901–2018. The average rate of sea level rise was 1.3 (0.6 to 2.1) mm $yr.^{-1}$ between 1901–1971, which increased to 1.9 (0.8 to 2.9) mm $yr.^{-1}$ between 1971–2006, and further increasing to 3.7 (3.2–4.2) mm $yr.^{-1}$ between 2006–2018 (high confidence). Human influence was very likely the main driver of these increases since at least 1971. Human-induced global warming has already caused multiple observed changes in the climate system (high confidence). Changes include increases in both land and ocean temperatures, as well as more frequent heatwaves in most land regions.

According to IPCC (2021), human-induced climate change is already affecting many weather and climate extremes in every region across the globe. Evidence of observed changes in extremes such as heatwaves, heavy precipitation, droughts, and tropical cyclones, and, in particular, their attribution to human influence, has strengthened since the release of Assessment Report 5. There is also high confidence that global warming has resulted in an increase in the frequency and duration of marine heatwaves. Most of the problems people have with weather and climate come from extremes. Extreme weather events—floods, violent storms, droughts, and forest fires (Fig. 2)—occurred on all inhabited continents in 2019. Annual losses from weather disasters such as hurricanes, hailstorms, or wildfires frequently run into hundreds of billions of dollars. In total, weather-related natural disasters have caused losses of some US $ 4,200 billion since 1980 and killed nearly a million people (Munich Re, 2019).

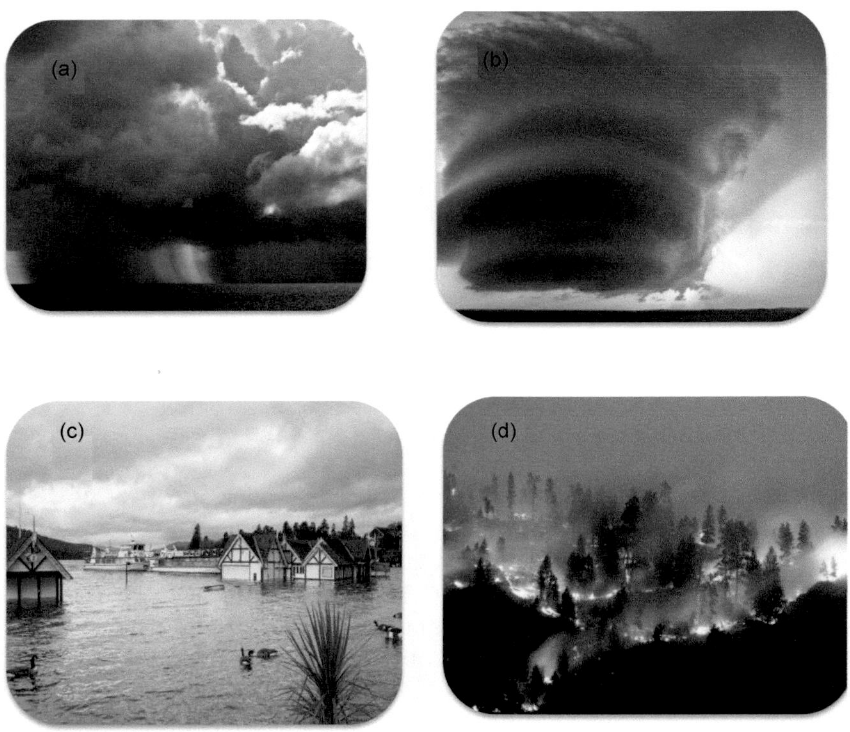

Fig. 2 Extreme weather events: (a) Hurricanes, (b) Storms, (c) Floods, and (d) Wildfires (*Source:* Munich Re 2019).

Scientific assessments have shown that over the past several decades, human activities, especially burning of fossil fuels for energy production and transportation, are changing the natural composition of the atmosphere. Proxy records indicate that for over at least the last 400,000 years, up to about AD 1800, the atmospheric concentration of carbon dioxide (CO_2) varied only by 1–3%. Since then, it has increased by more than 33%, and reached 376 parts per million (ppm) at the end of 2003 and by 2020, it reached 413 ppm. WMO's Global Atmosphere Watch, observing network monitoring atmospheric chemistry, show that today's atmospheric CO_2 concentration has not been exceeded during the past 420,000 years. More than half of that increase in CO_2 concentration has occurred since 1950.

The global atmospheric concentration of methane (CH_4) has increased from a pre-industrial value of about 715 ppb to 1732 ppb in the early 1990s, and was 1889 ppb in 2020. The global atmospheric nitrous oxide (N_2O) concentration increased from a pre-industrial value of about 270 ppb to 333 ppb in 2020. According to IPCC (2021), in 2019, atmospheric CO_2 concentrations were higher than at any time in at least 2 million years (high confidence), and concentrations of CH_4 and N_2O were higher than at any time in at least 800,000 years (very high confidence). Continued GHG emissions will cause further warming and changes in the climate system. Based on the assessment of multiple lines of evidence, global warming of 2°C, relative to 1850–1900, would be exceeded during the 21st century under the high and very high GHG emissions scenarios. The ocean has absorbed about 30% of the emitted anthropogenic CO_2, causing ocean acidification. Marine heatwaves have approximately doubled in frequency since the 1980s (high confidence), and human influence has very likely contributed to most of them since at least 2006.

It is also possible, even likely in some cases, that human-induced climate change will affect naturally occurring climate variability such as the frequency or intensity of El Niño/Southern Oscillation (ENSO) events. ENSO is a recurring cycle that refers to year-to-year variations in sea-surface temperatures, convective rainfall, surface air pressure, and atmospheric circulation that occur across the equatorial Pacific Ocean. The warm phase, in which the central and east-central equatorial Pacific Ocean warms, occurs at intervals of 2–7 years. (El Niño) and alternates with an opposite cold phase (La Niña) (Guilyardi et al., 2009). ENSO is known to affect weather events globally (teleconnection) such as, depending on location, warmer/colder or drier/wetter climate than normal conditions (potential droughts or floods), monsoon rainfall changes, and intensity and frequency of tropical cyclones. A growing number of extreme weather and climate events, some of which have been of unprecedented intensity, continue to be observed with associated degradation of the environment. This requires the global community to give urgent attention and high priority to addressing key issues related to climate change through appropriate measures and policies at national and regional levels.

Increased frequency of droughts and floods are projected to negatively affect local crop production, especially in subsistence sectors at low latitudes. Flash flood and drought may be long-lasting (Gommes and Nègre, 1992), but in Asia their shorter durations are more common. The disasters dealt with below are those relatively short-term extreme events chosen by Salinger et al., (2000): (i) high intensity rainfall and floods, (ii) tropical storms, tornadoes, and strong winds, (iii) extreme temperatures including heat waves and cold waves, (iv) droughts, and (v) wildfires and bushfires. Challenges to disaster risk mainstreaming and to mitigation practices need to be dealt with for each of these events while challenges to contingency planning and to basic preparedness have so much in common for these events that they can be dealt with in more general terms.

Tropical cyclones can form over a sea or ocean with a surface temperature of 27°C or more. The area where such temperatures are possible will definitely increase with warming. The year 1999 witnessed a super cyclone striking the eastern coast of India (Odhisha State). It was a major natural disaster affecting the subcontinent in recent years. On 20 May 2020, Supercyclone Amphan made a landfall, the most powerful storm to do so in the Bay of Bengal region in 20 years. Winds gusting up to 185 km hr^{-1} pounded the coast of the Indian state of West Bengal. Huge waves swept over the Indian and Bangladeshi coast. Trees were lifted out of the ground, city streets turned into rivers, and tens of thousands of people lost their homes. As of 27 May 2020, the number of deaths were about 100 overall.

Droughts of 1972 and 1987, the heat waves in 1995 and 1998, and the cold wave in 2003 killing several hundred people are still fresh in public memory. The drought and failed monsoon of 2002, in particular, an unusually dry July, is a matter of concern for scientists and planners (De et al., 2005). In 2020, during the Coronavirus Pandemic crisis, because of the abandonment of planes, trains and automobiles, the emissions of greenhouse gases decreased. Between January and March 2020, demand for coal dropped by 8% and oil by 5%, compared with the same period in 2019. According to International Energy Agency (IEA), by the end of 2020, the global industrial GHG emissions could be 8% lower than they were in 2019.

Climate Change: Future Projections

IPCC has projected that temperature increase by the end of this century is likely to be in the range of 1.8–4.0°C. For South Asia, the projections are 0.5–1.2°C rise in temperature by 2020, 0.88–3.16°C by 2050. and 1.56–5.44°C depending upon the scenario of future development (Table 1; IPCC, 2007b). Overall, the temperature increases are likely to be much higher in the rabi season than in the kharif season. Himalayan glaciers and snow cover are projected to contract in future due to rise in temperatures. It is very likely that hot extremes, heat waves, and heavy precipitation events will become more frequent. The projected sea level rise by the end of this century is likely to be 0.18–0.59 m. The absolute amount of precipitation is likely to increase in South Asia in future in all months except in the period December–February when this is likely to decrease (Table 1). This increase may, however, be associated by heavier precipitation events and fewer rainy days leading to increased frequency of floods and droughts in the region.

Table 1 Projected changes in surface air temperature and precipitation for South Asia under SRES A1FI (highest future emission trajectory) and B1 (lowest future emission trajectory) pathways for three time slices, namely 2020s, 2050s, and 2080s (from IPCC, 2007b).

Season	*2020*				*2050*				*2080*			
	Temperature, °C		*Precipitation, %*		*Temperature, °C*		*Precipitation, %*		*Temperature, °C*		*Precipitation, %*	
	AIFi	*B1*	*A1F1*	*B1*	*A1F1*	*B1*	*A1F1*	*B1*	*A1F1*	*B1*	*A1F1*	*B1*
Dec-Feb	1.17	1.11	–3	4	3.16	1.97	0	0	5.44	2.93	–16	–6
Mar-May	1.18	1.07	7	8	2.97	1.81	26	24	5.22	2.71	31	20
Jun-Aug	0.54	0.55	5	7	1.71	0.88	13	11	3.14	1.56	26	15
Sep-Nov	0.78	0.83	1	3	2.41	1.49	8	6	4.19	2.17	26	10

Producing enough food for the increasing demand against the background of reducing resources in a changing climate scenario, while also minimizing further environmental degradation, will be a challenging task. Ensuring food security in future at regional, national, and sub-regional levels remains a major challenge for South Asia in the coming decades. It is important to quantify the key sensitivities of food production systems in the region, and to identify options for increasing adaptive capacity.

According to IPCC (2023), global warming will continue to increase in the near term (2021–2040) due to increased cumulative CO_2 emissions in all considered scenarios and modeled pathways. In the near term, global warming *is more likely than not* to reach 1.5°C even under the a very low GHG emission scenario and *likely* or *very likely* to exceed 1.5°C under higher emissions' scenarios. With further warming, every region is projected to increasingly experience concurrent and multiple changes in climatic impact-drivers (IPCC, 2023). Compound heatwaves and droughts are projected to become more frequent, including concurrent events across multiple locations *(high confidence)*. Due to relative sea level rise, current 1-in-100-yr. extreme sea level events are projected to occur at least annually in more than half of all tide gauge locations by 2100 under all considered scenarios *(high confidence)*. Other projected regional changes include intensification of tropical cyclones and/or extratropical storms *(medium confidence)*, and increases in aridity and fire weather *(medium to high confidence)*.

It is reported that for any given warming level, the level of risk will also depend on trends in vulnerability and exposure of humans and ecosystems (IPCC, 2023). Future exposure to climatic hazards is increasing globally due to socio-economic development trends including migration, growing inequality, and urbanization. Human vulnerability will concentrate in informal settlements and rapidly growing smaller settlements. In rural areas, vulnerability will be heightened by high reliance on climate-sensitive livelihoods. Vulnerability of ecosystems will be strongly influenced by past, present, and future patterns of unsustainable consumption and production, increasing demographic pressures, and persistent unsustainable use and management of land, ocean, and water. Loss of ecosystems and their services has cascading and long-term impacts on people globally, especially for Indigenous Peoples and local communities who are directly dependent on ecosystems, to meet basic needs.

According to IPCC (2023), relative to 1995–2014, the *likely* global mean sea level rise under the SSP1-1.9 GHG emissions' scenario is 0.15–0.23 m by 2050, and 0.28–0.55 m by 2100; while for the SSP5-8.5 GHG emissions' scenario, it is 0.20–0.29 m by 2050, and 0.63–1.01 m by 2100 *(medium confidence)*. Over the next 2000 years, global mean sea level will rise by about 2–3 m if warming is limited to 1.5°C and 2–6 m if limited to 2°C. It is virtually certain that the land surface will continue to warm more than the ocean surface (likely 1.4–1.7 times more). It is also reported that the Arctic will continue to warm more than global surface temperature, with high confidence above two times the rate of global warming.

It is very likely that heavy precipitation events will intensify and become more frequent in most regions with additional global warming (IPCC, 2021). At the global scale, extreme daily precipitation events are projected to intensify by about 7% for each 1°C of global warming (*high confidence*). The proportion of intense tropical cyclones (categories 4–5) and peak wind speeds of the most intense tropical cyclones are projected to increase at the global scale with increasing global warming (*high confidence*).

According to AR6 Report of IPCC (2021), there is strengthened evidence since AR5 that the global water cycle will continue to intensify as global temperatures rise (*high confidence*), with precipitation and surface water flows projected to become more variable over most land regions within seasons (*high confidence*) and from year-to-year (*medium confidence*). Monsoon precipitation is projected to increase in the mid-to long-term at global scale, particularly over South and Southeast Asia, East Asia, and West Africa apart from the far west Sahel (*high confidence*). The monsoon season is projected to have a delayed onset over North and South America and West Africa (*high confidence*) and a delayed retreat over West Africa (*medium confidence*). An increase in temperature will trigger increased demand for water for evapotranspiration by crops and natural vegetation and will lead to more rapid depletion of soil moisture (FAO, 2013). Climate change is projected to reduce renewable surface water and groundwater significantly in most dry subtropical regions (Jiménez Cisneros et al., 2014). This will intensify competition for water use.

Impacts of Climate Change on Agriculture and Land Resources

IPCC (2007b) indicates that crop yields could decrease up to 30% in South Asia by the end of the century even if the direct positive physiological effects of CO_2 fertilization are taken into account. Several global studies indicate a probability of 10–40% loss in crop production in India with increases in temperature by 2080–2100 (Rosenzweig et al., 1994; Fischer et al., 2002; Parry et al., 2004). These long-time horizon estimates generally assume business as usual scenario, no new technology development, and nil or limited adaptation by all stakeholders. Recent studies in India suggest a 5–7% decline in wheat for every degree Celsius increase in temperature provided irrigation remains available in future at today's levels.

Easterling et al. (2007) analyzed that in mid-to high-latitude regions, moderate local increases in temperature (1–3°C), along with associated CO_2 increase and rainfall changes is likely to be beneficial for crop yields. As against this, in low-latitude regions a temperature increase of 1–2°C is likely to have negative yield impacts for major cereals. It implies that food security of most developing countries including South Asian countries, which happen to be in the low latitude regions, is likely to become vulnerable in near future. At the same time, differential impacts of climate change on food production in different parts of the world is likely to have consequences on international food prices and trade.

Analysis of the historical trends in yields of crops in the Indo-Gangetic plains using regional statistics, long-term fertility experiments, other conventional field experiments, and crop simulation models have shown that rice yields during the last three decades are showing a declining trend and this may be partly related to the gradual change in weather conditions during last two decades (Aggarwal et al., 2000; Pathak et al., 2003). Similar decline in rice yields and their association with rising nighttime temperatures in rice have also been noticed in the Philippines (Peng et al., 2004).

All countries of South Asia frequently experience natural climatic disasters. Bangladesh, parts of India and Sri Lanka are low lying coastal areas prone to cyclones, sea level rise, and floods. Bhutan, Nepal, India, and Pakistan have several mountain ecosystems that are vulnerable to glacier melt. Semi-arid and arid areas of India, Pakistan, and Bangladesh frequently experience heat and drought stress. Such climatic extremes are known to negatively impact agricultural production, and farmers' livelihood. The projected increase in these events could result in greater instability in food production and threaten livelihood security of farmers. Increased production variability could be perhaps the most significant impact of global impact change on south Asian economies.

Cold waves cause significant losses to crops such as mustard, mango, guava, papaya, brinjal, tomato, and potato in northern India and Pakistan. There are indications that such cold waves and frost events could decrease in future due to global warming and hence yield losses in these crops associated with frost damage is likely to decrease.

IPCC (2007b) has indicated significant increases in runoff in many parts of the world including South Asia due to global warming. This, however, may not be very beneficial because the increase was largely in the wet season and the extra water may not be available in the dry season unless storage infrastructure could be vastly expanded. This extra water in the wet season, on the other hand, may lead to increase in frequency and duration of floods. The increased melting and recession of glaciers associated with global climate change could further change the runoff scenario. In recent decades, Himalayan glaciers have recessed between 2.6–28 m/yr (Kulkarni and Bahugana, 2002). These glaciers are a major source of water for the rivers such as the Indus, Ganges, and Brahmaputra in the Indo-Gangetic plains. These rivers are crucial to millions of people in Pakistan, Nepal, Bhutan, India, and Bangladesh. Such increases in glacier melt in Himalayas could affect availability of irrigation especially in the Indo-Gangetic plains, and thus affect food and livelihood security of millions.

The nutritional quality of cereals may be moderately affected by climatic changes which, in turn, will have consequences for nutritional security of several developing countries where cereals are the primary diet. Research has indeed shown that the decline in grain protein content in cereals could partly be related to increasing CO_2 concentrations and temperature (Hocking and Meyer, 1991; Ziska et al., 1997).

Crop-pest interactions will change significantly with climate change leading to impact on pest distribution and crop losses. Diseases and insect populations are strongly dependent upon the temperature and humidity. Any increase in them, depending upon their base value, can significantly alter their population, which ultimately results in yield loss. With small changes, the virulence of different pests also changes. The swarms of locust produced in the Middle East usually fly eastward into Pakistan and India during summer season and they lay eggs during the monsoon period. Changes in rainfall, temperature, and wind speed pattern may influence the migratory behavior of locusts and other pests, thus threatening food and livelihood security.

Heat stress associated with global warming directly impacts forage quality, ingestion of food and feed, declines in physical activity, and ultimately reduces dairy milk yield. Increases in air temperature and/or humidity have the potential to affect the reproductive behavior of domestic animals. Global warming would further increase water, shelter, and energy requirement of livestock for meeting projected milk demands.

Increasing sea and river water temperature is likely to affect fish breeding, migration, and harvests. A rise in temperature as small as 1°C could have important and rapid effects on the mortality of fish and their geographical distributions. Oil sardine fishery did not exist before 1976 in the northern latitudes and along the east coast as the resource was not available/and sea surface temperature (SST) were not congenial. With warming of sea surface by 0.5–1.0°C, the oil sardine is able to find temperature to its preference especially in the northern latitudes and eastern longitudes, thereby extending the distributional boundaries and establishing fisheries in larger coastal areas.

Corals are affected by warm surface waters leading to bleaching due to losses of associated symbiotic algae. Heat stress in 1998 caused massive bleaching of coals in the Indian Ocean. Corals in Indian Ocean are likely to be soon exposed to summer temperatures that will exceed the thermal thresholds observed over the last 20 years. Annual bleaching of corals will become almost a certainty from 2050. Given the implication that reefs will not be able to sustain catastrophic events more than three times a decade, reef building corals are likely to start disappearing as dominant organisms on coral reefs between 2030–2040.

Agriculture is the economic sector that is most vulnerable to climate change, and it is directly responsible for about 25% of the world's total greenhouse gas emissions. Climate change is profoundly impacting the conditions in which agricultural activities are conducted. Cascading impacts of climate change can now be attributed along chains of evidence from physical climate through to intermediate systems and then to people (IPCC, 2014). A range of physical, biological, and biophysical impacts bear on agro-ecosystems and translate into impacts on agricultural production. This has quantity, quality, and price effects, with impacts on the income of farm households and on purchasing power of non-farm households. Climate change affects all four dimensions of food security and nutrition.

Increase of mean temperature, changes in rain patterns, increased variability both in temperature and rain patterns, changes in water availability, the frequency and intensity of 'extreme events'; sea level rise and salinization and perturbations in ecosystems will have intense impacts on agriculture. The relatively large inter-annual climate variability in the near term underlines the importance of managing risks to food security from current climate variability and extreme weather events. The extent of these impacts will depend not only on the intensity and timing of the changes but also on their combination, which are more uncertain, and on local conditions.

Climate change is projected to reduce renewable surface water and groundwater significantly in most dry subtropical regions (Jiménez Cisneros et al., 2014). This will intensify competition for water use. Climate change is adding significant uncertainty to the availability of water in many regions in the future. It will affect precipitation, runoff and snow/ice melt, with effects on hydrological systems as well as on water quality, water temperature and groundwater recharge. Climate change will also significantly impact sea level with potential impacts on the salinity of surface and groundwater in coastal areas.

The observed effects of past climate trends on crop production are evident in several regions of the world (Porter et al., 2014), with negative impacts more common than positive ones, including several periods of price spikes following climate extremes in key producing regions. There is evidence that climate change has already negatively affected wheat and maize yields in many regions and also at global level (Lobell et al., 2011). The increased frequency of unusually hot nights in most regions is damaging for most crops, with observed impacts on rice yields and quality.

According to the United Nations (2015), there are still 836 million people in the world living in extreme poverty (less than US $1.25/day). And according to the International Fund for Agricultural Development (IFAD), at least 70% of the very poor live in rural areas, most of them depending partly or completely on agriculture for their livelihoods. It is estimated that 500 M smallholder farms in the developing world are supporting almost 2 billion people, and in Asia and sub-Saharan Africa these small farms produce about 80% of the food consumed (IFAD, 2011). The rural poor often depend partly on forests for their livelihoods (World Bank, 2002).

Climate change and increased climate variability are already impacting the smallholder agriculture and food security and will make the challenge of ending hunger and malnutrition even more difficult. These existing pressures will demand the development and implementation of appropriate methods to address issues of vulnerability to weather and climate. These will be needed to assist farmers to further develop their adaptive capacity with improved planning and better management decisions.

Agriculture is the sector most vulnerable to climate change due to its high dependence on climate and weather. Climate change and increasing climate variability affect food security in all of its 4 dimensions: availability, accessibility, utilization, and stability (FAO, 2008). More frequent and intense extreme weather events and increasing uncertainty in rainy season patterns are already having significant impacts on food production, food distribution infrastructure, livelihood assets, human health, and food emergencies, in both rural and urban areas (FAO, 2008). Battisti and Naylor (2009) concluded that projected changes in the frequency and severity of weather and climate events, including higher growing season temperature, have significant consequences for food production, farm income, and food security.

Climate change can have both direct and indirect impacts on agricultural production systems. The direct impacts are those that are directly caused by a modification of physical characteristics such as temperature levels and distribution along the year and water availability on a specific agricultural production. Indirect effects are those that affect production through changes in other species such as pollinators, pests, disease vectors, invasive species.

In their article on climate change in South Asia, Sivakumar and Stefanski (2011) enlisted the following climate change impacts in South Asia: impacts of enhanced temperatures; impacts of precipitation variability and water resources; impacts of increased frequency of extreme events and natural disasters; impacts of crop, pasture and forest productivity; impacts on crop pests and diseases; impacts on fisheries; and impacts of sea level rise.

Impacts in the mid to high latitudes are expected to be more mixed, especially at lower levels of warming. Some high-latitude regions are expected to benefit, sometimes substantially, from warmer temperatures and longer growing seasons; however, other environmental conditions, such as soil quality issues in the far north, will likely constrain expansion (Porter et al., 2014). Spatial differences are also observed at regional and subregional scales, particularly where there are substantial differences in elevation. Contrasted impacts between high- and low-latitude regions indicate that climate change is likely to exacerbate existing imbalances between the developed and developing world (Elbehri et al., 2015). Overall climate change will also increase variability in crop yields in many regions.

The risks of climate change impacts on agricultural production directly translate into additional risks for the food security and nutrition of the people who directly depend on agriculture for their food and livelihood. They can also have an impact on the food security and nutrition of distant populations through price volatility and disrupted trade. As shown in Fig. 3, there is thus a cascade of risks from climate changes to agro-ecosystems, to agricultural production, to economic and social consequences and finally to food security and nutrition (FAO, 2016).

The effects of climate change on the ecosystems are already severe and widespread, and ensuring food security in the face of climate change is among the most daunting challenges facing humankind. Extreme weather and climate events such as severe droughts, floods, or temperature shocks often strongly impede sustainable farming development, particularly in the tropics and sub-tropics. Analysis of drought trends in different locations confirm that there is an increasing frequency and intensity of droughts, which can lead to increased land degradation. Factors such as climate variability and change contribute to the vulnerability of individual farms, as well as on whole rural communities.

Developing countries are likely to face the highest reductions in agricultural potential in the world due to climate change. As a result, climate change will place an additional burden on efforts to meet long-term sustainable development goals. Slow agricultural productivity growth, declining income growth, and problems of maintaining food security already pose challenges to many developing countries in the world. The impacts on the territories of indigenous people and marginalized populations are both early and severe due to their location in vulnerable environments, including low rainfall areas, drylands, and desert margins. Climate change affects livestock production in multiple ways, both directly and indirectly. The most important impacts are experienced in animal productivity, yields of forages and feed crops, animal health and biodiversity, as summarized in Table 2.

Consistent warming trends and more frequent and intense extreme weather events have been observed across the different regions in the world in recent decades. Of the total annual crop losses in world agriculture, many are due to direct weather and climatic effects such as droughts, flash floods, untimely rains, frost, hail, and severe storms (Hay, 2007). Chattopadhyay and Lal (2007) estimated that around 28% of the land in India is vulnerable to droughts, 12% to floods, and 8% to cyclones. But in the year 1918, which was ranked as the worst drought year of the last century in India, about 68.7% of the total area of the country was affected by drought (Chowdhury et al., 1989).

Farmers and farming communities throughout the world have, in most instances, survived and developed by mastering the ability to adapt to widely varying weather and climatic conditions. However, the dramatic growth in human population is imposing enormous pressure on existing farming production systems. Agricultural production systems are expected to produce food for a global population that will

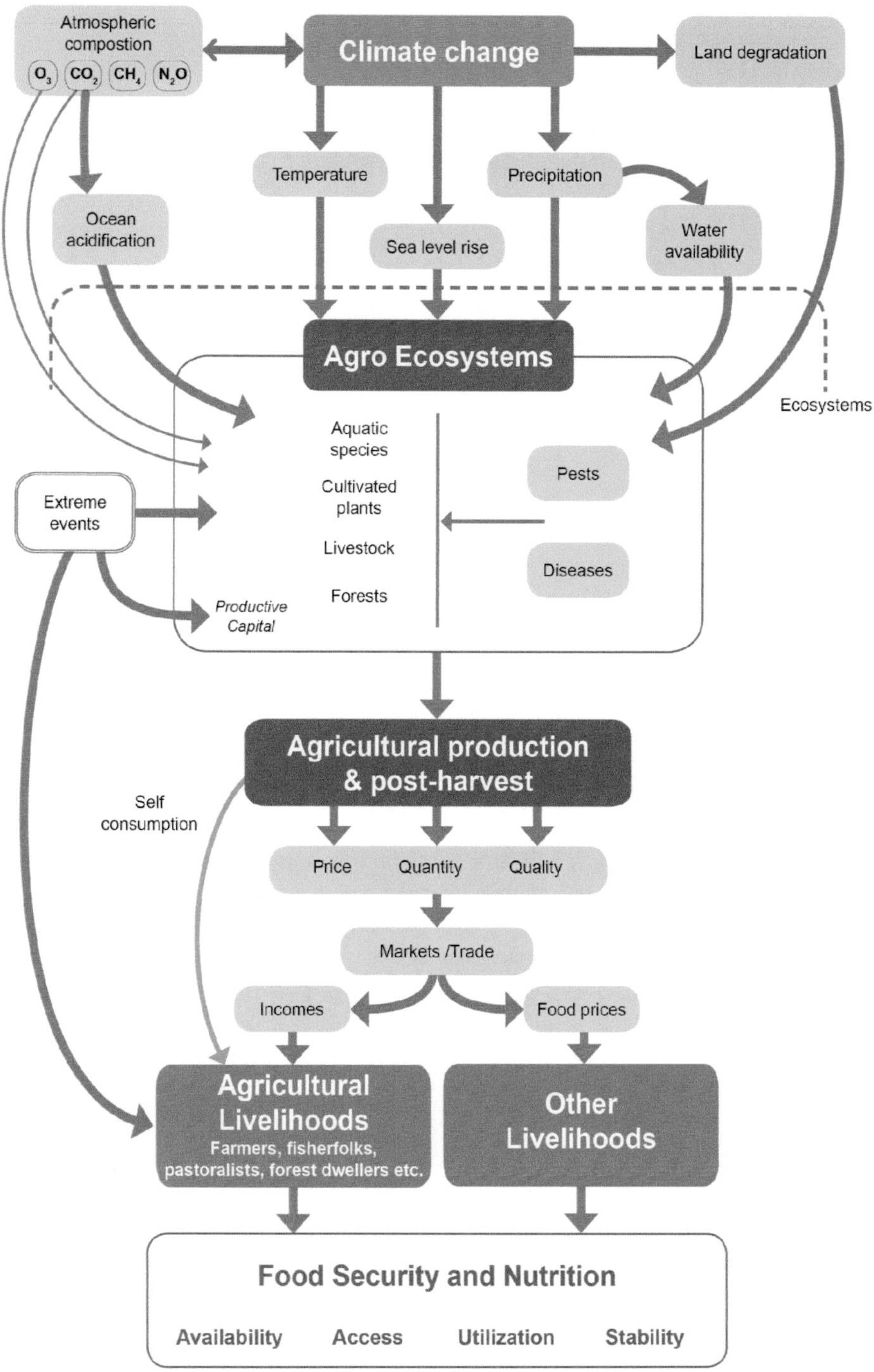

Fig. 3 Schematic representation of the climate change impacts on food security and nutrition (*Source:* FAO 2016).

Table 2 Pathways of impacts of climate change on livestock (***Source***: Thornton et al., 2009)

	Animals	*Forage and Feed Crops*	*Labor Force and Capital*

Variability in rainfall	• Shortages in drinking and servicing water.
	• Diseases
	  ▪ Increased pathogens, parasites, and vectors.
	  ▪ Changed distribution and transmission.
	• New diseases
		• Decreased yields.
		• Decreased forage quality.
		• Changes in pasture composition (species, communities).
		Changes in production system (e.g., from mixed crop-livestock to rangelands)
			• Altered human health and resources allocation to livestock.
			• Decreased productivity.
			• Migrations
			• Conflicts

Temperature	• Heat stress
	  ▪ Decreased feed intake and livestock yields
	  ▪ Decreased conception rates
	  ▪ Altered metabolism and increased mortality.
	• Diseases
	  ▪ distribution and transmission through pathogens, parasites, and vectors
	  ▪ Decreased resistance of livestock
	  ▪ New diseases
	• Domestic biodiversity losses
		• Decreased yields
		• Decreased forage quality
		  ▪ Change in pasture composition	

CO_2 in the atmosphere		▪ Partial stomata closure and reduced transpiration ▪ Change in pasture composition	

amount to 9.1 billion people in 2050 and over 10 billion by end of the century (UNFPA, 2011). Most of the additional billions of people will live in developing countries. At the same time, more people will be living in cities (70% against the current 50%). Population and income increase as well as urbanization are driving increased and changing food and feed demand. Given these trends, FAO estimates that production will have to increase by 60% by 2050 to satisfy the expected demands for food and feed (FAO, 2015).

Slow agricultural productivity growth, declining income growth, and problems of maintaining food security already pose challenges to many developing countries in the world. In future, the increase of mean temperature, changes in rainfall patterns, increased variability both in temperature and rainfall patterns, changes in water availability, the frequency and intensity of extreme events, sea level rise, and salinization and perturbations in ecosystems will have intense impacts on agriculture. Developing countries are likely to face the highest reductions in agricultural potential in the world due to climate change.

Climate change is adding significant uncertainty to the availability of water in many regions in the future. It will affect precipitation, runoff and snow/ice melt, with effects on hydrological systems as well as on water quality, water temperature and groundwater recharge. According to the IPCC (2012), there is "medium confidence" that "droughts will intensify in the twenty-first century in some seasons and areas, due to a combination of more variable precipitation and/or increased evapotranspiration". Climate change will also significantly impact sea level with potential impacts on the salinity of surface and groundwater in coastal areas.

The observed effects of past climate trends on crop production are evident in several regions of the world (Porter et al., 2014), with negative impacts more common than positive ones, including several periods of price spikes following climate extremes in key producing regions. There is evidence that climate change has already negatively affected wheat and maize yields in many regions and also at global level (Lobell et al., 2011). The increased frequency of unusually hot nights in most regions is damaging for most crops, with observed impacts on rice yields and quality.

A Framework for Adapting Agriculture to Climate Change

In view of the projected large-scale adverse effects of global warming on food production in South Asia, we need to analyze the possible options that could assist in increasing the regions adaptive capacity. There is considerable traditional wisdom in the region for adapting to climatic risks that is valuable even today. Sharing such experiences accumulated over centuries across South Asia could be useful at the household, community as well as regional level. The region has earlier adapted to climatic stresses by resorting to mixed cropping, changing varieties and planting times, by diversifying sources of income for farmers, and maintaining buffer stocks of food for managing periods of scarcity. These management strategies would also help in the future climate change scenarios but may not be enough in view of the increasing intensity of climatic risks and pressure on land to produce more food and also with much higher efficiency. Some of the possible adaptation options, which are relevant for adapting to current climatic risks as well as to climate change, are discussed below.

- It is important to promote the research needed to continue international efforts to develop, disseminate, and apply climate forecast information that will increase sustainable agricultural, fisheries, and forestry productivity and be of particular benefit to developing countries.
- Maintain, promote, and establish the preparedness strategies and mechanisms, including development and application of climate forecast information for surveillance and early warning, drought, flood, other natural disasters, pest, and disease alertness.
- Promote judicious application of meteorological, including climatological and hydrological, knowledge and information, and seasonal and long-range forecasts, to assist the agricultural community in developing and operating sustainable agricultural systems and increasing production in an environmentally sustainable manner. Hence, improve collection and dissemination of weather-related information by improving weather station networks to strengthen monitoring of extreme events and their impacts on food production and availability.
- Create a Climate Change and Food Security Network in each region and establish a Climate Outlook Forum. This network will share information on management of climate change and related science, data, tools and methodologies in the region and develop seasonal climate predictions to assist farmers to optimally adjust their planting dates, crop varieties, and management practices to reduce agricultural vulnerability to hydro-meteorological hazards.
- Stimulate multidisciplinary research on topics related to climate change and food security in the region which include hydrological cycles of major river systems and deltas; improved germplasm for tolerance to biotic and abiotic stresses exacerbated by climate change; and promotion of carbon credits through Clean Development Mechanisms (CDM) of the Kyoto Treaty and other national and international organizations.
- Initiate and strengthen cooperation among academic, research, and policy institutions, international organizations, and NGOs to provide opportunities for human resource development and relevant training needs on issues related to mitigation and adaptation of climate change.
- Improve national agricultural extension services in order to secure farmers' self-reliance by providing better information about effective weather and climate risk management and the sustainable use of natural resources for agricultural production.
- Develop innovative financial mechanisms to scale up technical and financial support for adaptation efforts in developing countries and identify mainstream perspectives about current development policies to enhance integrative and cost-effective responses to adaptation and development needs.
- South Asian countries have limited resources, which are targeted towards attaining food security and poverty alleviation, and other developmental targets. Addressing future climate change concerns at the cost of current priority of development cannot be an option. However, integrating perspectives on climatic risks in their current policies and programs in different sectors such as disaster management, water resources management, land use, biodiversity conservation, and agricultural development will lead to increased adaptive capacity to current as well as future climatic variability.

Conclusions

There is an alarming increase in the number of countries affected by natural disasters, which rose from 28–46, mostly in the developing world, where disasters have exacerbated the problems of food security, poverty, and land degradation. As per available scientific evidence, climate extremes are on the rise, and they continue to target the world's poorest and least-developed and there must be greater investment in the reduction of the risks from climate extremes. Despite a long history of climate extremes affecting agriculture, rangelands, and forestry, comprehensive documentation of these climate extremes at the national, regional, and international levels has been weak and a comprehensive assessment of their impacts on agriculture, forestry, and fisheries, and strategies for their mitigation is critical for sustainable development, especially in the developing countries. To gain a better understanding of the climate variability and climate change and to cope with the impacts of climate extremes, improved agrometeorological databases are critical to develop more appropriate levels and forms of mitigation and preparedness. Programs for improving prediction methods and dissemination of warnings should be expanded and intensified. Agricultural experts, especially agrometeorologists, should make more efficient use of the improvements in the information and communication technologies over the past two decades to provide timely and efficient agrometeorological information and products to the farming communities to help them cope with the impacts of climate extremes and improve the agricultural productivity.

References

Aggarwal, P.K., Bandyopadhyay, S.K., Pathak, H., Kalra, N., Chander, S. and Sujith Kumar, S. (2000). Analysis of the yield trends of rice-wheat system in northwestern India. *Outlook on Agric.*, *29*(4): 259–68.

Battisti, D.S. and Naylor, R.L. (2009). Historical warnings of future food security with unprecedented seasonal heat. *Science*, *323*: 240–44.

Bryant C.R. and Johnston, T.R.R. (1992). *Agriculture in the City's Countryside*. Toronto: University of Toronto Press.

Chattopadhyay, N. and Lal, B. (2007). Agrometeorological Risk and Coping Strategies: Perspective from Indian Subcontinent. *In*: M.V.K. Sivakumar, R.P. Motha (Eds.), *Managing Weather and Climate Risks in Agriculture*, 83–98. Springer, Berlin Heidelberg.

Chowdhury, A., Dandekar, M.M. and Raut, P.S. (1989). Variability of drought incidence over India: A Statistical Approach. *Mausam*, *40*: 207–14.

De, U.S., Dube, R.K. and Prakasa Rao, G.S. (2005). Extreme Weather Events over India in the last 100 years, I. *Ind. Geophys. Union*, *9*: 173–87.

Easterling, W.E., Aggarwal, P.K., Batima, P., Brander, K.M., Erda, L., Howden, S.M., Kirilenko, A., Morton, J., Soussana, J-F.,. Schmidhuber, J. and Tubiello, F.N. (2007). Food, fiber, and forest products. *Climate Change 2007: Impacts, Adaptation, and Vulnerability*. Contribution of Working Group II to the Fourth Assessment Report of the Intergovernmental Panel on Climate Change, (Eds., M.L. Parry, O.F. Canziani, J.P. Palutikof, P.J. van der Linden, C.E. Hanson) 273–13. Cambridge University Press, Cambridge, UK.

Elbehri, A., Elliott, J. and Wheeler, T. (2015). Climate change, food security and trade: An overview of global assessments and policy insights. *In*: A. Elbehri (Ed.), *Climate Change and Food Systems: Global Assessments and Implications for Food Security and Trade*. Food and Agriculture Organization of the United Nations, Rome, Italy.

FAO. (2008). Climate change adaptation and mitigation: Challenges and opportunities for food security. Information document prepared for the high-level *Conference on World Food Security: The Challenges of Climate Change and Bioenergy*, 3–5 June 2008. Food and Agriculture Organization of the United Nations, Rome, Italy.

FAO. (2013). *Resilient Livelihoods: Disaster Risk Reduction for Food and Nutrition Security Framework Programme*. Food and Agriculture Organization of the United Nations, Rome, Italy.

FAO. (2015). *The Impact of Natural Hazards and Disasters on Agriculture and Food and Nutrition Security: A Call for Action to Build Resilient Livelihoods*. Food and Agriculture Organization, Rome, Italy.

FAO. (2016). *Climate Change and Food Security: Risks and Responses*. Food and Agriculture Organization of the United Nations. Rome, Italy.

Fischer, G., Shah, M. and van Velthuizen, H. (2002). *Climate Change and Agricultural Vulnerability*. International Institute for Applied Systems Analysis. Luxemburg, Austria.

Gommes, R. and Nègre, T. (1992). The role of agrometeorology in the alleviation of natural disasters. *Agrometeorology Series Working Paper No. 2.* FAO, Rome.

Guilyardi, E., Wittenberg, A., Fedorov, A., Collins, M., Wang, C., Capotondi, A., van Oldenborgh, G.J. and Stockdale, T. (2009). Understanding El Niño in ocean-atmosphere general circulation models: Progress and challenges. *Bulletin American Meteorological Society*, *90*: 325–40.

Hay, J. (2007). Extreme Weather and Climate Events, and Farming Risks, *In*: M.V.K. Sivakumar, R. Motha (Eds.), *Managing Weather and Climate Risks in Agriculture*, 1–19. Springer, Berlin Heidelberg.

Hocking, P.J. and Meyer, C.P. (1991). Carbon dioxide enrichment decreases critical nitrate and nitrogen concentrations in wheat. *Journal of Plant Nutrition*, *14*: 571–84.

IFAD. (2011). Rural poverty report 2011. New realities, new challenges: New opportunities for tomorrow's génération (available at http://www.ifad.org/rpr2011/report/e/rpr2011.pdf.

IPCC. (2007b). *Climate Change 2007: Climate Change Impacts, Adaptation and Vulnerability*. Summary for Policymakers. Inter-Governmental Panel on Climate Change.

IPCC. (2012). *Managing the Risks of Extreme Events and Disasters to Advance Climate Change Adaptation*, 582 pp. (Eds., C.B. Field, C. Barros, T.F. Stocker, D. Qin, D.J. Dokken, K.L. Ebi, M.D. Mastrandrea, K.J. Mach, G.-K. Plattner, S.K. Allen, M. Tignor, P.M. Midgley). Cambridge University Press, Cambridge, UK and New York, NY, USA.

IPCC. (2014). Summary for policymakers. *In*: IPCC. *Climate Change 2014: Impacts, Adaptation, and Vulnerability*, 1–32. Contribution of Working Group II to the Fifth Assessment Report of the Intergovernmental Panel on Climate Change. Cambridge University Press, Cambridge, UK and New York, USA.

IPCC. (2021). Summary for Policymakers. *In*: *Climate Change 2021: The Physical Science Basis*. Contribution of Working Group I to the Sixth Assessment Report of the Intergovernmental Panel on Climate Change. (Eds., V. Masson-Delmotte, P. Zhai, A. Pirani, S.L. Connors, C. Péan, S. Berger, N. Caud, Y. Chen, L. Goldfarb, M.I. Gomis, M. Huang, K. Leitzell, E. Lonnoy, J.B.R. Matthews, T.K. Maycock, T. Waterfield, O. Yelekçi, R. Yu, B. Zhou). Cambridge University Press, Cambridge, UK and New York, NY, USA.

IPCC (2023). Summary for Policymakers. *In*: *Climate Change 2023: Synthesis Report*. A Report of the Intergovernmental Panel on Climate Change. Contribution of Working Groups I, II, and III to the Sixth Assessment Report of the Intergovernmental Panel on Climate Change (Core Writing Team, H. Lee and J. Romero (Eds.)). IPCC, Geneva, Switzerland, p. 36. (In press).

Jiménez Cisneros, B.E., Oki, T., Arnell, N.W., Benito, G., Cogley, J.G., Döll, P., Jiang, T. and Mwakalila, S.S. (2014). Freshwater resources. *In*: *Climate Change 2014: Impacts, Adaptation, and Vulnerability. Part A: Global and Sectoral Aspects*, 229–69. Contribution of Working Group II to the Fifth Assessment Report of the Intergovernmental Panel on Climate Change. (Eds., C.B. Field, V.R. Barros, D.J. Dokken, K.J. Mach, M.D. Mastrandrea, T.E. Bilir, M. Chatterjee, K.L. Ebi, Y.O. Estrada, R.C. Genova, B. Girma, E.S. Kissel, A.N. Levy, S. MacCracken, P.R. Mastrandrea, L.L. White). Cambridge University Press, Cambridge, UK and New York, NY, USA.

Kulkarni, A.V. and Bahuguna, I.M. (2002). Glacial retreat in the Baspa basin, Himalaya, monitored with 17 satellite stereo data (Correspondence). *Journal of Glaciology*, *48*: 171–72.

Lal, R. (1991). Soil structure and sustainability. *Journal Sustainable Agriculture, 1*: 67–92.

Lobell, D.B., Schlenker, W. and Costa-Roberts, J. (2011). Climate trends and global crop production since 1980. *Science, 333*: 616–620.

Munich Re. (2019). *Natural Disasters Overview*. Munich Re, Munich, Germany. https://www.munichre.com/topics-online/en/climate-change-and-natural-disasters/natural-disasters.html.

Olmstead, C.W. (1970). The phenomena, functioning units, and systems of agriculture. *Geographica Polonica, 19*: 31–41.

Parry, M.L., Rosenzweig, C., Iglesias, A., Livermore, M. and Fischer, G. (2004). Effects of climate change on global food production under SRES emissions and socio-economic scenarios. *Global Environmental Change, 14*: 53–67.

Pathak, H., Pathak, J.K., Ladha, P.K., Aggarwal, S., Peng, S., Das, Yadvinder Singh, Bijay Singh, S.K. Kamra, B., Mishra, A.S.R.A.S. Sastri, H.P., Aggarwal, Das, D.K. and Gupta, R.K. (2003). Trends of climatic potential and on-farm yields of rice and wheat in the Indo-Gangetic Plains. *Field Crops Research, 80*: 223–34.

Peng, S., Huang, J., Sheehy, J.E., Laza, R.C., Visperas, R.M., Zhong, X., Centeno, C.S., Khush, G.S., and Cassman, K.G. (2004). Rice yields decline with higher night temperature from global warming. *Proceedings of the National Academy of Science, 101*(27): 9971–75.

Porter, J.R., Xie, L., Challinor, A.J., Cochrane, K., Howden, S.M., Iqbal, M.M., Lobell, D.B. and Travasso, M.I. (2014). Food security and food production systems. *In*: *Climate Change 2014: Impacts, Adaptation, and Vulnerability. Part A: Global and Sectoral Aspects*, 485–33. Contribution of Working Group II to the Fifth Assessment Report of the Intergovernmental Panel on Climate Change. (Eds., C.B. Field, V.R. Barros, D.J. Dokken, K.J. Mach, M.D. Mastrandrea, T.E. Bilir, M. Chatterjee, K.L. Ebi, Y.O. Estrada, R.C. Genova, B. Girma, E.S. Kissel, A.N. Levy, S. MacCracken, P.R. Mastrandrea, L.L. White. Cambridge University Press, Cambridge, UK, and New York, USA.

Rosenzweig, C. and Parry, M.L. (1994). Potential impact of climate change on world food supply. *Nature, 367*: 133–38.

Salinger, M.J., Desjardins, R.L., Janzen, H., Karing, P.H., Veerasamy, S. and Zipoli, G. (2000). *Climate Variability, Agriculture, and Forestry: Towards Sustainability*, 41 p. Technical Note No. 200 (WMO No. 928). Geneva: World Meteorological Organization.

Sivakumar, M.V.K. and Stefanski, R. (2011). Climate change in South Asia. *In*: R. Lal, M.V.K. Sivakumar, S.M.A. Faiz, A.H.M. Mustafizur Rahman, K.R. Islam (Eds.), *Climate Change and Food Security in South Asia*, 13–30. Springer, New York and London.

UNFPA. (2011). *The State of World Population 2011*. United Nations Population Fund, New York, USA.

United Nations. (2015). *Sendai Framework for Disaster Risk Reduction 2015–2030*. Agenda Item 19 (c). United Nations General Assembly, 69th Session, New York, USA.

World Bank. (2002). *Sustaining forests. A World Bank Strategy*. Available at http://siteresources.worldbank.org/Intforests/214573-113990657527/20632625/ Forest_Strategy_Booklet.pdf).

Ziska, L.H., Namuco, O., Moya, T. and Quiland, J. (1997). Growth and yield response of field grown tropical rice to increasing carbon dioxide and air temperature. *Agronomy Journal, 89*: 45–53.

CHAPTER 10

Evolving Landraces of the Coast and their Utilization in Salt Tolerance Breeding

Zeba I. Seraj[1]*

Introduction

More than two million hectares of the coastal region of Bangladesh is affected by varying degrees of salinity depending upon the season and its remoteness from the major rivers flowing through the south of Bangladesh. Lands in close proximity to the rivers usually have their accumulated salts washed out every year during the monsoon season from the river catchments originating above Bangladesh and flowing out into the Bay of Bengal. Lands further away can have low to moderate salt in the monsoon season but become highly saline in the dry winter months. The resource poor farmers of these regions can plant only one crop of rice in the monsoon season. The rice they produce is not enough to last them throughout the year and their capacity to buy rice from the market is limited. In contrast, these areas are home to many rice landraces which have adapted and evolved over hundreds of generations to survive the saltiness but only produce one third or less rice yield compared to modern day commercial but sensitive rice. These rice landraces are usually photo period sensitive and flower only once during the year (Lisa et al., 2004). One such farmer-admired rice landrace is called Horkuch. This traditional rice shows lower loss in shoot biomass under salt stress, better Na^+ partition into older leaves, and higher retention of Ca^{2+} and Mg^{2+} in flag leaves (Lisa et al., 2011).

We mapped the genetic loci responsible for the salt tolerance of Horkuch using a reciprocal cross-population with sensitive but high-yielding IR29 as its partner, followed by identification of 16 reproductive and 8 seedling stage QTLs. Two RILs, I-14 and I-71 had the highest number of QTLs from Horkuch. Root Length and K^+ content in I-14, while Saltol, Filled Grain Number (FGN), and Spikelet Fertility (SF) in I-71 (Elias et al., 2020; Haque et al., 2022). Saltol is a previously identified QTL

---

[1] Department of Biochemistry and Molecular Biology, University of Dhaka, Bangladesh.
* Corresponding author: zebai@du.ac.bd

which confers salinity tolerance at seedling stage and is responsible for 64–80% of the phenotypic variance. The Saltol QTL maintains low Na^+/K^+ ratio, an important phenomenon in providing salt tolerance (Thomson et al., 2010). Many studies have successfully introgressed Saltol QTL in commercial lines to improve their tolerance. The total potassium trait was mapped in chromosome 2 and 3 in the RILs of Horkuch and IR29. The QTL region of chromosome 2 co-localized with Root Length QTL. The K^+ QTL on chromosome 3 harbors various transmembrane transporters like vacuolar two pore potassium channel and HAK type transporter which maintains K^+ homeostasis (Haque et al., 2022, 2023). The Root length QTL is also a seedling stage parameter which was found to have a high heritability (72.45%), making it suitable for incorporation in hybrid population (Noor et al., 2019).

In the current work, we have designed fluorescent markers based on the SNPs in the QTLs between salt tolerant Horkuch and sensitive but high-yielding IR29. The fluorescent markers are based on a technique called KASP (Kompetitive Allele Specific PCR). KASP is a fluorescence-based SNP genotyping technique based on extension of allele specific primers having a fixed tail of sequences, one for each alternate allele. Subsequently the amplified sequences with allele A or B sequence-tails hybridize with either of two fluorescent labelled oligonucleotides which emits signal on amplification. Thus, reading the fluorescence signal is a simple way to determine the genotype (Kumptala et al., 2012). KASP has gained popularity due to its low cost, high throughput, gel-free system and excellent specificity and sensitivity which is ideal for large breeding applications (Zhao et al., 2017).

Methods

Crossing Scheme

The recombinant inbred lines I-14 and I-71 were chosen as the donor parents of the three salt tolerant QTLs namely Saltol, Root Length, and K^+ content as explained in the introduction. BRRI dhan 63, 67, and 74 were chosen as the recipient (recurrent) parents and are popular high yielding dry winter season developed by BRRI. They have excellent yield related features and are widely cultivated across the country. BRRI dhan 67 and BRRI dhan 74 apart from being high yielding are moderately salt tolerant and blast resistant, respectively. At first, crossing between each of the commercial varieties with I-14 and I-71 was done to produce X_{BD1} and X_{BD2}. These were crossed with each other to produce a hybrid rice Y having all the QTLs which were confirmed by fluorescent markers.

The hybrid rice Y was back crossed into the respective commercial cultivar and progenies selected with fluorescent markers on the basis of the presence of all three QTLs and maximal high yielding background genotype. This step-wise selection reduced the number of desired progenies from 1,500 to 1–2. The selected plant then went through another round of backcrossing with the same stepwise selection to produce 3 types of plants BD63, BD67, and BD74 having additional 3 loci containing the salt tolerance QTLs. These plants are referred to as BC_2F_1 for BD63, 67, and 74. These lines were selfed to produce BC_2F_2 lines and tested for their physiological performance at BRRI and DU and in the fields in Satkhira. Further selfing produced

BC_2F_3 plants which were more uniformly homozygous and were also tested for the physiological performance under salinity stress. The crossing and field trials were done by BRRI, whereas the plant selection by markers was done at DU.

KASP, Plant Selection and their Validation by Phenotypic Selection

The SNPs between Horkuch and IR29 derived from DArT sequencing were compared with the reference genome using NCBI Blast. A 101 bp DNA sequence around the SNP with alternative SNPs at the 50th position was submitted to 3CR group (https://3crbio.com/) for designing the desired SNP markers. A unique tail sequence with the capacity to bind either of the dyes FAM or HEX was added to allele specific primers. The KASP Master Mix contained all components for the thermal cyclic amplification reaction which included Universal FRET cassette (for fluorescence), ROX™ passive reference dye, Taq polymerase, free nucleotides, and $MgCl_2$ in an optimized buffer solution. 2 forward primers with alternate SNPs at the 3' end and the unique tail sequence at 5' end and a common reverse primer were used. The SNP markers for KASP assay were designed targeting Saltol, Root Length and K^+ content QTLs of Chromosome 1 (10.8–16.4 Mb), 2 (26–29 Mb) and 3 (30–31.9 Mb), respectively. Primers homologous to DNA sequence within the QTLs called foreground markers were designed for the 3 QTL regions. Recombinant markers were designed for selection of background genotype at the 2 ends of the QTL regions to restrict the non-desired donor allele introgression. Markers for selection of background or recurrent parent (BD63, 67, or 74) were designed for the region without the QTL on either side of the specific chromosome with the QTL and for all over for the rest of the chromosomes. This was so that we could select maximal amount of the genotypes BD63, 67, or 74 but including the 3 QTLs. DNA was isolated from the population at various time frames during the breeding process. A modified CTAB protocol was used to isolate 96 DNA samples in a DNA grinder (Doyle, 1991). The KASP assay was performed using the standard protocol of LGC genomics in BIO-RAD CFX96 Real Time PCR machine (https://www.biosearchtech.com/support/education/kasp-genotyping-reagents/how-does-kasp-work). After completion of the thermal cycle, FAM and HEX signals that binds to alternative oligos (or are amplified) are read and plotted on x- and y-axes respectively to generate a graph. Selected plants containing all 3 QTLs were screened for salinity tolerance (Amin et al., 2012) and the best lines further screened by BRRI in the Satkhira zone at Debhata.

Results

We obtained between 900–1300 BC_2F_2 plants having the genetic loci with the foreground or QTL-specific markers for each of the BD63, 67, and 74 genetic backgrounds. After recombinant and background genotype selection, various lines had between 64–90% genetic background recovery for the 3 commercial cultivars BD63, 67, and 74, respectively. Seventeen BD63 plants, 11 BD67, and 16 BD74 plants were chosen all of which were homozygous for all the QTLs. All the lines were re-tested and confirmed for the presence of all the QTLs. These lines were subjected to seedling stage salinity tolerance tests.

Physiological screening under salinity stress in hydroponic culture at 10 dS/m showed that 75% of BRRI dhan63 and BRRI dhan67 progenies and 50% of BRRI dhan74 progenies showed a better tolerance score than their respective recipient parents. Sixty-five lines selected having better score from the mass screening were re-evaluated for seedling stage salinity tolerance under a higher salinity stress (12 dS/m) in hydroponics culture. A total of 9, 2, and 1 progenies of BRRI dhan 63, BRRI dhan 67, and BRRI dhan 74 background, respectively, having better score than their parents were found. These 12 best lines were selected and screened at 12 dS/m salinity again. The root length of the best lines was significantly higher than their respective parent under stress. Almost all the selected lines had higher mean shoot length compared to their respective recipient parents, but which was lower than their donor parents. Fifty percent of the selected lines of BD63 background had higher root length than its recipient parent, while most of BD67 and BD74 had a higher mean root length. Therefore, it is apparent that the root QTL in all the selected progenies resulted in better performance under saline stress compared to their recipient parents.

The Percent Yield reduction over control was assessed in the selected progenies of the three populations, their recipient parents, donor parents and tolerant and sensitive checks. The yield reduction over control of the BD63 and BD74 was slightly higher than their respective tolerant parents and tolerant check but significantly lower than the recipient parents and the sensitive checks. However, the yield reduction of the BD67 population was lower than their parents and comparable to tolerant check IR58443.

In the field trials conducted by BRRI at Debhata, the salinity levels varied between 6 dS/m –12 dS/m. However, the salinity was > 6dS/m throughout the trial period in the Boro season of 2022–2023. Here two lines of BD63 and one line each of BD67 and BD74 did significantly better than parental high yielding material and their performance was similar to the tolerant checks.

Discussion and Summary

This work describes a farmer-favorite low-yielding rice, molecular technologies to discover QTLs with the help of an advanced laboratory in the US, further breeding and molecular work accomplished by BRRI rice breeders and molecular biologists at DU. The entire work took 2.5 years and was the result of the use of efficient visible marker technology like KASP. These markers are efficient, simple and cost-effective. Although SNP markers can have a wide range of applications, but their use is limited by their high expense to develop. KASP could be used easily to generate a moderate number of markers sufficient for plant breeding and screen a large breeding population. We will hopefully be able to select 2–3 lines for eventual release for the saline coast of Bangladesh.

Acknowledgements

This work was funded by the World Bank through the National Agricultural Technical Project (NATP), The International Fund for Agricultural Development (IFAD) and Government of Bangladesh through Ministry of Agriculture. The research fund was supervised by the Program Based Research Grant (PBRG) of the Bangladesh Agricultural Research Council (BARC). Research Scientists from BRRI included, M. Sazzadur Rahman and Ruhul Amin Sarker. From DU, the researchers were, GMNA Jewel, Tomalika Azim and Yeamin F. Ria. Technical help included Nazrul Islam of BRRI and Rabin Sarker of DU,

References

Amin, M., Elias, S.M., Hossain, A., Ferdousi, A., Rahman, M.S., Tuteja, N. and Seraj, Z.I. (2012). Over-expression of a DEAD-box helicase, PDH45, confers both seedling and reproductive stage salinity tolerance to rice (*Oryza sativa* L.). *Mol. Breed.*, *30*: 345–54.

Doyle, J. (1991). DNA Protocols for Plants. *In*: Hewitt, G.M., Johnston, A.W.B., Young, J.P.W. (Eds.), *Molecular Techniques in Taxonomy*. NATO ASI Series, Vol. 57. Springer, Berlin, Heidelberg.

Elias, S.M., Rahman, M.S., Khan, S.F., Biswas, S., Haque, T., Razzaque, S. and Seraj, Z.I. (2020). Combination of traits at two developmental stages under salt stress as a measure of tolerance in a reciprocally crossed rice (*Oryza sativa*) population. *Crop Pasture Sci.*, *71*: 334–48.

Haque, U.S., Elias, S.M., Jahan, I. and Seraj, Z.I. (2023). Functional genomic analysis of K^+ related salt-responsive transporters in tolerant and sensitive genotypes of rice. *Front. Plant Sci.*, *13*: 1089109.

Haque, T., Elias, S.M., Razzaque, S., Biswas, S., Khan, S.F., Jewel, G.M.N.A., Rahman, M.S., Juenger, T.E. and Seraj, Z.I. (2022). Salt tolerance QTLs of an endemic rice landrace, Horkuch at seedling and reproductive stages. *Scientific Reports*, *12*: 17306.

Kumpatla, S., Buyyarapu, R., Abdurakhmonov, I. and Mammadov, J. (2012). *Genomics-Assisted Plant Breeding in the 21st Century: Technological Advances and Progress*. https://doi.org/10.5772/37458.

Lisa, L.A., Seraj, Z.I., Elahi, C.M.F., Das, K.C., Biswas, K., Islam, M.R., Salam, M.A. and Gomosta, A.R. (2004). Genetic variation in microsatellite DNA, physiology, and morphology of coastal saline rice (*Oryza sativa* L.) landraces of Bangladesh. *Plant Soil*, *263*: 213–28.

Lisa, L.A., Elias, S.M., Rahman, M.S., Shahid, S., Iwasaki, T., Hasan, A.M., Kosuge, K., Fukami, Y. and Seraj, Z.I. (2011). Physiology and gene expression of the rice landrace Horkuch under salt stress. *Functional Plant Biol.*, *38*: 282–92.

Noor, A.U.Z., Jewel, G.M.N.A, Haque, T., Elias, S.M., Rahman, M.S. and Seraj, Z.I. (2019). Validation of QTLs in Bangladeshi rice landrace Horkuch responsible for salt tolerance in seedling stage and maturation. *Acta. Physiol. Plant*, *41*: 173.

Thomson, M.J., de Ocampo, M., Egdane, J. et al. (2010). Characterizing the Saltol Quantitative Trait Locus for Salinity Tolerance in Rice. *Rice*, *3*: 148–60.

Zhao, S., Li, A., Li, C., Xia, H., Zhao, C., Zhang, Y., Hou, L. and Wang, X. (2017). Development and application of KASP marker for high throughput detection of AhFAD2 mutation in peanut. *Electronic J. Biotech.*, *25*: 9–12.

Chapter 11

Climate-Smart Management Practices for Crop Production in Bangladesh

S. Akhter[1], *R. Sen*[2], *M.K. Alam*[3], *M.M. Masud*[2] and *J.C. Biswas*[4*]

Introduction

Increasing agricultural productivity to meet global food demand is essential, but it must be achieved with consideration for the agricultural ecosystem. Agriculture is one of the sources of major greenhouse gases (GHGs) such as methane (CH_4), nitrous oxide (N_2O), and carbon dioxide (CO_2). Currently, agriculture contributes about 20% of GHG emissions, and this figure could rise to 30% by 2050 (FAO, 2014).

Various soil management associated challenges have been identified, including soil fertility depletion, organic matter loss, micronutrient deficiencies, marginal soils, poor water retention, compaction, secondary salinization, acidification, erosion, and pollution. Proactive soil management techniques are crucial for sustainable crop production with enhanced agroecosystem services. However, frequent tillage practices accelerate CO_2 emissions and adversely impact soil health and sustainability (Lal and Stewart, 2013). While soil management can contribute to soil carbon (SOC) sequestration, several of these techniques require time, labor, and financial investments, and can contribute to increased GHG emissions. Therefore, the adoption of climate-smart agricultural technologies is essential to conserve resources and improve crop yields to address food security.

In Bangladesh, N fertilizers especially urea alone accounts for approximately 80% of the total fertilizers used (FRG, 2018). However, excessive dependence on chemical fertilization has caused soil and water degradation, compromising their quality over time (Stamatiadis et al., 1999). Moreover, a low N-use efficiency of crops contributes to GHG gas emissions into the atmosphere. The manufacturing,

[1] Tuber Research Center, Bangladesh Agricultural Research Institute, Bangladesh.
[2] Soil Science Division, Bangladesh Agricultural Research Institute, Bangladesh.
[3] Commonwealth Scientific and Industrial Research Organization, Canberra, Australia.
[4] Krishi Gobeshona Foundation.
* Corresponding author: http://kgf.org.bd.

transportation, and use of chemical fertilizers also lead to substantial GHG emissions, especially chemical N fertilization, which contributes to significant N_2O emissions from agricultural soils (Bouwman, 1996). Maintaining and/or accumulating SOC by sustainable agricultural productivity and reducing CO_2 emissions is crucial (Alam et al., 2017). Indiscriminate use of chemical fertilizers also leads to soil, water, and environmental pollution with increased ecosystem disservices. Hence, climate-smart fertilizer management approaches are necessary in Bangladesh to address these issues.

Irrigation is vital for crop production, with different crops requiring varying quantities. However, flood irrigation, commonly practiced in Bangladesh, results in significant water loss through evapotranspiration, percolation, drainage, and seepage. Moreover, irrigation management contributes to the emission of N_2O and CH_4, both potent GHGs. To ensure sustainable agricultural production, adopting climate-smart water management technologies that mitigate GHG emissions becomes essential. Therefore, a study was conducted to summarize climate-smart soil, fertilizer, and water management practices for crop production associated with food security in Bangladesh.

Climate Smart Agriculture Practices

Climate Smart Agriculture (CSA) is a regenerative approach that seeks to achieve sustainable agricultural development by increasing productivity, building resilience, and reducing GHG emissions. It is based on three main pillars:

1. ***Increasing Productivity*:** The primary focus of CSA is to enhance agricultural productivity in an environmentally viable and environmentally compatible manner. The goal is to ensure long-term food security by boosting production levels while preserving the health of the ecosystem.
2. ***Enhancing Resilience*:** CSA aims to strengthen agricultural systems' ability to adjust and withstand the impacts of climate change. By building resilience against factors like changes in temperature and rainfall patterns, it ensures that agriculture remains viable and productive under changing climatic conditions.
3. ***Reducing Mitigation*:** Another crucial aspect of CSA is mitigating GHG emissions from agricultural activities. By adopting appropriate practices and technologies, CSA aims to minimize the agricultural contribution to climate change and environmental degradation.

Generally, CSA involves various techniques, including conservation agriculture, improved soil management, optimized fertilizer use, efficient water management, and other sustainable practices. By implementing these approaches, CSA contributes to the overall goal of achieving climate resilience and sustainable agricultural development.

Climate-smart Soil Management

Benefits of climate-smart soil management are numerous like require less quantity of energy, conserve soil moisture, sequester SOC, and emitting less GHGs. To combat the climate change effect on agriculture and soil health, conservation agriculture (CA) practices like minimum tillage performed well in establishing crops along with yield, economics, soil health and environment. While CA is a type of climate-smart agriculture that focuses on the efficient conservation and improvement of natural resources, it is characterized by three key practices:

1. ***Minimum Disturbance of Soil***: CA aims to minimize soil disturbance through reduced tillage or no-till practices. This helps to enhance biodiversity, increase SOM content, improve soil structural stability, control erosion, and regenerate soil health.
2. ***Crop Rotation***: The practice of crop rotation is an important aspect of CS. By alternating different crops in a sequence, it helps to improve biodiversity, disrupt pest and disease cycles, influence nutrient cycling, and increase soil fertility.
3. ***Incorporation of Crop Residue***: CA encourages the incorporation of crop residues into the soil. This acts as a mulch to control evaporation and erosion, increase organic matter content, improve moisture retention, and foster nutrient cycling.

The adoption of CA offers several benefits including reduced tillage and labor costs by up to 70%, conserved moisture by 35%, and reduced CO_2 emissions by 80%. Furthermore, CA has been shown to increase crop yields by approximately 40%. These advantages make CA an effective approach for sustainable and productive farming systems.

1. ***Minimum disturbance of soil***: CA practices minimize soil disturbance through minimum tillage, no-till, or strip tillage practices. Conventional tillage plays a crucial role in crop production, but plowing requires significant amounts of energy, labor, money, and time and adversely affect soil and water ecosystems. CA is helpful in reducing organic carbon loss from soils, which may result in improved physical, chemical and biological properties as well as improved crop productivity. It can reduce CO_2 emissions by 44% compared to the conventional tillage (Table 1). There are two types of minimum or CA practices in Bangladesh.

Table 1 Use of diesel fuel in conventional and conservation tillage system (Grace et al., 2003; Hossain et al., 2009a).

Tillage System	*Diesel Used ($L\ ha^{-1}\ yr^{-1}$)*	*CO_2 emission* ($kg^{-1}\ ha^{-1}\ yr^{-1}$)*	*Fuel Saved ($L\ ha^{-1} yr^{-1}$)*	*CO^2 Emission Reduced ($kg^{-1}\ ha^{-1}\ yr^{-1}$)*
Conservation	119	309.4	94	244.4
Conventional	213	553.8	–	–

a. Zero or minimum tillage: In specific regions of Bangladesh like Rajshahi, Pabna, Dinajpur, and Jessore, farmers practice zero or minimum tillage techniques. This approach has proven effective for various crops, including wheat, mustard, mung bean, lentil, and chickpea, successfully reducing crop establishment costs by 30%. Notably, in Pabna's Chabanel area, garlic cultivation is carried out without any tillage. To support these sustainable practices, Bangladesh Agricultural Research Institute (BARI) has developed a zero-tillage planter, which is conveniently operated using a power tiller. This innovative planter enables farmers to continue reaping the benefits of zero tillage while streamlining their planting processes and enhancing overall agricultural productivity.

b. ***Strip tillage***: In the northern and northwestern regions of Bangladesh, farmers adopt strip tillage as a common agricultural practice. This method entails plowing pits, sowing seeds, applying fertilizer, and finally covering the seeds. By implementing strip tillage, farmers can save up to 25% in crop establishment costs. Recognizing the significance of this technique, BARI has developed a specialized strip till planter, tailor-made to efficiently utilize this approach, further enhancing agricultural productivity in the region.

2. ***Contour ploughing:*** In the hilly areas of Bangladesh, farmers have adopted contour ploughing as a climate-smart practice to replace traditional slash and burn (Jhum) cultivation. This method proves beneficial by effectively reducing soil erosion in the challenging terrain while conserving valuable organic matter and nutrients. To support the adoption of contour ploughing, BARI has developed a dibbler, simplifying and promoting the use of this technique.

   By embracing climate-smart soil management practices such as conservation tillage and contour ploughing, farmers can achieve multiple benefits. Following these practices efficiently contribute to improved soil health, reduce environmental degradation, and result in sustainable agricultural systems.

3. ***Crop rotation:*** Diversified crop rotation especially including legume crops help to enhance biodiversity, extended land cover, diverse quality residue inputs, SOC sequestration, disrupt pest and disease cycles, minimize erosion and converse moisture, and improve soil health.

4. ***Mulching:*** CA includes an incorporation of crop residues or the addition of mulching materials into the soil. Mulching practice proves particularly advantageous for drought-prone and saline soils. For drought-prone soils, mulching plays a vital role in reducing evaporation and preserving soil moisture, thereby enhancing water availability for plants. In saline soils, mulch helps limit the capillary rise of saline water by minimizing evaporation. As a result, mulching practices aid in effectively conserving soil moisture and mitigating the adverse effects of water scarcity and salinity. By adopting mulching techniques, farmers can promote soil management, improve crop resilience, and achieve agricultural sustainability.

Climate-smart Fertilizer Management

Efficient use of fertilizers is vital in crop production due to their essential role and cost. To ensure optimal crop growth and reduce environmental impacts, adopting climate-smart fertilizer management practices is crucial. Several techniques fall under this category:

Urea Deep Placement (UDP)

Placing large urea granules deep in the soil to improve N availability to crops and decrease edge-of-field loss are the approach of UDP. By keeping N closer to the plant roots, the risk of leaching due to rainfall or flooding is reduced. This environmentally friendly technique utilizes urea super granules (USGs) that have shown significant yield increases compared to the traditional methods (IFDC, 2017). The use of USGs with UDP is cost-effective and aids in mitigating N_2O emissions, contributing to environmental sustainability. BARI has developed a specialized USG applicator for precise deep placement of urea in paddy fields. By adopting climate-smart fertilizer management practices like USG application, farmers can improve nutrient-use efficiency, minimize environmental pollution, and achieve sustainable crop production. Notably, studies have shown lower nitrous oxide (N_2O) emissions in USG/urea briquette application, further highlighting the benefits of this approach (Fig. 1).

Neem Oil Coated Urea

A specialized type of urea fertilizer that has been treated with neem seed oil is the neem oil coated urea (NCU). This unique coating imparts a slow-release mechanism for N, leading to more efficient N utilization. Neem seed oil contains triterpenes, which act as inhibitors of the nitrification process in urea. Consequently, the utilization of neem oil coated urea enhances N- use efficiency (Hala et al., 2014). In comparative analysis, N-use efficiency varies among different forms of urea, with values ranging from 16.4–,20.9 kg kg^{-1} (Table 2). Particularly noteworthy is the highest N-use efficiency observed in plots treated with urea super granules (USG), which showed a comparable efficiency level to that of neem oil coated urea-treated plots. The lowest N-use efficiency of 16.4 kg^{-1} was found in plots treated with Prilled urea. To enhance N-use efficiency in agriculture, the introduction of neem oil coated urea presents a promising approach, further contributing to sustainable and efficient crop production.

Organic Manure Management

***Bio-slurry*:** Bio-slurry, a valuable byproduct of biogas plants that generate renewable and clean energy, plays a crucial role in sustainable agriculture. While CH_4 produced in these plants serves as fuel, the bio-slurry uses various beneficial applications. As an organic recyclable material, it enriches crop fields, promoting healthy plant growth. Additionally, it can serve as fish feed, organic pesticide, and seed coat filler, diversifying its utility. In vegetable production, it is recommended to apply bio-slurry, derived from cow dung, at a rate of 5 t ha^{-1} or bio-slurry from poultry litter at a rate of 3 t ha^{-1}, alongside the full dose of chemical fertilizer. This integration of bio-slurry allows farmers to effectively recycle organic waste, enhance soil fertility, and reduce reliance on synthetic fertilizers. Embracing bio-slurry in agricultural practices fosters a sustainable approach that benefits both the environment and crop yields.

***Biochar amendments*:** Biochar, a pyrolyzed C-enriched material, results from the slow and controlled combustion of organic matter in an environment with a limited oxygen concentration. It can be derived from various sources such as wood chips, bark, crop residues, straw, rice husks, grass, paper mill residues, sugarcane husks, chicken manure, dried dung, and other waste materials. Being rich in C and metals oxides, biochar serves as a valuable soil amendment. Its C content typically exceeds 60%, making it an effective means of SOC sequestration due to its stability that can endure for a period of long time. As a soil amendment, biochar plays multiple roles, serving as a liming material, aiding in water retention, enhancing soil fertility, and the reduction of GHG emissions. Its diverse benefits make it a promising tool for sustainable agriculture and environmental stewardship.

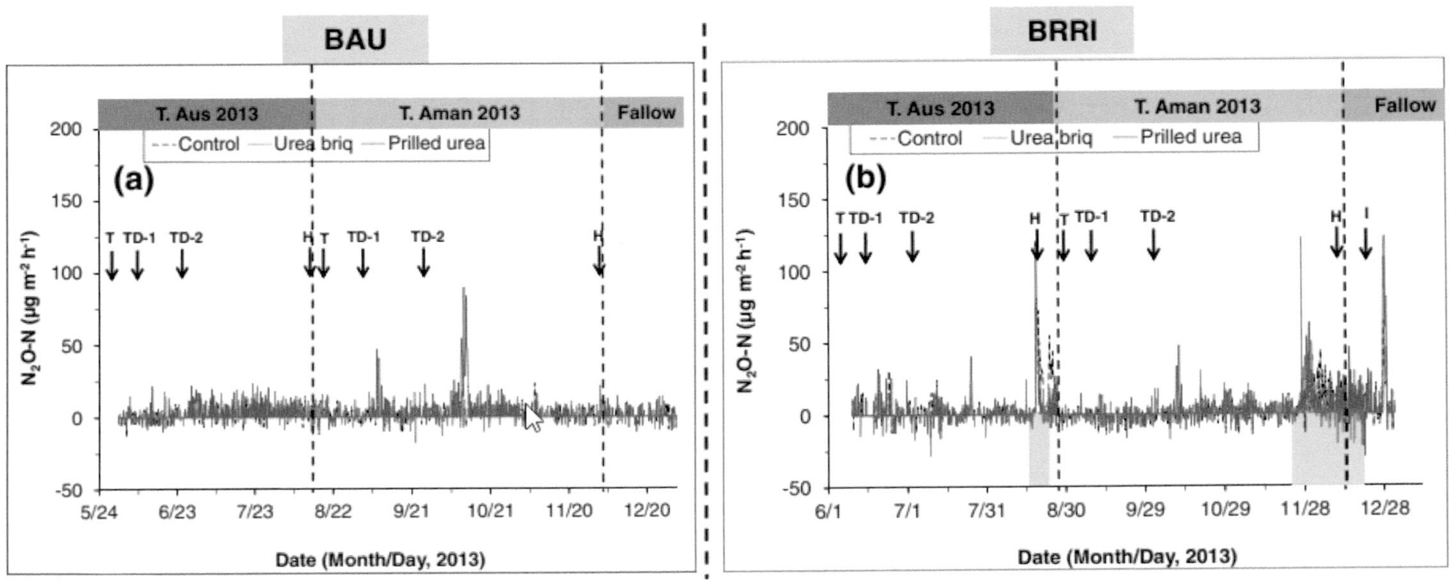

Fig. 1 Nitrous oxide (N_2O) emission rates under different fertilizer treatments during the Aus, Aman, and Fallow seasons, 2013 at BAU and BRRI. T, TD-1, TD-2, H represent transplanting, first and second topdressing and harvesting, respectively. The drying period represented by the shaded areas (Gaihre, Y.K., et al., 2015).

Table 2 Nitrogen use efficiency of different form of urea (Sen et al., 2021).

Treatment	*Amount of N Applied*	*Increase N over N-control*	*Amount of Grain Yield Obtained*	*Increase Grain Yield over N Control*	*N-use Efficiency*
	(kg ha^{-1})				
N-Control	0	–	6260	–	–
Prilled Urea	225	225	9960	3700	16.44
USG	200	200	10450	4190	20.95
NCU	205	205	10400	4140	20.20

When biochar is incorporated into the soil, it exhibits the remarkable capacity to absorb and retain carbon, effectively sequestering it within the soil matrix. This application of agricultural waste-derived biochar as a soil amendment brings about numerous benefits, including heightened soil fertility and productivity, improved utilization of chemical fertilizers, and decreased nutrient leaching and runoff into surface and groundwater (Lehmann et al., 2006). As a consequence of these advantages, a consistent and ample supply of nutrients becomes available to plants, leading to augmented crop yields (Table 3).

Table 3 Influence of biochar on yield of maize, *okra,* and cassava (Frimpong, et al., 2021).

Treatment	*Maize (t ha^{-1})*	*Okra (t ha^{-1})*	*Cassava (kg ha^{-1})*
Control	1.62 ± 0.03a	4.99 ± 1.45a	6.91 ± 1.53a
Biochar	1.65 ± 0.10a	5.56 ± 0.69ab	11.85 ± 0.85b
Compost	2.75 ± 0.19c	11.77 ± 2.63c	13.78 ± 0.74c
Biochar + Compost	2.84 ± 0.15cd	9.05 ± 1.03bc	18.84 ± 0.42e
NPK	2.53 ± 0.15b	9.30 ± 2.21c	17.49 ± 1.09de
Biochar + NPK	3.97 ± 0.02e	8.66 ± 3.67bc	16.20 ± 1.19d
Compost + NPK	3.03 ± 0.02d	10.15 ± 1.66c	17.68 ± 0.74de
Biochar + Compost + NPK	5.17 ± 0.05f	12.23 ± 0.50c	23.90 ± 1.70f
P value	<0.001	0.003	<0.001

Additionally, the implementation of biochar can play a pivotal role in mitigating greenhouse gas emissions, thereby helping to combat potential climate change and global warming effects (Yang et al., 2022). Biochar significantly reduces the release of greenhouse gases by sequestering carbon in the soil. In summary, utilizing biochar as a soil amendment offers a multifaceted approach for improving soil fertility, boosting crop productivity, enhancing nutrient efficiency, and potentially mitigating the effects of climate change (Table 4).

***Compost amendments*:** Composts are organic fertilizers that are derived from decomposing plant and animal residues. Sources of waste compost are various parts of plants, industrial waste and household waste. Common sources of waste manure are household sweeping waste, dried leaves, green leaves and stems, weeds, cow calf

Table 4 Effect of biochar on the yield of different crops in climate vulnerable areas (*Source:* Project Completion Report, *Improvement of soil health and crop productivity in climate vulnerable and polluted areas through organic amendments project*, NATP, Phase II).

Area	*Cropping Pattern*	*Crop*	*Yield in Biochar + RD over Control (RD)*
Acidic (Norsingdi)	Cauliflower-ladies finger-Indian spinach	Cauliflower (Snow White)	28%
		Ladies Finger (Green finger)	25%
		Indian spinach (BARI Puishak-1)	20%
Drought (Rajshahi)	Mustard-Boro rice-T. Aus	Mustard (BARIShorisha-14)	28%
		Boro rice (BRRI dhan28)	19%
		T. Aus rice (BRRI dhan88)	13%
Saline (Khulna)	Sweet gourd-fallow-T. Aman	Sweet gourd (Monica)	25%
		T. Aman rice (BRRI dhan78)	13%

scraps, farm abandoned haystacks, stubble, paddy husks, husks, sugarcane husks, chickpeas/peas, vegetable peel, fruit peel, fish scales, fish washing water, blood of slaughtered animal, animal entrails, eggshell etc. To reduce GHG emissions from rice-rice cropping systems, applying compost can be a good management option (Seung et. al., 2018).

***Vermicompost*:** Vermicompost, also known as earthworm compost, is a type of organic fertilizer. It is produced when earthworms consume plant or animal waste and excrete nutrient-rich feces mixed with chemicals from their bodies. It contains approximately 8–10% more nutrients compared to other types of compost. It increases the rate of microbial activities in the soil and helps to enhance nutrient availability for plants. Using vermicompost is more convenient than other composts, and it can reduce the reliance on chemical fertilizers by up to 50%. Vermicompost increases SOM content and can be used for SOC sequestration to mitigate CO_2 emissions in future climate change scenarios (Dominguez and Edwards, 2011).

***Tricho-compost*:** Tricho-compost. is a natural organic fertilizer produced by the decomposition of various organic matter by the fungus Trichoderma. It is rich in organic C, N, phosphorus, and other essential plant nutrients (Table 5). Crop residues, animal waste, and food waste can be used as substrates for Tricho-compost production. Tricho-compost enhances plant growth, suppresses plant diseases, improves soil porosity and microbial activity, and enhances water retention and aeration in the soil. The leachate derived from Tricho-compost known as troche-leachate, is a nutrient-rich liquid fertilizer. The recommended application rate of Tricho-compost is 3 ton/ha for field and vegetable crops, while fruit trees require 4–8 kg per tree, depending on their age. Using 3 ton/ha of Tricho-compost (no fertilizer) can reduce the reliance on chemical fertilizers by 20–25%. Tricho-compost fosters the proliferation of beneficial microorganisms, such as Trichoderma, in the soil, leading to increased soil fertility and the suppression of harmful fungi. By enhancing crop production and quality, Tricho-compost can help increase farmers' income.

Table 5 Plant nutrient content in different organic sources (*Fertilizer Recommendation Guide*, 2018, BARC, Farmgate, Dhaka, Bangladesh).

Fertilizer	*Nitrogen (%)*	*Phosphorus (%)*	*Potassium (%)*
Bio-dung (dung)	1.01±0.01	0.59±0.06	0.28±0.03
Bio-slurry(chicken excrement)	1.48±0.01	0.69±0.07	0.36±0.04
Compost (rural)	0.75±0.07	0.6±0.06	1.0±0.1
Compost (urban)	1.5±0.15	0.6±0.06	1.5±0.5
Compost	1.5±0.15	0.8±0.08	0.3±0.3
Vermicompost	2.04±0.12	1.16±0.05	1.06±0.5
Trichocompost	2.66±0.07	1.33±0.17	1.62±0.16
Biochar (Crop straw)	2.29±0.65	0.55±0.09	1.94±0.72

***Green manure and crop residues*:** Certain crops have a higher demand for N, which results depletion of N in soil; however, this deficit can be effectively addressed by incorporating leguminous crops into the crop rotation. Legumes have a unique symbiotic relationship with bacteria called Rhizobium, which reside in their root nodules. These bacteria have the ability to fix atmospheric N and lead to enhance soil fertility. Cultivating crops like dhaincha (Sesbania), mung beans, lentils, peas, and Sun hemp and incorporating them into the soil as green manure or cover crop can significantly improve soil fertility. Green manure refers to the practice of growing specific crops in between main crops and then incorporating them into the soil while they are still green and actively growing. This process enriches the soil with labile organic matter and nutrients, including N, released from the decomposing residues.

It is advisable to incorporate green manure into the soil at least once a year or two years to maintain soil fertility. Additionally, leaving crop residues, such as straw, on the land, after harvesting can contribute to soil fertility when mixed into the soil before the next cultivation. The nutrient content in different organic sources used in the soil can be found in Table 6, indicating their respective contributions to soil nutrient levels.

Table 6 Influence of biofertilizer on the yield of legume crops (Ali et al., 2009).

Leguminous Crops	*Suitable Strain*	*Yield Increase over Control*
Lentil	BARI RLc-103, BARI RLc-104 BARI RLc-107, BARI RLc-138	15–45%
Mung bean	BARI RVr-401, BARI RVr-403 BARI RVr-404, BARI RVr-408	17–35%
Cowpea	BARI RVs-605, BARI RVs-405 Hathazari	15–45%
Soybean	BARI RGm-901, Ishurdi, Kalapara	30-40%
Groundnut	BARI RAh-TAL-1002, BARI RAh-TAL-803, Noakhali	20–65%
Chickpea	BARI RCa-202, BARI RCa-Ic-59,	10–50%
French bean	BARI RPv-701, BARI RPv-702 BARI RPv-703	20–48%

Biofertilization: Biofertilizers are fertilizers that contain living microorganisms. They are typically applied to seeds, soil, or plants. The microorganisms in biofertilizers reproduce in the roots of the plants and form nodules in the roots of legumes. This process enables them to store atmospheric N, increase availability of phosphorus, and enhance plant growth, which occur naturally. By using biofertilizers, the reliance on chemical fertilizers can be reduced while improving soil fertility. For example, in tea cultivation, the application of organic manure has led to a reduction in N_2O emissions by 33.3–71.8% and increased tea yields by 16.2–62.2% compared to the use of chemical N fertilizers (Ali et al., 2009). The beneficial bacteria are also referred to as growth-promoting rhizobacteria.

Leaf Color Chart for Rice Production

Among the nutrition management strategies, the use of N fertilizer is one of particular concern in rice production. Applying N fertilizer based on leaf N content by measuring leaf greenness is a good method. Fertilizer application by measuring leaf greenness using leaf color chart (LCC) is becoming very popular (Ravi et al., 2007). N determines the appropriate time of application which is related to the N status of the leaf. This method increases N utilization efficiency, in addition to reasonably high yields.

Climate-smart Water Management Practices

Water is a major input for crop production. Without water crops cannot be grown. Irrigation requirement for crops varies crop to crop, climate and seasons, soil types, and management practices. For wetland crop like rice, irrigation requires 5–6 times than irrigation requirement for upland crops, viz. pulses, oilseeds, etc. Conventional method of irrigation, i.e., under flood irrigation, a large volume of water is lost. There are some climate-smart water management technologies have introduced in some areas of Bangladesh. These are:

1. ***Alternate Wetting-drying of Soil*** The escalating global warming and dwindling freshwater resources pose significant threats to rice production and global food security. Traditional rice farming practices necessitate substantial water inputs and result in GHG emissions, which have adverse effects on the environment. In response, several water-saving approaches have been tested to enhance rice production sustainability. Among these techniques, the alternative wetting-drying (AWD) of irrigation lands has emerged as a promising and environmentally friendly irrigation approach. The implementation of AWD can yield positive environmental outcomes by reducing groundwater extraction. By adopting AWD, a reduction of 25–40% in total water consumption can be achieved, thereby resulting in a decrease of 11–45% in CH_4 emissions.
2. ***Fertigation or Chemigation through Drip Irrigation*** Fertigation refers to the practice of applying fertilizers and other nutrient chemicals along with irrigation. Drip irrigation, which involves the root-adjacent delivery of water drop by drop, is commonly employed in high-value horticultural or specialty crops. Fertigation through drip irrigation is considered a climate-smart technology due to its advantages. By localizing the application of fertilizers, nutrient loss is minimized, and leaching is significantly reduced. When managed correctly,

drip irrigation exhibits high water application efficiency. In Bangladesh, drip irrigation has been successfully implemented in crops such as guava, orange, and other high-value fruit trees. This practice not only conserves fertilizer and water resources but also reduces energy costs associated with irrigation.

3. ***Rainwater Harvesting*** Rainwater harvesting is recognized as a climate-smart approach that plays a crucial role in crop production, particularly in drought-prone and salinity-affected areas. In regions susceptible to drought and salinity, the establishment of Kuni/Mini ponds is vital for small-scale irrigation. These mini ponds are also prevalent in the drought-prone Barind region of Bangladesh. By capturing and storing rainwater, farmers can mitigate the adverse effects of water scarcity and salinity, thus ensuring sustainable crop production. Rainwater harvesting serves as a valuable strategy to enhance water availability for agricultural purposes, contributing to climate resilience and adaptation efforts.

Conclusions

The prospect of climate-smart soil, fertilizer, and water management practices in Bangladesh is promising. Climate-smart agriculture is crucial for enhancing agricultural production while safeguarding natural resources and mitigating the climate change effects. With increasing awareness about GHG gas emissions, global warming, and environmental protection, the adoption of climate-smart technologies is expected to rise steadily to address food security for increasing population in the South Asian countries.

References

Alam, M., Salahin, N., Islam, S., Begum, R., Hasanuzzaman, M., Islam, M. and Rahman, M. (2017). Patterns of change in soil organic matter, physical properties, and crop productivity under tillage practices and cropping systems in Bangladesh. *The Journal of Ag. Sci.*, *155*(2): 216–38.

Ali, M.E., Khanom, D., Khatun, M.R., Bhuiyan, M.A.H. Noor, S. (2009). Soil microorganisms in crop production (in Bangla). *Strengthening of Soil Science*, *16*. Laboratory and Research projectSoil Science Division, BARI, Gazipur-1701.

Bouwman, A.F. (1996). Direct emission of nitrous oxide from agricultural soils. *Nutr. Cycl. Agroecos.*, *46*: 53–70.

Dominguez, J. and Edwards, C.A. (2011). Biology and ecology of earthworm species used for vermicomposting. *In*: Edwards, C.A., Arancon, N.Q., Sherman (Eds.), *Vermiculture Technology: Earthworms, Organic Wastes, and Environmental Management*. CRC Press, Taylor & Francis Group, Boca Raton, FL, USA.

FAO (2014). *Agriculture, Forestry and Other Land Use Emissions by Sources and Removals by Sinks*. ESS working paper 2. Climate, Energy, and Tenure Division, FAO.

Frimpong , K.A., Phares, C.A., Boateng, I., Baidoo , E.A. and Apure, L. (2021). One-time application of biochar influenced crop yield across three cropping cycles on tropical sandy loam soil in Ghana. *J. Helion.*, *7*(2): e06267.

FRG. (2018). *Fertilizer Recommendation Guide: 2018*, 48pp. Bangladesh Agricultural Research Council (BARC), Farmgate, Dhaka-1215.

Gaihre, Y.K., Singh, U., Islam, S.M.M., Huda, A., Islam, M.R., Satter, M.A., Sanabria, J., Islam, M.R. and Shah, A.L. (2015). Impacts of urea deep placement on nitrous oxide and nitric oxide emissions from rice fields in Bangladesh. *Geoderma*, 259–60, 370–79.

Grace, P.R., Jain, M.C. and Hamington, L.W. (2003). Environmental concerns in Rice-wheat systems. *In*: *Proc. International Workshop on Development of action Program for Farm Level Impact in Rice-wheat System of the Indo-Gangetic Plains*, 99–111; 25–27.

Hala, Y., Jumasi, O., Muis, A., Hartati, H. and Inubushi, K. (2014). Development of Urea Coated with Neem (*Azadirachta indica*) to Increase Fertilizer Efficiency and Reduce Greenhouse Gases Emission. *Jurnal Teknologi*, *69*(5).

Hossain, M.I., Islam, M.S., Meisner, C.A., Bodruzzaman, M. and Ilias Hossain. (2009a). Minimum tillage one pass seeder for sustaining cropping intensity and profitability in rice-wheat system. *Int. J. Sustain. Agric. Tech. 5*(6): 32–37.

International Fertilizer Development Center (IFDC). (2017). *Rapid introduction and market development for urea deep placement technology for lowland transplanted rice: A reference guide*. International Fertilizer Development Center. Muscle Shoals, AL.

Lal, R., Stewart, B.A. (2013). *Principles of Sustainable Soil Management in Agroecosystems*. Boca Raton, FL: CRC Press.

Lehmann, J., Gaunt, J., Rondon, M. (2006). Bio-char sequestration in terrestrial ecosystems: A review. *Mitigation and Adaptation Strategies for Global Change*, *11*: 403–27.

Ravi, S., Ramesh, S. and Chandrasekaran, B. (2007). Exploitation of hybrid vigour in rice hybrid (*Oryza sativa* L.) through green manure and Leaf Colour Chart (LCC) based on N application. *Asian Journal of Plant Sciences*, *6*(2): 282–87.

Sen, R., Masud, Barman, M.M., Akhter, A. and Naser, H.M. (2021). *Efficacy of Different Form of Urea on Nitrogen Availability and Yield of Maize*, 144–51. Annual Res. Report,

Seung, T.J., Gil, W.K., Hyun, Y.H., Pil, J.K. and Sang, Y.K. (2018). Beneficial effect of compost utilization on reducing greenhouse gas emissions in a rice cultivation system through the overall management chain. *Science Total Environ.*, *613–14*: 115–22.

Stamatiadis, S., Werner, M. and Buchanan, M. (1999). Field assessment of soil quality as affected by compost and fertilizer application in a broccoli field (San Benito County, California). *Appl. Soil Ecol.*, *12*: 217–25.

Yang, W., Feng, G., Miles, D. Gao, L., Jia, Y., Li, C. and Qu, Z. (2020). Impact of biochar on greenhouse gas emissions and soil carbon sequestration in corn grown under drip irrigation with mulching. *Sci. Total Environ.*, *729*: 138752.

Yang, X., Vancov, T., Peñuelas, J., Sardans, J., Singla, A., Alrefaei, A.F., Song, X., Fang, Y. and Wang, W. (2022). Optimal biochar application rates for mitigating global warming and increasing rice yield in a subtropical paddy field. *J. Exp. Agric.*, *57*: 283–99.

CHAPTER 12

Soil Microorganisms for Sustainable Crop Production in Bangladesh

A.R.M Solaiman[1]*

Introduction

By the year 2050, the global population is estimated to increase 10 billion (FAO, 2009; FAO, 2018); therefore, the demand of food security will increase. To feed the world population, food production would require to increase by 25–70% between now and 2050 (FAO, 2018; Hunter et al., 2017). The sustainable intensification of agricultural production is needed to meet food demand by maintaining, functioning, and healthy ecosystems.

Soil contains most of the nutrients essential for plant growth. Due to intensive cropping with little or no organic amendments, the fertility of soils of Bangladesh is declining day by day. Concern is growing about the long-term sustainability of agriculture to address food security for the increasing population. The overall strategy for increasing crop yields and sustaining them to high level must include an integrated approach focusing on soil nutrient management by integrating inorganic, organic, and biofertilizers. To complement traditional crop management practices, the use of biofertilizers or bioinoculants containing beneficial microorganisms, is gaining increasing attention (Santoyo et al., 2021). Biofertilizers synthesized with *Rhizobium, Azotobacter, Azospirillum, arbuscular mycorrhiza*, and blue green algae are being used in Bangladesh in a limited scale for a long time. These types of biofertilizers are gaining popularity due to the increasing emphasis on maintaining soil health, minimizing environmental pollution, and decreasing the use of reactive chemicals in agriculture.

Biofertilizers or bioinoculants are eco-friendly that, when applied to seeds, plant surfaces, or soil, have the capability to promote plant growth and development. Soil

[1] Bangabandhu Sheikh Mujibur Rahman Agricultural University, Bangladesh.
Present address: Mawlana Bhashani Science and Technology University, Bangladesh.
* Corresponding author: arm_solaiman@yahoo.com

microorganisms support plant growth and development using different mechanisms, such as (i) increasing nutrient availability; (ii) enhancing root growth; (iii) reducing the effects of toxic compounds in soil; and (iv) increasing resistance to both biotic and abiotic stresses (Glick and Gamalero, 2021; Santoyo et al., 2021; Lugtenberg and Kamilova, 2009; de los Santos-Villalobos et al., 2021). According to the mechanisms of action, beneficial microorganisms can be categorized as nitrogen fixers, phosphate solubilizers, phyto-stimulators, plant resistance inducers, and antagonists of phytopathogens, among other groups (Parnell et al., 2016; Köhl et al., 2019). The best known and the most ancient beneficial microorganisms used in biofertilizers are the rhizobia, which are to able inhabit the roots of leguminous plants and provide them with fixed nitrogen via biological symbiosis.

Beneficial soil microorganisms can act as free-living bacteria, symbiotic (establish specific symbiotic relationships with plants, i.e., *rhizobia* and *Frankia*), and endophytes (colonize the interior plant tissues). Within them, a group of soil bacteria called plant growth-promoting bacteria (PGPB) stimulates the growth, yield, and health of the plants through different mechanisms (Glick and Gamalero, 2021). A thorough understanding of the microbial diversity perspectives in agriculture is important and useful to select measures that can act as indicators of soil quality and plant productivity. Beneficial microbial diversity especially bacteria and fungi and their potentiality for sustainable crop production are discussed.

Diversity of Microorganisms in Soil

Bacteria and fungi are the dominant microorganisms present in soil. Among the microorganisms, some are beneficial to soil-plant system and others are detrimental. A large number of existing species of soil microorganism especially beneficial bacteria and fungi has not been described to date. Microbial populations in soil and their behavior in agroecosystem often change in response to agricultural management practices. Agrochemicals, cropping diversity, and irrigation used for crop production influence the soil microbes to a great extent.

Diversity of Beneficial Bacteria

Bacteria are the most dominant group of microorganisms in soil. Rhizosphere population of bacteria may be more or less 10^6/g soil. They can be categorized into two groups: autotrophic and heterotrophic, depending on the methods of obtaining and utilizing nutrition and energy. Autotrophic bacteria are of two types: (i) photoautotrophs: obtain energy from sunlight and carbon as energy from CO_2, e.g., *Chromatium*, and (ii) chemoautotrophs: utilize inorganic materials such as nitrogen, iron, or sulphur for energy and carbon from CO_2, e.g., *Thiobacillus*. Heterotrophic bacteria are of two types: (i) photoheterotrophs: obtain energy from sunlight and carbon from organic compounds, e.g., *Rhodopseudomonas*, and (ii) chemoheterotrophs: obtain energy and nutritive carbon from organic compounds, e.g., *Rhizobium*.

Bacteria can also be categorized into aerobic, e.g., *Bradyhizobium* and anaerobic, e.g., *Clostridium* depending on oxygen requirement. Depending on biochemical properties, bacteria are of two types: Gram positive, e.g., *Bacillus* and Gram negative, e.g., *Pseudomonas*. Bacteria living in the soil are called free-living as they do not depend on root exudates for their survival. Several bacteria have the ability to attach to the root surfaces (rhizoplane) allowing these to derive maximum benefit from root exudates. Some of these are more specialized, as they possess the ability to penetrate inside the root tissues (endophytes) and have direct access to organic compounds present in the apoplast.

Role of Microorganisms in Agriculture

Soil contains different types of microorganisms and play various roles in soil-plant ecosystems. The role of microorganisms ranges from atmospheric N fixation, bioavailability of plant nutrients, plant growth promotion, decomposition of organic matter, biodegradation and bioremediation of organic pollutants in soil, and biocontrol of plant diseases, etc. Some of the microorganisms perform their role through symbiotic association with higher plants and some other perform their work in non-symbiotic manner.

Nitrogen-fixing Bacteria

Biological N fixation is a process by which N in the atmosphere is converted into ammonium compounds via a series of chemical reactions. Nitrogen fixation takes place by both symbiotic and non-symbiotic manner of which 80% comes from symbiotic associations and the rest from free-living or associative systems.

Symbiotic N fixing bacteria: More than 30 species belonging to genera, viz., *Rhizobium, Bradyrhizobium*, *Allorhizobium*, *Azorhizobium*, *Mesorhizobium*, and *Sinorhizobium* fix atmospheric N (Sahgal and Johri, 2003). Nodules are normally formed in the root, but a few aquatic legumes bear stem nodules in addition to the normal root nodules. This type of nodule formation is restricted to 15 of the 250 species of *Aeschynomene*, one out of 15 species of *Neptunia* (*N. oleracea*), and one out of 70 species of *Sesbania* (*S. rostrata*) (Subba Rao et al., 1981, 1991; Tomekpe, et al., 1991). Root nodules of more than 280 species of woody plants from eight different families are formed by *Frankia* (Schwintzer, R. and Tjepkema, 1990). *Frankia* forms effective symbiosis with species of *Alnus* and *Casuarina.*

Effects of *Rhizobium/Bradyrhizobium* Inoculation on Crops

Among 10 *Rhizobium* strains isolated from chickpea grown in AEZ 11, AEZ 12, AEZ 27, and AEZ 28 of Bangladesh, the BSMRAUCR6, BSMRAUCR7, and BSMRAUCR11 performed better compared to other strains at pH 4.0 in the presence of 100 μM aluminum in recording nodule number, N content, and N uptake in chickpea (Islam, 2007). Roots with nodules of different leguminous crops such as lentil (*Lens esculentus*), grass pea (*Lathyrus sativus*), chickpea (*Cicer arietinum*), and rhizosphere soils were collected from Kushtia (AEZ-11), Rajbari (AEZ-11), Faridpur (AEZ-12), and Patuakhali (AEZ-13) districts of Bangladesh for isolation and characterization of *Rhizobium* strains. Strains isolated from lentil were: Le 1, Le 2, Le 3, Le 4, Le 6, Le 7, and Le 8, chickpea: Ca 1, Ca 2, Ca 3, Ca 4, and Ca 5, and grass pea: Ls 1, Ls 2, Ls 3, Ls 4 Ls 6, Ls 7, Ls 8, and Ls 12. Some of the strains increased biomass yield in rice (Solaiman et al., 2011).

Rhizobium strains, such as CR1, CR2, CR3, and CR4 isolated from nodules of chickpea grown in Faridpur and Rajbari (AEZ 12) increased nodulation, dry-matter yield, and N fixation in chickpea. Among the strains, the CR2 and CR3 performed better than the CR1and CR4 (Solaiman et al., 2010). Six *Rhizobium* strains, such as R1, R2, R3, R4, R_5, and R_6 were isolated from nodules of pea grown in AEZ 11 (Jessore and Ishurdi), AEZ 12 (Faridpur and Rajbari), and AEZ 13 (Barisal). Strains R3 and R_5 performed better compared to other strains on nodulation, dry-matter yield, and N fixation in pea (Solaiman et al., 2008). Inoculation of chickpea with *Rhizobium* strains increased nodulation, dry-matter yield, and N fixation (Solaiman et al., 2007). *Rhizobium/Bradyrhizobium* inoculant BINA- LT-18, BINA CP-2, BINA-MB-1, BINA-COP-7, BINA-GN-2, BINA-SB-4, BINA –DG-1, and BINA-DC-9 developed for inoculation of lentil, chickpea, mung bean, cowpea, groundnut, soybean, black gram, and dhaincha (*Sesbania bispinosa*), respectively, increased grain yield ranging from 25–50%, when compared to the control (Anonymous, 2012a). *Rhizobium/ Bradyrhizobium* inoculants, such as BARI RLc-105, BARI RLc-107, BARI RLc-108, and BARI RLc-140 developed for inoculation of lentil; BARI RVr-402 for mung bean; BARI RCa-220, BARI RCa-259, BARI RCa-204, and BARI RCa-205 for chickpea; BARI RVm-301 for black gram; BARI RPv-701 for bush bean, and BARI RGm-901 for soybean increased crop yield ranging from 40–50% compared to the control (Anonymous, 2011). The influence of inoculation of crops with different free living and symbiotic bacterial isolates are shown in Tables 1–3 and Figs. 1–9.

Table 1 Effect of diazotrophic bacteria on growth and yield of rice.

Treatment	*Tiller Hill^{-1}*	*SPAD Value*	*Leaf Area Index (cm^2)*	*Plant Height (cm)*	*Root Length (cm)*	*Root Vol. (cm^3)*	*Dry-Atter Yield (g)*
T_1	5.5f	27.7ab	1.37 d	42.5e	15.63 c	4.00 g	4.0f
T_2	8.7ef	27.7ab	1.85cd	50.4bcd	23.00 b	8.00 ef	6.9def
T_3	9.7def	28.7ab	2.78bcd	52.1abcd	25.75 ab	10.25 de	10.4cde
T_4	14.2c	29.1ab	3.55ab	53.5ab	26.33 ab	12.25 cde	11.7bcde
T_5	16.5abc	28.0ab	4.01 ab	54.5ab	27.63 ab	16.25 bc	13abc
T_6	15.5bc	31.6a	4.16 ab	53.7ab	27.13 ab	19.75 ab	16.7ab
T_7	20.5a	31.6a	4.41 a	56.0a	29.88 a	22.50 a	17.7a
T_8	18.7ab	31.4a	4.51 a	52.7abc	27.25 ab	16.25 bc	13.8abc
T_9	12.2cde	28.4ab	3.55 ab	47.2de	28.88 a	15.50 bcd	12.5abcd
T_{10}	13.7cd	28.7ab	2.87 bc	51.6abcd	27.13 ab	13.00 cde	10.1cde
T_{11}	14.2c	25.0bc	4.53 a	52.1abcd	27.50 ab	14.75 bcd	9.0cdef
T_{12}	6.5f	26.0bc	1.50 cd	47.7d	26.13 ab	8.750 ef	9.7cdef
T_{13}	8.7ef	22.13 c	1.57 cd	48.4cd	25.63 ab	12.00 cde	6.4ef
LSD (0.05)	9.3	4.5	1.46	12.0	11.90	13.72	5.9

Notes: T_1: Control (No fertilizer), T_2: 25 % N + Full PKSZn, T_3: 50% N + Full PKSZn, T_4: 75% N + Full PKSZn, T_5: Recommended Fertilizer Dose (RFD), T_6: 50% N + Full PKSZn + *Rhizobium sp.* –1 (Le 6), T_7: 50% N + Full PKSZn + *Rhizobium sp.* –2 (Ls 6), T_8: 50% N + Full PKSZn + *Rhizobium sp.* –3 (Ca 4), T_9: 50% N + Full PKSZn + *Klebsiella pneumoneae,* T_{10}: 50% N + Full PKSZn + *Azospirillum brasilense*, T_{11}: 50% N + Full PKSZn + *Azospirillum sp. (Azs. 1)*, T_{12}: 50% N + Full PKSZn + *Azotobacter sp.* 2 (Azt 2) and T_{13}: 50% N + Full PKSZn + *Azotobacter sp.* 1 (Azt 1).

Table 2 Influence of *Rhizobium* in combination with urea and poultry manure on growth and yield of rice.

Treatment	*Plant Ht. (cm)*	*Tiller No. Hill^{-1}*	*Root Length (cm)*	*Root Vol. (cm^3)*	*Biomass Yield (t ha^{-1})*	*Yield Hill^{-1} (t ha^{-1})*
T_1	64.07 h	10.00 j	10.21 g	12.67 h	2.39 g	2.04 h
T_2	75.07 fg	16.87 gh	14.08 f	16.00 gh	3.50 ef	3.36 g
T_3	82.27 cd	22.40 e	18.48 c	22.00 d	4.37 cde	3.79 fg
T_4	88.53 ab	22.20 cd	25.94 b	26.67 c	5.41 abc	4.53 de
T_5	80.22 de	21.87 ef	19.01 c	21.67 d	4.32 de	4.10 ef
T_6	73.73 g	17.93 g	14.21 f	15.67 gh	3.33 efg	3.37 g
T_7	89.33 ab	32.00 a	25.88 b	31.20 a	5.72 ab	5.77 ab
T_8	90.47 a	33.00 a	30.08 a	31.67 a	6.43 a	5.96 a
T_9	73.04 g	18.63 g	13.37 f	16.33 fg	3.43 efg	3.45 g
T_{10}	81.00 d	25.53 d	17.48 cd	20.67 de	3.86 def	3.79 fg
T_{11}	76.57 efg	19.40 fg	14.48 ef	17.67 efg	3.58 ef	3.54 fg
T_{12}	89.71 ab	31.80 ab	26.61 b	30.67 ab	5.80 ab	5.72 ab
T_{13}	86.13 bc	28.53 c	25.54 b	26.67 c	4.93 bcd	4.75 cd
T_{14}	78.07 def	13.87 i	15.48 def	18.00 efg	3.06 fg	3.26 g
T_{15}	80.20 de	14.33 hi	16.94 cde	19.67 def	3.39 efg	3.67 fg
LSD (0.05)	4.255	2.908	2.587	3.445	1.069	0.582

Notes: T_1: Absolute Control (No N fertilizer), T_2: 50% N from Urea , T_3: 75% N from Urea, T_4: 100% N from Urea , T_5: 100% N from PM, T_6: 25% N from urea + Rhizobium sp., T_7: 50% N from urea + Rhizobium sp., T_8: 75% N from urea + Rhizobium sp., T_9: 25% N from Urea + 25% N from PM, T_{10}: 25% N from Urea + 25% N from PM + Rhizobium sp., T_{11}: 50% N from Urea + 25% N from PM, T_{12}: 50% N from Urea + 25% N from PM + Rhizobium sp., T_{13}: 25% N from Urea + 50% N from PM + Rhizobium sp., T_{14}: 50% N from PM + Rhizobium sp. and T_{15}: 75% N from PM + Rhizobium sp.

Table 3 Combined effect of *rhizobacteria* and urea on growth and yield of rice.

Treatments	*Plant Ht. (cm)*	*SPAD Value*	*Tiller Number*	*Root Length (cm)*	*Root Vol. (cm^3)*	*1000-Grain Wt. (g)*	*Yield (t ha^{-1})*
T_1	55.57	28.7	12.5	16.53	14.9	17	2.4
T_2	62.3	37.2	19.3	22.89	20.5	19.5	3.05
T_3	67.9	42.1	24	24.8	23.5	20.33	3.66
T_4	71.3	46.14	28	28.5	27.3	22.2	4
T_5	69.9	43.12	27	25	26.67	21	3.5
T_6	66.01	39.1	26.31	27.9	22.95	20.66	3.7
T_7	73	48.91	31.11	30.14	28.13	22.97	4.23
T_8	69.5	42.59	26.93	26.97	25.1	20.98	3.8

Contd.

Table 3 *Contd.*

Treatments	*Plant Ht. (cm)*	*SPAD Value*	*Tiller Number*	*Root Length (cm)*	*Root Vol. (cm^3)*	*1000-Grain Wt. (g)*	*Yield ($t\ ha^{-1}$)*
T_9	68.1	42.15	24.1	24.95	21.3	20.82	3.58
T_{10}	70.1	40.9	25	25.25	24.5	21.5	3.8
T_{11}	67	39.11	21.36	24	22.1	20.23	3.44
T_{12}	68.1	38.53	22.19	23.9	21.99	20	3.4
CV	10.03	6.74	9.62	10.18	12.05	5.64	5.20

Notes: T_1 = Absolute control (No nitrogen fertilizer), T_2 = 50% urea, T_3 = 75% urea, T_4 = 100% urea, T_5 = 50% urea + *Rhizobium sp.* (LS-2), T_6 = 75% urea + *Rhizobium sp.* (LS-2), T_7 = 50% urea + *Rhizobium sp.* (LS-6), T_8 = 75% urea + *Rhizobium sp.* (LS-6), T_9 = 50% urea + *Rhizobium sp.* (LE-6), T_{10} = 75% urea + *Rhizobium sp.* (LE-6), T_{11} = 50% urea + combination. T_{12} = 75% urea + combination.

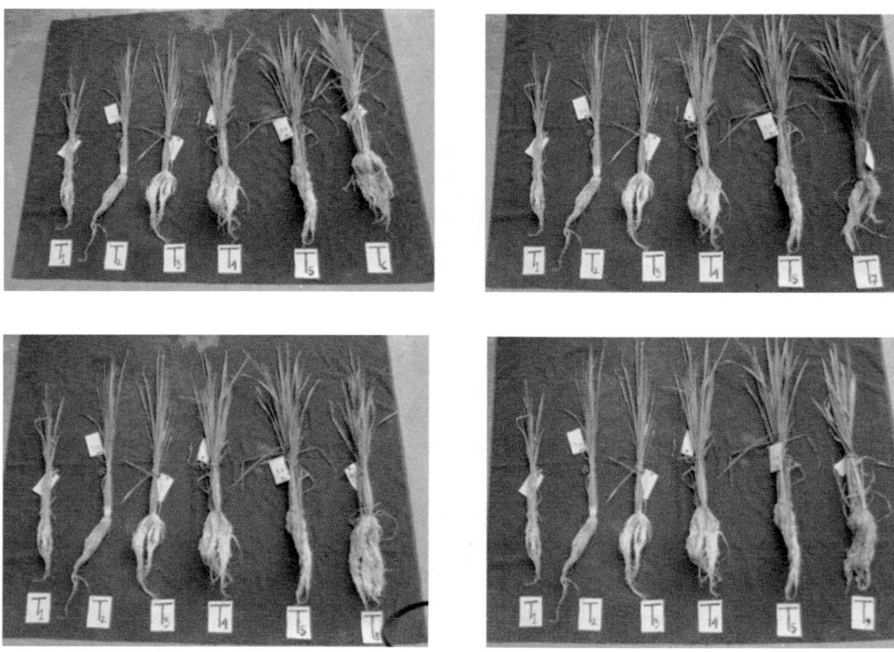

Fig. 1 Effect of inoculation with different microorganisms on root growth of rice.
Notes: T_1: Control (No fertilizer), T_2: 25 % N + Full PKSZn, T_3: 50% N + Full PKSZn, T_4: 75% N + Full PKSZn, T_5: Recommended Fertilizer Dose (RFD), T_6: 50% N + Full PKSZn + *Rhizobiun sp.* -1 (Le 6), T_7: 50% N + Full PKSZn + *Rhizobium sp.* -2 (Ls 6), T_8: 50% N + Full PKSZn + *Rhizobium sp.*-3 (Ca 4), T_9: 50% N + Full PKSZn + *Klebsiella pneumoneae,* T_{10}: 50% N + Full PKSZn + *Azospirillum brasilense*, T_{11}: 50% N + Full PKSZn + *Azospirillum sp. (Azs. 1)*, T_{12}: 50% N + Full PKSZn + *Azotobacter sp.* 2 (Azt 2) and T_{13}: 50% N + Full PKSZn + *Azotobacter sp.* 1 (Azt 1).

Non-symbiotic N-fixing bacteria: Biological N-fixation through non-symbiotic manner takes place by both free living and associative bacteria. Important genera of bacteria that fix N non-symbiotically are *Achromobacter*, *Acetobacter*, *Alcaligenes*,

Fig 2 Effect of *diazotrophic* bacteria on growth of rice.

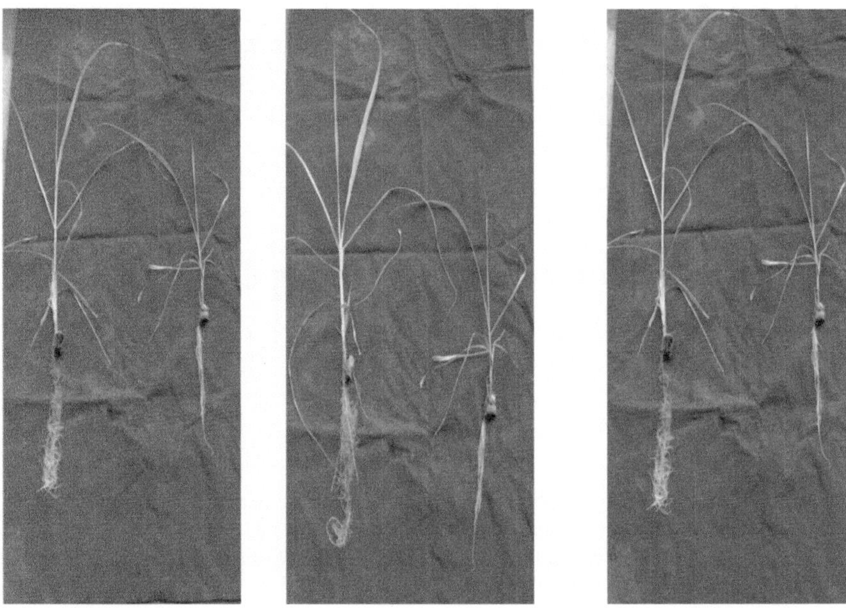

Fig 3 Effect of inoculation with different bacteria on root growth of rice.

Arthrobacter, *Azospirillum*, *Azotobacter*, *Azomonas*, *Bacillus*, *Beijerinckia*, *Clostridium*, *Corynebacterium*, *Derxia*, *Enterobacter*, *Herbaspirillum*, *Klebsiella*, *Pseudomonas*, *Rhodospirillum, Rhodopseudomonas,* and *Xanthobacter* (Saxena and Tilak, 1998).

(i) *Azotobacter*: The genus *Azotobacter* comprises of six species (Tchan and New, 1984), namely, *A. chroococcum, A. vinelandii, A. beijerinckii, A. nigricans, A. armeniacus*, and *A. paspali.* Species of *Azotobacter* are regarded as free-living aerobic N-fixing bacteria. Among the non-symbiotic bacteria species, the

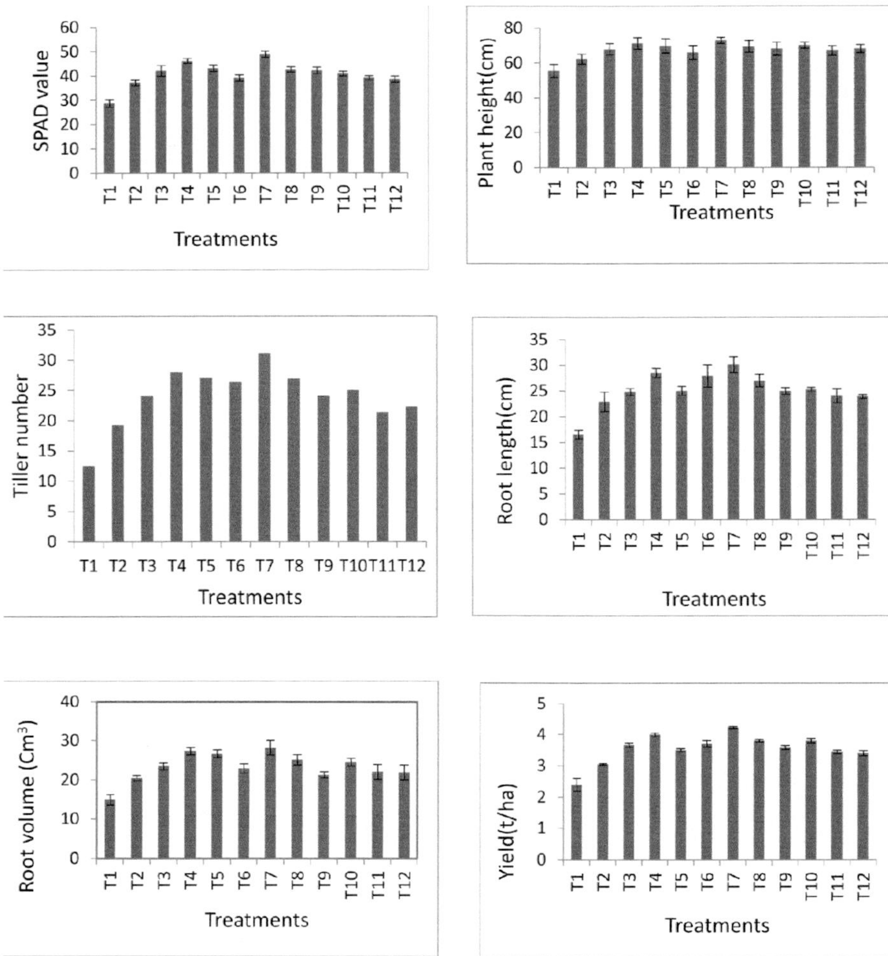

Fig. 4 Effect of Rhizobium inoculation on different parameters of rice.

Notes: T_1 = Absolute control (No nitrogen fertilizer), T_2 = 50% urea, T_3 = 75% urea, T_4 = 100% urea, T_5 = 50% urea + *Rhizobium sp.* (LS-2), T_6 = 75% urea + *Rhizobium sp.* (LS-2), T_7 = 50% urea + *Rhizobium sp.* (LS-6), T_8 = 75% urea + *Rhizobium sp.* (LS-6), T_9 = 50% urea + *Rhizobium sp.* (LE-6), T_{10} = 75% urea + *Rhizobium sp.* (LE-6), T_{11} = 50% urea + combination. T_{12} = 75% urea + combination.

Azotobacter was studied thoroughly. It is reported that inoculation of *Azotobacter* and *Azospirillum* improved seed germination of *Cicer arietinum, Phaseolus mungo, Vigna catjung*, and *Zea mays* (Saikia and Bezbaruah, 1995).

(ii) *Azospirillum*: The species of the genus *Azospirillum* frequently associated with root and rhizosphere of a large number of agriculturally important crops. The increase in crop yield through inoculation of *Azotobacter* is attributed to improved root development due to the production of growth-promoting substances, and consequently increased water and mineral uptake for plant growth (Fallik, et al., 1994). There are seven species of *Azospirillum* – *Azospirillum brasilense, A.*

Fig. 5 Effect of *Rhizobium* inoculation and urea on growth of rice.

Fig. 6 Effect of *Rhizobium* inoculation on growth of rice (*Source:* Solaiman et al., 2011).

Fig. 7 Vegetative growth (a) and harvesting stage (b) of rice (BRRI dhan29) as influenced by plant growth promoting bacteria.

Fig. 8 Influence of different bacterial isolates on Brinjal.

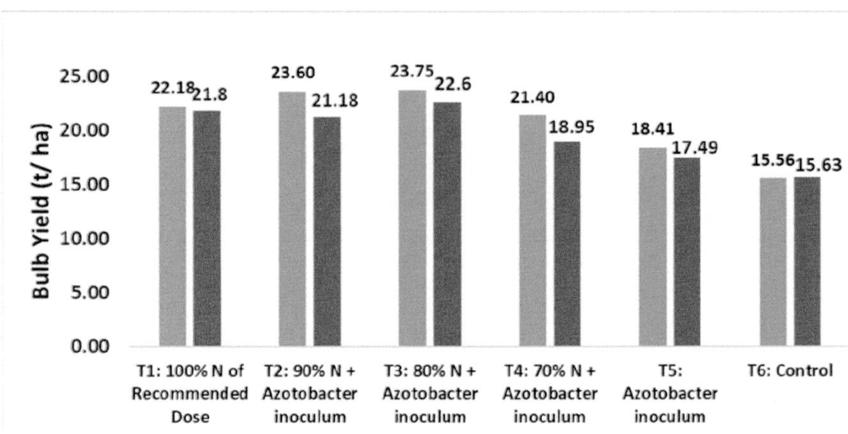

Fig. 9 Influence of *Azotobacter* inoculation on bulb yield of onion.

lipoferum, *A. amazonense*, *A. halopraeferens*, *A. irakense*, *A. dobereinerae*, and *A. largimobile* (Tilak et al., 2005).

(iii) *Acetobacter*: One member of *Acetobacter*, viz., *A. diazotropicus* isolated from sugarcane fall under the family Acetobacteriaceae associated with N-fixation. Important genera of this family are: *Acetobacter, Gluconobacter*, *Gluconoacetobacter*, and *Acidomonas* (Yamada et al., 1997). The name *Acetobacter diazotrophicus* has been changed to *Gluconoacetobacter diazotrophicus* (Yamada et al., 1998). In addition to *G. diazotrophicus*, two more diazotrophs, *G. johannae* and *G. azotocaptans* have also been reported (Fuentez-Ramirez et al., 1997).

(iv) *Azoarcus*: *Azoarcus* is an aerobic/microaerophilic N-fixing bacterium. This was isolated from surface-sterilized tissues of kaller grass (*Leptochloa fusca* (L.) Kunth) (Reinhold, et al., 1986), and can also infect roots of rice plants. The genus *Azoarcus* has two species, *A. indigens* and *A. communis*. *Azoarcus* can fix N efficiently.

Effect of Non-symbiotic Bacteria on Crops

Combined effects of three diazotroph strains, such as *Bacillus cereus*, *Acinetobacter calcoaceticus* and *Azospirillum brasilense* along with three levels of urea-N, i.e., no N, 50% N, and 100% N was assessed in sugarcane. Treatment receiving 50% N along with *Bacillus cereus* inoculation significantly increased the growth parameters, juice quality, sugar, and cane yield (Hossain, 2014). The effects of diazotrophic bacterial inoculation on sugarcane are presented in Tables 4–5 and Figure 10. Forty-five diazotrophs that were isolated from research fields of BRRI, the population of diazotrophs was higher in the rhizosphere than in the soil and root. Diazotrophs fixed 10–37 kg N ha^{-1} and produced 4–7 mgl^{-1} of indoleacetic acid. Root endophyte fixed the highest concentration of N (Anonymous, 2012b).

Table 4 Effect of *diazotrophic* bacterial inoculation on growth and yield of sugarcane.

Treatment	*Tiller No. Hill^{-1}*	*Millable Cane Hill^{-1}*	*Cane Diameter (cm)*	*Cane Ht. (m)*	*Leaf Area Index*	*Biomass Yield (kg hill^{-1})*	*Cane Yield (kg hill^{-1})*
T_1	7.25 d	5.25 d	7.52 e	2.64 e	3.76 f	2.36 d	7.83 d
T_2	8.25 c	6.25 bcd	8.17 cd	3.49 cd	4.32 e	2.93 bc	8.32 cd
T_3	9.50 a	7.25 ab	8.75 ab	3.81 a	5.80 ab	3.59 a	10.93 a
T_4	8.50 bc	6.50 abc	8.52 bc	3.67 abc	5.01 cd	3.06 bc	9.02 bc
T_5	9.25 ab	6.75 abc	8.67 ab	3.75 ab	5.17 c	3.16 b	9.20 bc
T_6	9.75 a	7.50 a	8.98 a	3.85 a	6.29 a	3.69 a	11.13 a
T_7	8.25 c	5.75 cd	8.47 bc	3.59 bc	4.52 de	2.76 c	8.72 cd
T_8	8.50 bc	6.00 cd	8.55 bc	3.65 abc	5.10 c	2.95 bc	8.83 c
T_9	9.00 abc	6.50 abc	8.57 b	3.71 ab	5.41 bc	3.11 b	9.83 b
T_{10}	9.50 a	7.25 ab	8.82 ab	3.83 a	5.93 ab	3.60 a	10.98 a
T_{11}	8.25 c	5.76 cd	7.94 d	3.35 d	4.94 cd	2.98 bc	8.92 bc
CV (%)	6.48	12.64	3.15	4.18	7.42	7.41	7.22

Notes: T_1: Control (no nitrogen), T_2: 50% N, T_3: 100 % N, T_4: 50% N + *Acinetobacter calcoaceticus* (BUSo 9), T_5: 50% N + *Azospirillum Brasiliense* (SP7), T_6: 50% N + *Bacillus cereus* (BUSo 13), T_7: 100% N + *Acinetobacter calcoaceticus* (BUSo 9), T_8: 100 % N+ *Azospirillum Brasiliense* (SP7), T_9: 100% N + *Bacillus cereus* (BUSo 13), T_{10}: 50% N + Mixed inoculum, T_{11}: 100% N + Mixed Inoculum.

Table 5 Effect of *diazotrophic* bacterial inoculation on sugarcane juice quality parameters.

Treatment	*Brix (%)*	*Pol % Cane*	*Purity (%)*	*Recoverable Sucrose (RS) (%)*
T_1	18.60 f	13.85 c	86.59 b	7.22 a
T_2	19.80 e	14.14 bc	87.66 ab	7.32 a
T_3	21.70 ab	14.43 abc	88.79 a	7.44 a
T_4	20.50 d	14.30 abc	87.67 ab	7.35 a
T_5	21.00 c	14.34 abc	87.89 ab	7.42 a
T_6	21.90 a	14.99 a	88.91 a	7.62 a
T_7	20.00 e	14.28 abc	87.47 ab	7.25 a
T_8	20.80 cd	14.32 abc	87.57 ab	7.27 a
T_9	21.50 b	14.39 abc	87.83 ab	7.33 a
T_{10}	21.80 ab	14.90 ab	88.82 a	7.55 a
T_{11}	20.00 e	13.86 c	87.25 b	7.24 a
CV (%)	1.05	3.79	1.20	6.58

T_1: Control (no nitrogen), T_2: 50% N, T_3: 100% N, T_4: 50% N + *Acinetobacter calcoaceticus* (BUSo 9), T_5: 50% N + *Azospirillum Brasiliense*(SP7), T_6: 50% N + *Bacillus cereus* (BUSo 13), T_7: 100% N + *Acinetobacter calcoaceticus* (BUSo 9), T_8: 100% N+ *Azospirillum Brasiliense* (SP7), T_9: 100% N + *Bacillus cereus* (BUSo 13), T_{10}: 50% N + Mixed inoculum, T_{11}: 100% N + Mixed Inoculum. *Source:* Hossain, 2014.

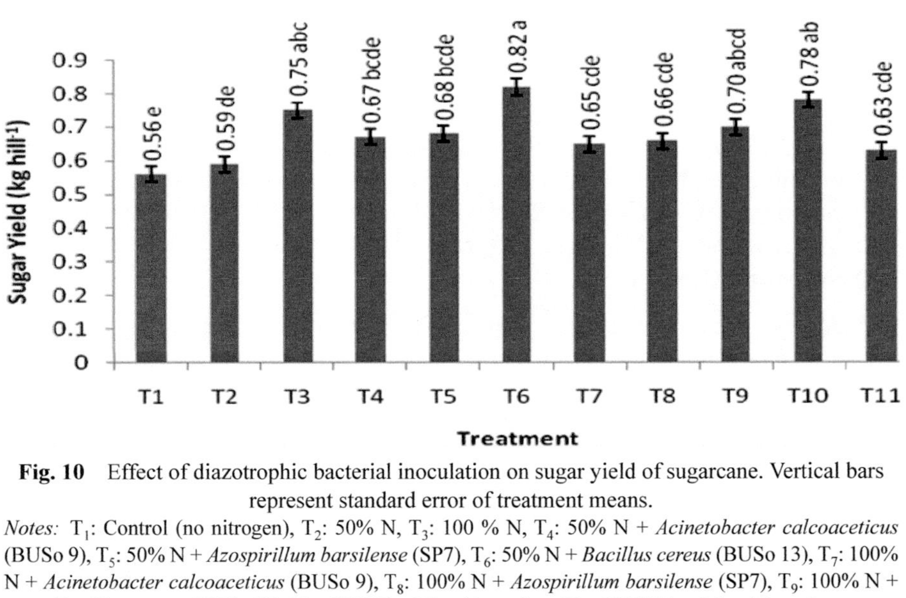

Fig. 10 Effect of diazotrophic bacterial inoculation on sugar yield of sugarcane. Vertical bars represent standard error of treatment means.

Notes: T_1: Control (no nitrogen), T_2: 50% N, T_3: 100 % N, T_4: 50% N + *Acinetobacter calcoaceticus* (BUSo 9), T_5: 50% N + *Azospirillum barsilense* (SP7), T_6: 50% N + *Bacillus cereus* (BUSo 13), T_7: 100% N + *Acinetobacter calcoaceticus* (BUSo 9), T_8: 100% N + *Azospirillum barsilense* (SP7), T_9: 100% N + *Bacillus cereus* (BUSo 13), T_{10}: 50% N + Mixed inoculum, T_{11}: 100% N + Mixed Inoculum.

Nitrogen transforming bacteria: These bacteria are responsible for the breakdown of complex proteinaceous and nitrogenous organic compounds to produce amine and ammonium through ammonification process. Microorganisms responsible are *Actinomycetes, Bacillus, Pseudomonas*, etc. Ammonia is converted to nitrite, and finally nitrate by *Nitrosomonas* and *Nitrobacter*, respectively.

Denitrifying bacteria: Denitrifying organisms are *Pseudomonas, Bacillus,* and *Paracoccus* which reduced nitrates to oxides of N and gaseous N.

Cellulose decomposing bacteria or cellulolytic bacteria: Decomposition of cellulose, a major constituent of plant material or organic residues takes place by the bacteria of the genus *Cytophaga, Angiococcus, Polyangium*, etc.

Phosphate solubilizing microorganisms: Phosphorus is one of the major essential plant nutrients. When applied as phosphorus fertilizer it gets fixed with Fe, Al, and Mn at low pH and with Ca at high pH of soil. The fixed phosphorus are then solubilized by a group of heterotrophic bacteria and fungi excreting various organic acids and enzymes. *Bacillus* and *Pseudomonas* amongst the bacteria and *Aspergillus* and *Penicillium* amongst the fungi are the most important phosphate solubilizing microorganisms. The important spp. of bacilli and pseudomonas are *B. brevis*, *B. cereus*, *B. circulans*, *B. firmus*, *B. licheniformis*, *B. megaterium*, *B. mesentricus*, *B. mycoides*, *B. polymyxa*, *B. pumilis*, *B. pulvifaciens B. subtilis*, *Pseudomonas striata*, *P. cissicola*, *P. fluorescens*, *P. pinophillum*, *P. putida*, *P. syringae*, *P. aeruginosa*, *P. putrefaciens*, and *P. stutzeri*. Microorganisms such as *Escherichia freundii, E. intermedia, Serratia phosphaticum*, and species of *Achromobacter, Brevibacterium*, *Corynebacterium*, *Erwinia, Micrococus, Sarcina*, and *Xanthomonas* are also involved in phosphorus solubilization (Tilak et al., 2005). Blue green algae, viz., *Anabaena* sp., *Calothrix brauni*, *Nostoc* sp., *Scytonema* sp., and *Tolypothrix ceylonica* can also solubilize phosphates (Gupta et al., 1998).

Phosphate solubilizing fungi include *Aspergillus niger*, *A. flavus*, *A. nidulans*, *A. awamori*, *A. carbonum*, *A. fumigatus*, *A. terreus*, and *A. wentii* (Prerna et al., 1997). *Paeciliomyces fusisporus*, *Penicillium digitatum*, *P. simplicissimum*, *P. aurantiogriseum, Sclerotium rolfisii*, and species of *Cephalosporium*, *Alternaria*, *Cylindrocladium*, *Fusarium*, and *Rhizoctonia* are also involved in solubilization of inorganic phosphates (Tilak et al., 2005). Yeasts such as *Torula thermophila*, *Saccharomyces cerevisiae*, and *Rhodotorula minuta* can also solubilize inorganic phosphate. An amount of 30–35 kg P_2O_5 ha^{-1} can be released by phosphate solubilizing microorganisms (Gaur et al., 2004).

Arbuscular mycorrhiza (*AM*): Arbuscular mycorrhizal fungus is a type of mycorrhiza in which the fungus penetrates the cortical cells of the roots of vascular plants. AM fungi help the plant to capture nutrients such as phosphorus, sulphur, nitrogen, and micronutrients from the soil. Among the 29 crops surveyed in four AEZs of Bangladesh, maize, wheat, sorghum, grass pea, chickpea, cowpea, groundnut, soybean, and onion were found highly mycotrophic. Average colonization was the highest (43.3%) at AEZ-9 (Jamalpur) and the lowest (38.8%) at AEZ-28 (Joydebpur). The maximum average spore number (157 per 100 g soil) was recorded at AEZ-23 (Hathazari) and the minimum (99 per 100 g soil) at AEZ-28. *Glomus, Gigaspora*, and

Acaulospora were the most commonly distributed genera. Edaphic factors such as soil texture, moisture, pH, organic matter, and available nutrient contents influenced the colonization and sporulation of AM fungi (Khanam, 2002; Khanam et al., 2006).

Grass pea, chickpea, cowpea, lentil, mung bean, garden pea, black gram, and soybean were found highly mycotrophic. Root colonization ranged from 20–76, 20–80, 14–69, 20–76, and 20–60 and that of total number of spores ranged from 89–495, 40–390, 89–390, 97–298, and 100–295 per 100 g soil at AEZs 3, 11, 12, 18, and 28, respectively. The AM fungal structures varied among leguminous crops. *Glomus* was the most common genus while *Sclerocystis* was the least prevalent genus in the rhizosphere (Molla and Solaiman, 2009). Eight pulse crops, i.e., grass pea, chickpea, cowpea, lentil, mung bean, garden pea, black gram, and bush bean and two oil seed crops, soybean and groundnut were found highly dependent on mycorrhizal inoculation (Molla et al., 2010). AM fungi inoculants increased dry-matter yield of lentil and chickpea significantly and minimized foot and root diseases of these crops (Anonymous, 2011).

Combined Effect of *Rhizobium* and Arbuscular Mycorrhiza (AM)

Co-inoculation of *Rhizobium* and AM improved root colonization and spore population, growth, yield, nutrient content, and uptake by crops, such as mung bean, chickpea, and pea (Khanam et al., 2008; Rahman et al., 2010; Molla et al., 2011; Solaiman et al., 2012). The combined effect of *Rhizobium* and AM are presented in Tables 6–8.

Table 6 Effect of different combinations of arbuscular mycorrhiza, *Rhizobium* inoculant, nitrogen, and phosphorus on growth of chickpea.

Treatment	*Nodule No. Plant^{-1}*	*Nodule Dry-Wt. (g) Plant^{-1}*	*Plant Ht. (cm)*	*Root Length (cm)*	*Branch No. Plant^{-1}*	*Shoot Dry-Wt. (g) Plant^{-1}*	*Root Dry-Wt. (g) Plant^{-1}*
*T*1: Control	1.33 ft	0.013e	19.17 f	20.60g	3.33 e	1.13 i	1.00 e
*T*2: Arbuscular mycorrhiza(AM)	4.50 e	0.021de	22.83 de	37.95e	3.60 d	1.50 g	1.27 d
*T*3: *Rhizobium* (R)	8.00 c	0.023cd	21.00 e	46.83b	3.67 cd	1.36 h	1.30 d
*T*4:AM+R	10.33 b	0.031be	21.00 e	41.00d	4.23 b	1.70 f	1.57 c
Ts:AM+R+N	11.00 b	0.033b	24.33 cd	34.67f	4.00 be	2.07 e	1.63 c
*T*6: AM + R + P	12.50 a	0.054a	30.33 a	52.67a	4.83 a	3.23 a	2.17 a
*T*7: AM+N+P	4.00 e	0.019de	25.33 be	40.00d	3.67 cd	2.40 d	1.80 b
Ts:AM+N	6.00d	0.023cd	25.67 bc	44.33c	4.87 a	2.83 c	1.87 b
*T*9:AM+P	5.00 de	0.037ab	26.00 be	37.36e	3.63 d	2.80 c	1.53 c
TIO : AM + R + N + P	11.40 ab	0.044a	26.67 b	46.46b	4.67 a	2.90b	2.06 a
CV(%)	10.40	17.10	4.50	2.40	4.60	1.70	5.80

Note: Means followed by common letter(s) in a column are not significantly different at 5% level by DMRT.

Table 7 Interaction effect of AM and pulses and oil seed crops on number of branches plant per plant and number of pods per plant at pre-flowering and 50% flowering stages.

Treatment	*No. of Branches Plant^{-1}*		*No. of Pods Plant^{-1}*
	Pre-flowering	*50% Flowering*	*50% Flowering*
T_1: Black gram x U	2.00 f-i	5.00 cd	1.60 f-g
T_2: Black gram x I	3.67 cd	7.00 b	3.70 ab
T_3: Bush bean x U	1.67 g-i	4.00 de	1.30 gh
T_4: Bush bean x I	3.67 cd	6.00 bc	3.30 a-c
T_5: Chickpea x U	2.33 e-h	4.00 de	1.00 h
T_6: Chickpea x I	5.00 ab	7.33 b	1.70 gh
T_7: Cowpea x U	1.00 i	2.30 f	1.00 h
T_8: Cowpea x I	1.33 hi	3.33 ef	2.30 d-f
T_9: Garden pea x U	2.67 d-g	2.33 f	1.00 h
T_{10}: Garden pea x I	3.33 c-e	3.33 ef	2.00 ef
T_{11}: Grasspea x U	1.00 i	4.00 de	1.30 gh
T_{12}: Grasspea x I	2.67 d-g	6.00 bc	3.30 a-c
T_{13}: Mung bean x U	4.00 bc	5.00 cd	1.30 gh
T_{14}: Mung bean x I	6.00 a	9.00 a	4.00 a
T_{15}: Lentil x U	2.00f-i	3.00 ef	1.00 h
T_{16}: Lentil x I	3.00e-f	6.33 bc	3.00 bcd
T_{17}: Groundnut x U	1.00 i	3.00 f	1.00 h
T_{18}: Groundnut x I	2.00 f-i	4.33 de	2.70 c-e
T_{19}: Soybean x U	1.00 g	2.00 g	1.00 h
T_{20}: Soybean x I	2.33 e-h	2.33 f	2.30 def
CV %	19.4	19.00	19.40
SE (±)	0.34	0.45	0.25

Notes: Means followed by common letter are not significantly different at 5 % level by DMRT.
U = Uninoculated, I = Inoculated.

Table 8 Distribution of *Arbuscular Mycorrhiza* in the rhizosphere soil of different pulses and oil seed crops.

Treatments (Legume Crops)	*Glomus (%)*	*Acaulo. (%)*	*Gigaspora (%)*	*Scutell. (%)*	*Entphos. (%)*	*Sclero. (%)*	*Sporo. (%)*	*Uniden. (%)*
T1: Blackgram	37	14	5	1	–	–	–	43
T2: Bushbean	43	23	4	–	1	–	–	29
T3: Chickpea	52	27	8	5	–	–	–	8
T4: Cowpea	56	–	17	–	5	–	–	22
T5: Gardenpea	40	19	–	7	–	–	–	34
T6: Grasspea	45	25	10	1	–	–	–	19

Contd.

Table 8 *Contd.*

Treatments (Legume Crops)	*Glomus (%)*	*Acaulo. (%)*	*Gigaspora (%)*	*Scutell. (%)*	*Entphos. (%)*	*Sclero. (%)*	*Sporo. (%)*	*Uniden. (%)*
T7: Mungbean	80	10	4	–	–	2	–	4
T8: Lentil	64	14	–	8	2	4	2	10
T9: Groundnut	59	17	8	–	1	–	1	14
T10: Soybean	56	10	9	5	–	–	–	20

Notes: Acaulo. = *Acaulospora*, Giga. = *Gigaspora*, Scutel. = *Scutellispora*, Ent. = *Entrophosphora*, Sclero. = *Sclerocystis*, Sporo. = Sporocarp and Uniden. = Unidentified.

Soil Microbial Biomass

Soil microbial biomass is the main driving force in the decomposition of organic matter, and is frequently used as an early indicator of changes in soil properties resulting from management practices and environment stresses. Addition of different organic materials, glucose, or cellulose increased the amount of microbial biomass C, N, and S in soils (Howlader et al., 2008 a, b; Howlader et al., 2010 a, b; Paul and Solaiman, 2004). The amount of microbial biomass C and N were assessed in soils collected from Bilashchara Tea Estate under BTRI, Srimangal of Moulavibazar district, and Sripur Tea Garden of Sylhet districts in Bangladesh, respectively. Soil microbial biomass C and N in Bilashchara Tea Estate varied from 90–144 and 20–29 mg per kg soil and that of Sripur Tea Garden varied from 121–362 and 27–59 mg per kg soil, respectively (Abdur Rahman and Solaiman, 2008). The effect of long-term nutrient management practices on soil microbial populations is shown in Fig. 11.

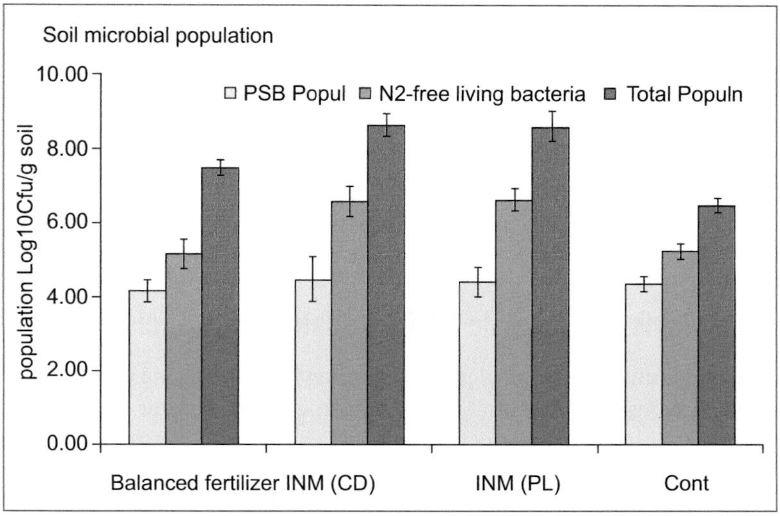

Fig. 11 Effect of long-term nutrient management practices on soil microbial population.
Notes: CD-Cow-dung; PL-Poultry litter; Cont-Control; INM-Integrated nutrient management.

Fungi: They are next only to bacteria in dominance. They are eukaryotic, heterotrophic, aerobic organisms (referred as molds, mildews, rusts, smuts, yeasts, mushrooms). Fungi are responsible for decomposition of organic residues, mobilization of P and N. They dominate in acid soils as bacteria and actinomycetes cannot survive well. Their average cell number is 918×10^3/ g of soil. There are four classes of fungi: Phycomycetes, Ascomycetes, Basidiomoycetes and Fungi imperfecti. Important fungi that decompose plant materials are: *Penicillium, Trichoderma, Aspergillus, Fusarium, Mucor, Citromycetes*, *Allternaria, Botrytis, Rhizopus, Pythium* and *Rhizoctonia.*

Blue green algae: They belong to the family Cyanophyceae and are photoautotrophic microorganisms that have the capability to fix atmospheric N via nitrogenase enzyme. Numerically, they are not as many as fungi, bacteria, or actinomycetes. Common genera of blue green algae are *Chroococcus, Oscillatoria, Nostoc, Anabaena, Scytonema, Fischerella, Calothrix*, and *Tolypothrix.*

Actinomycetes: They are morphologically alike bacteria and their characteristics are transitional between bacteria and fungi. Some Actinomycetes are morphologically and physiologically the same as *Mycobacterium* and *Corynebacterium*. They are intolerant to acidity with an optimum pH range between 6.5–8. They are aerobic and heterotrophic organisms; however, waterlogging is detrimental for their growth. Common genera of Actinomycetes are *Streptomyces, Nocardia*, and *Micromonospora.*

Protozoa: They are unicellular and larger than the bacteria. This group of organisms exist in all arable soils but smaller in numbers (10,000 to more than a million/g soil). They are abundant in the surface soil and can tolerate adverse environmental conditions. Protozoa feed on bacteria and actinomycetes, thus help in balancing microflora diversity in soils.

Plant Disease Controlling Microorganisms

Microorganisms are used to control plant disease. They used to eliminate or reduce pathogens or to decrease damage caused by them are called biocontrol agents. The mode of action of biocontrol agents are competition for food and nutrients, space or infection sites, antibiosis, parasitism, production of cell-wall degrading enzymes, degradation of pathogenicity factors, induced resistance, and plant growth promotion. Biopesticides are prepared with bacteria such as *Bacillus, Pseudomonas, Burkholderia, Agrobacterium, Erwinia*, and *Streptomyces* and fungi such as *Trichoderma, Coniothyrium, Gliocladium*, and *Phlebiopsis* to control plant disease.

Microorganisms Responsible for Degradation of Pesticides

Biodegradation refers a biological process that converts an organic compound to CO_2, water, and/or inorganic salt(s) or, sometimes, to an intermediate or metabolite that could be higher or lower in toxicity than the parent compound. Pesticides are common, non-point source environmental pollutants. Some pesticides are readily degradable by soil microorganisms, such as aerobic and anaerobic bacteria and chemo-lithotrophic and photosynthetic microorganisms. Amongst others, the *Alcaligenes, Bacillus,*

Pseudomonas, Burkholderia, Arthrobacter, Enterobacter, and *Flavobacterium* degrade pesticides.

Constraints

Some of the major constraints in relation to microbial resources used in soil-plant system are as follows:

- Lack of information on the status of microbial resources in soils of Bangladesh.
- A lack of trained professionals for collection, identification, preservation, and documentation of the beneficial soil microorganisms.
- Inadequate laboratory facilities for studying microbial genetic resources.
- Lack of systematic research for assessing the effect of beneficial microorganisms.
- Lack of suitable carriers to prepare multi strain inoculants for nutrient management.
- Difficult to maintain suitable population of beneficial microorganisms in soil applied through bioinoculants.
- Proper storage facilities of bioinoculants are inadequate.
- Gene bank for preservation of microbial genetic resources is not available.
- Lack of awareness of farmers about bioinoculants and their beneficial effects.
- Government regulations to control quality and use of bioinoculants.

Future Outlook

- Detailed studies on the status of soil microbial diversity in soils of Bangladesh.
- Microbial bio-functional activities need to be assessed using molecular techniques.
- Suitable bioinoculants need to be developed which are reliable, ecofriendly, and act as suitable alternatives to reactive chemicals, leading to improve soil fertility and plant growth and biocontrol against phytopathogens.
- Need fine-tuning of the desired effects of the beneficial microbes on soil-plant system and the mechanisms involved therein.
- Development of improved carriers that will provide longer shelf-life, higher number of viable cells, protection against environment, convenient use, and are cost-effective.
- Farmers need trained, so that they have comprehensive knowledge about the properties, application, and the advantages of the bioinoculants.
- Different beneficial microorganisms can be combined into a single preparation that may help plants against a wide range of biotic and abiotic stresses.
- The performances of bioinoculants need to be evaluated under different climatic and agronomic conditions.
- Technologies generated to be transferred from laboratory to field, as well as technical, regulatory, and marketing issues to be addressed.
- Data base of beneficial soil microorganisms need to be developed and a gene bank should be established to preserve them.
- Government outreach activities are needed to transfer science-based knowledge on various aspects of bioinoculants using different media.

Conclusions

Soil microorganisms contribute to the growth and development of plants and thereby increase crop productivity to address food security. Research on the role of soil microorganisms in crop production should be intensified and sustainable novel and holistic technologies should be developed for improving agricultural production. It is the need of the hour to rapidly expand the technologies developed in the laboratory to the field level, resolve technical issues, make regulations, and market the bio-products.

References

Abdur Rahman, S.M. and Solaiman, A.R.M. (2008). Assessment of microbial biomass carbon and nitrogen in some tea soils of Bangladesh. *Bangladesh J. Microbiol.*, *25*(1): 21–25.

Anonymous. (2011). *Annual Report, 2010–2011*. Bangladesh Agricultural Research Institute, Joydebpur, Gazipur.

Anonymous. (2012a). *BINA A PROFILE* (4th Edn.), September 2012. Bangladesh Institute of Nuclear Agriculture, Mymensingh.

Anonymous. (2012b). *Annual Report of Bangladesh Rice Research Institute 2009–2010*. BRRI, Gazipur 1701.

de los Santos-Villalobos S., Díaz-Rodríguez A.M., Ávila-Mascareño M.F., Martínez-Vidales A.D. and Parra-Cota F.I. (2021). COLMENA: A Culture Collection of Native Microorganisms for Harnessing the Agro-Biotechnological Potential in Soils and Contributing to Food Security. *Diversity*, *13*: 337. doi: 10.3390/d13080337.

Fallik, E., Sarig, S. and Okon, Y. (1994). Morphology and physiology of plant roots associated with *Azospirillum*. *In*: Okon, Y. (Ed.), *Azospirillum–Plant Associations*, 77–84. CRC Press, Boca Raton.

FAO. (2009). How to feed the World in 2050? *Proceedings of the Expert Meeting on How to Feed the World in 2050*; 12–13 October. Rome, Italy.

FAO. (2018). *The Future of Food and Agriculture-Alternative Pathways to 2050. Summary Version*, FAO, Rome, Italy. 60.

Fuentez-Ramirez, L.E., Cabellaro-Mellado, J., Sepulveda-Sanchez, J. and Martinez-Romero, E. (1997). Location of *Acetobacter diazotrophicus* in inoculated sugarcane by GUS detection. *11th Intl. Congress on Nitrogen Fixation*, Abs, 18 p.

Gaur, R., Shani, N., Kawaljeet, Johri, B.N., Rossi, P. and Aragno, M. (2004). Diacetyl phloroglucinol producing *Pseudomonas* do not influence AM fungi in wheat rhizosphere. *Curr. Sci.*, *86*: 453–57.

Glick, B.R. and Gamalero, E. (2021). Recent developments in the study of plant microbiomes. *Microorganisms*, *9*: 1533.

Gupta, R.P., Vyas, M.K. and Pandher, M.S. (1998). Role of phosphorus solubilizing microorganisms in P-economy and crop yield. *In*: Kaushik, B.D. (Ed.), *Soil–Plant–Microbe Interaction in Relation to Nutrient Management*, 95–101. Venus Printers & Publishers, New Delhi.

Hossain, G.M.A. (2014). *Influence of Diazotrophic Bacteria on Growth and Yield of Sugarcane* (*Saccharum officinarum l.*). PhD Dissertation (Unpublished). Bangabandhu Sheikh Mujibur Rahman Agricultural University, Gazipur, Bangladesh.

Howlader, M.A.R., Solaiman, A.R.M. and Chowdhury, M.A.H. (2008a). Biodynamics of microbial biomass nitrogen and sulfur in organic matter amended soil. *Bull. Inst. Trop. Agr. Kyushu Univ.*, *31*: 19–29.

Howlader, M.A.R., Solaiman, A.R.M. and Chowdhury, M.A.H. (2008b). Biodynamics of microbial biomass nitrogen and sulfur in glucose and cellulose amended soil. *Bull. Inst. Trop. Agr. Kyushu Univ.*, *31*: 9–18.

Howlader, M.A.R., Solaiman, A.R.M. and Chowdhury, M.A.H. (2010a). Biodynamics of microbial biomass nitrogen and sulfur in soil amended with organic matter and fertilizer (1): soil biomass. *Bull. Inst. Trop. Agr. Kyushu Univ.*, *33*: 37–47.

Howlader, M.A.R., Solaiman, A.R.M., Islam, M.S. and Chowdhury, M.A.H. (2010b). Biodynamics of microbial biomass nitrogen and sulfur in soil amended with organic matter and fertilizer (2): Their roles in the yield of wheat. *Bull. Inst. Trop. Agr. Kyushu Univ.*, *33*: 49–62.

Hunter, M.C., Smith R.G., Schipanski M.E., Atwood L.W. and Mortensen D.A. (2017). Agriculture in 2050: Recalibrating targets for sustainable intensification. *Bioscience*, *67*: 386–91. doi: 10.1093/biosci/bix010.

Islam, M.R. (2007). *Influence of Soil Acidity Factors on Rhizobium sp* (*Cicer*). PhD Dissertation (Unpublished), Bangabandhu Sheikh Mujibur Rahman Agricultural Univ., Gazipur.

Khanam, D., Mridha, M.A.U. and Solaiman, A.R.M. (2006). Comparative study of *Arbuscular mycorrhizal* association with different agricultural crops among four AEZs of Bangladesh. *J. Agric. Res.*, *44*: 147–59.

Khanam, D. (2002). *Biodynamics of Arbuscular Mycorrhizal Fungi in Agricultural Crops and Their Interaction with Rhizobium on Chickpea* (*Cicer arietinum* L.). PhD Dissertation (Unpublished). Bangabandhu Sheikh Mujibur Rahman Agric. Univ. Gazipur, Bangladesh.

Khanam, D., Mridha, M.A.U. and Solaiman, A.R.M. (2008). Effect of combined inoculation of *Arbuscular mycorrhiza* and *Rhizobium* on nutrient content and uptake by chickpea. *Bangladesh J. Agric. and Environ.*, *4*: 1–11.

Köhl J., Kolnaar R. and Ravensberg W.J. (2019). Mode of action of microbial biological control agents against plant diseases: Relevance beyond efficacy. *Front. Plant Sci.*, *10*: 845. doi: 10.3389/fpls.2019.00845.

Lugtenberg B. and Kamilova F. (2009). Plant-growth-promoting rhizobacteria. *Annu. Rev. Microbiol.*, *63*: 541–56. doi: 10.1146/annurev.micro.62.081307.162918.

Molla, M.N. and Solaiman, A.R.M. (2009). Association of *arbuscular mycorrhizal* fungi with leguminous crops grown in different agroecological zones of Bangladesh. Archives *Agron. Soil Sci.*, *55*: 233–45.

Molla, M.N., Solaiman, A.R.M., Mridha, M.A.U. and Khanam, D. (2010). The dependence of some pulses and oil seeds on *arbuscular mycorrhizal* fungi. *Bull. Inst. Trop. Agr., Kyushu Univ.*, *33*: 27–36.

Molla, M.N., Solaiman, A.R.M., Jahiruddin, M., Mridha, M.A.U. and Khanam, D. (2011). Influence of different doses of phosphorus in the presence of *arbuscular mycorrhiza* and *Rhizobium* on the growth and yield of mung bean. *Bull. Inst. Trop. Agr., Kyushu Univ.*, *34*: 49–67.

Parnell J., Berka R., Young H., Sturino J., Kang Y., Barnhart D. and Dileo M. (2016). From the lab to the farm: An industrial perspective of plant beneficial microorganisms. *Front. Plant Sci.*, *7*: 1110.

Paul, G.C. and Solaiman, A.R.M. (2004). Changes of microbial biomass carbon and nitrogen in upland sugarcane soil amended with different organic materials. *Commun. Soil Sci. Plant Anal.*, *35* (17&18): 2433–47.

Prerna, A., Kapoor, K.K. and Akhaury, P. (1997). Solubilization of insoluble phosphate by fungi isolated from compost and soil. *Environ. Ecol.*, *15*(3): 524–27.

Rahman, M.M., Solaiman, A.R.M., Khanam, D., Karim, A.J.M.S. and Karim, M.A. (2010). Effects of inoculation with *Rhizobium* and *arbuscular mycorrhiza* and phosphorus on growth, yield and nutrient uptake by pea grown in soil. *Bangladesh J. Microbiol.*, *27*(1): 22–27.

Reinhold, B., Hurek, T., Niemann, E.G. and Fendrik, I. (1986). Close association of *Azospirillum* and *diazotrophic* rods with different root zones of Kallar grass. *Appl. Environ. Microbiol.*, *52*: 520.

Saikia, N. and Brezbaruah, B. (1995). Iron-dependent plant pathogen inhibition through *Azotobacter* RRLJ 203 isolated from iron-rich acid soils. *Indian J. Exptl. Biol.*, *33*: 571–75.

Sahgal, M. and Johri, B.N. (2003). The changing face of rhizobial systematics. *Curr. Sci.*, *84*: 43–48.

Santoyo G., Guzmán-Guzmán P., Parra-Cota F.I., de los Santos-Villalobos S., Orozco-Mosqueda M.D.C. and Glick B.R. (2021). Plant Growth Stimulation by Microbial Consortia. *Agronomy*, *11*: 219.

Saxena, A.K. and Tilak, K.V.B.R. (1998). Free-living nitrogen fixers: Its role in crop production. *In*: Verma, A.K. (Ed.), *Microbes for Health, Wealth, and Sustainable Environment*, 25–64 Malhotra Publ. Co., New Delhi.

Schwintzer, R. and Tjepkema, J.D. (1990). *The Biology of Frankia and Actinorhizal Plants*, 9.(Run on) Academic Press Inc., San Diego, USA.

Solaiman, A.R.M., Hossain, GMA and Mia, MAB. 2011. Effect of *Rhizobium* on Growth and Biomass Production of Rice. *Bangladesh J. Microbiol., 28*(2): 64–68.

Solaiman, A.R.M., Talukder, M.S. and Rabbani, M.G. (2010). Influence of *Rhizobium* strains on chickpea: Nodulation, dry matter yield, and nitrogen uptake. *Bang J. Microb., 27*: 61–64.

Solaiman Talukder, M., Solaiman, A.R.M., Khanam, D. and Rabbani, M.G. (2008). Characterization of some *Rhizobium* isolates and their effectiveness on pea. *Bangladesh J. Microbiol., 25*: 45–48.

Solaiman, A.R.M., Hossain, D. and Rabbani, M.G. (2007). Influence of *Rhizobium* inoculant and mineral nitrogen on some chickpea varieties. *Bangladesh J. Microbiol., 24*(2): 146–50.

Solaiman, A.R.M., Rabbani, M.G., Hossain, D., Hossain, G.M.A. and Alam, M.S. (2012). Influence of phosphorus and inoculation with *Rhizobium* and *AM* fungi on growth and dry matter yield of chickpea. *Bangladesh J. Sci. Res., 25*(1): 23–32.

Subba Rao, N.S., Tilak, K.V.B.R. and Singh, C.S. (1981). Root nodulation studies in *Aschenomene aspera. Plant Soil, 56*: 491–94.

Subba Rao, N.S., Gaur, Y.D. and Murthy, A. (1991). Biology of root and stem nodules of *Aeschynomene aspera* and *A. indica* potential green manure plants. *In*: Datta, S.K., Sloger, C. (Eds.), *Biological Nitrogen Fixation Associated with Rice Production*, 31–38. Oxford & IBH, New Delhi.

Tchan, Y.T. and New, P.T. (1984). Genus I. *Azotobacter beijerinck. In*: *Bergey's Manual of Systematic Bacteriology*, 220 pp. Williams and Wilkins, Baltimore.

Tilak, K.V.B.R., Ranganayaki, N., Pal, K.K., De, R., Saxena, A.K., Shekhar Nautiyal, C., Shilpi Mittal, Tripathi, A.K. and Johri, B.N. (2005). Diversity of plant growth and soil health supporting bacteria. *Current Science, 89*: 136–50.

Tomekpe, K, Dommergues, Y., Gillis, M., Holsters, M. and Dreyfus, B. (1991). Stem nodule forming rhizobia and Sesbania green manure crop. *In*: Datta, S.K., Sloger, C. (Eds.), *Biological Nitrogen Fixation Associated with Rice Production*, 17–24. Oxford & IBH, New Delhi.

Yamada, Y., Hoshino, K. and Ishikawa, T. (1997). The phylogeny of acetic acid bacteria based on the partial sequences of 16S ribosomal RNA: The elevation of the subgenus *Gluconoacetobacter* to generic level. *Biosci. Biotechnol. Biochem., 61*: 1244–51.

Yamada, Y., Hoshino, K. and Ishikawa, T. (1998). Taxonomic studies of acetic acid bacteria and allied organisms. XII, the phylogeny of acetic acid bacteria based on the partial sequences of 16S ribosomal RNA. *Int. J. Syst. Bacteriol., 48*: 3270–80.

Chapter 13

Integrated Effects of Vermicompost with Inorganic Fertilizer on Mustard–rice Cropping Systems

Md. Azizul Haque[1*], *Md. Mohsin Ali*[1] and *Kamrun Nahar*[1]

Introduction

Bangladesh is divided into 30 diverse agro ecological zones (AEZ), which vary in area, land and soil, climate, and cropping intensity (FAO UNDP, 1988). Several AEZs have great potential for agriculture and soil fertility, while a few are exhibiting poor fertility due to intensive and cropping under conventional management practices. Increasing cropping intensity to meet the demands of food security for an ever-rising population has led to mining out the natural soil fertility to support crop production, thereby soil fertility severely decreased in Bangladesh (Ali et al., 1997; Khan et al., 2008).

Integrated nutrient management with diverse cropping systems is critical to sustain crop productivity and soil fertility. Integrated use of different sources of nutrients is congenial for soil health and sustainable for long-term crop production (Dekhane et al., 2014). While organic sources of nutrients applied to the preceding crop benefits the succeeding crop (Hedge and Dwivedi, 1992), the various implications of chemically reactive fertilizers are associated with decreasing the native soil fertility and crop productivity (Moral et al., 2009). The ever-increasing cost of chemical fertilization compels stakeholders to think of the use of organic manures (Moral et al., 2009). Greater reliance on the use of chemical fertilizers is responsible for decreased ecosystem services associated with soil-water pollution and accelerated greenhouse gas emissions into the atmosphere. Hence, integrated use of both organic and inorganic sources of nutrients might be an appropriate option for improving soil fertility to sustain crop productivity.

[1] Soil Science Division, Bangladesh Institute of Nuclear Agriculture (BINA), Mymensingh.

[*] Corresponding author: azizul_bina@yahoo.com

All nutrient sources are considered in an integrated plant nutrient system, which supports economic viability and environmental compatibility with the reduced use of chemical fertilizers. A previous study indicated that about 60% of the cultivable lands in Bangladesh are deficient in N, P, and K contents (Miah et al., 2008). Moreover, soil organic matter content is low, the majority of soils had below the threshold level (1.5%), and it gradually depleted by 5–36% during the green revolution period of 1967–1995 (Ali et al., 1997). Currently, it is well accepted that depleted soil fertility is a major constraint for improving crop production in Bangladesh and indeed that crop yields, including rice and mustard, have been reduced.

Maintenance and/or regeneration of soil fertility is a prerequisite for climate-smart agriculture, and it is certain that organic manures like vermicompost can play a vital role in improving soil fertility and crop productivity (Bejbaruah, 2013). Vermicomposting is a biological process of composting in which certain species of earthworms are used to enhance the process of waste conversion into a nutrient-enriched product more so than conventional composting (Camberdella et al., 2003). The resulting earthworm castings (worm manure) are rich in NPK nutrients with narrow C:N (Goyal et al., 2005).

While vermicompost provides nutrients in readily available form that are well synchronized for plants' nutrients demand (Gaur, 2005; Alam et al., 2016), the present study was undertaken to investigate the integrated effects of different rates of vermicompost with chemical fertilizers on the growth and yield of mustard-Boro-Aman rice and their consequent effects on soil fertility at different locations in Bangladesh.

Materials and Methods

Experiment and Cultural Practices

Field experiments, with eight treatment combinations of vermicompost and chemical fertilization, were conducted in randomized complete block design at three locations, viz. (i) BINA farm Mymensingh (AEZ-9); (ii) farmer's field at Khoddo Tampat, Rangpur Sadar (AEZ-3); and (iii) farmer's field at Ista, Ishurdi, Pabna (AEZ-11). The values of initial soil properties are shown in Table 1. The treatments were T_1: native soil fertility; T_2: 100% chemical fertilizer (CF); T_3: 75% CF; T_4: 75% CF +

Table 1 Initial soil properties at different experimental fields.

Soil Properties	*BINA Farm Mymensingh*	*Rangpur Sadar (Farmer's Field)*	*Ishurdi, Pabna (Farmer's Field)*
Texture	Silt loam	Silt loam	Silty clay loam
pH	6.85	5.9	7.1
Organic carbon (%)	0.93	0.91	0.84
Total N (%)	0.1	0.09	0.1
Available P (mg kg^{-1} soil)	15.0	13.0	14.0
Exchangeable K (cmol + kg^{-1})	0.151	0.141	0.171
Available S (mg kg^{-1} soil)	15.0	14.0	13.0

vermicompost (VC) @ 4 t ha^{-1}; T_5: 85% CF; T_6: 85% CF + VC @ 4 t ha^{-1}; T_7: 75% CF +VC @ 2 t ha^{-10}; and T_8: 85% CF+VC @ 2 t ha^{-1}. Each treatment combination was replicated thrice. Fertilizers were applied based on soil test results and crop nutrient requirements (Table 2). The amount of NPK nutrients added from VC to support crop production was determined (Tables 3, 4).

Table 2 Calculated fertilization rates (kg ha^{-1}) based on soil test results for mustard, Boro, and Aman rice at different locations.

Location	*Mustard*				*Boro Rice*				*T. Aman Rice*			
	N	*P*	*K*	*S*	*N*	*P*	*K*	*S*	*N*	*P*	*K*	*S*
BINA farm, Mymensingh	87	18	47	10	131	9.8	60.4	10.2	87	8	40	12
Rangpur Sadar (Farmer's field)	120	20	49	11	136	11.9	64	11.2	120	8	42	10
Ishurdi, Pabna (Farmer's field)	87	19	49	12	131	11	63.6	11.5	87	8	35	10

Table 3 Nutrient concentration in vermicompost.

N (%)	*P (%)*	*K (%)*	*S (%)*
1.3	0.6	1.4	0.38

Table 4 Amount of nutrients added to soil from vermicompost amendments.

Vermicompost (t ha^{-1}) Application Rate	*Equivalent Amount of Nutrients (kg ha^{-1})*			
	N	*P*	*K*	*S*
2	26	12	28	7.6
4	52	24	56	15.2

After harvest of mustard (Binasharisha-10), Binadhan-14, and Binadhan-17 were transplanted in the Boro and Aman seasons during 2017–2018, respectively. Mustard was sown in the last week of November 2017. Boro and Aman rice were transplanted in the third week of February 2018 and the second week of August 2018, at all the locations. Yield of mustard, Boro, and Aman rice were recorded during the harvest. Post-harvest soil was collected at the end of the cropping cycle (i.e., after harvesting of Aman rice). Plant samples were collected after harvesting of each crop. Collected soil and plant samples were processed and analyzed following standard methods.

Soil Analyses

Particle size analysis was carried out using the hydrometer method (Black, 1965) and textural class was determined as per Marshall's Triangular Coordinates of the USDA system. Soil pH and electrical conductivity (ECe) were measured by a glass electrode pH and EC meter using soil: water suspension (1:2.5) as described by Jackson (1967). Soil organic carbon was determined by wet oxidation method (Black, 1965). To obtain the soil organic matter content, the organic carbon concentration was multiplied with the Van Bemmelen factor 1.73. Total N content was determined following the micro-Kjeldahl method (Jackson, 1967) followed by distillation and acid-base titration. Available phosphorus was extracted from the soil samples by shaking with 0.5 M $NaHCO_3$ solutions at pH 8.5 (Olsen and Sommers, 1982), and the extracted phosphorus was determined spectrophotometrically at 660 nm wavelength and the readings were calibrated to the standard P curve. Exchangeable K was extracted from the soil samples with 1N NH_4OAC (pH 7) and cations were determined from the extract by flame photometer (Black, 1965) and calibrated with a standard curve. Available sulfur was determined by extraction with 0.15% $CaCl_2$ solution (1:5 soil extractant ratio) and estimated according to the turbidimetric method using spectrophotometer at 535 nm wavelength (Williams and Steinbergs, 1959).

Plant Analysis

Random samples of both grain and straw biomass were oven-dried at 65°C for about 72 h until a constant was obtained, followed by grinding using a grinder, then stored prior to analysis. Total N was analyzed using the micro-Kjeldahl method (Jackson, 1967) followed by distillation and acid-base titration. Total phosphorus concentration in the micro-Kjeldahl digestion aliquot was analyzed spectrophotometrically at 660 nm wavelength and the readings were calibrated to the standard phosphorus curve. Likewise, total potassium concentration in the micro-Kjeldahl digestion aliquot was determined from the extract by flame photometer (Black, 1965) against a standard curve. Total sulfur concentration in the micro-Kjeldahl digestion aliquot was determined according to the turbidimetric method using a spectrophotometer at 535 nm wavelength (Williams and Steinbergs, 1959).

Nutrient Uptakes by Crops

The N, P, and K concentrations were used for the calculation of nutrient uptake in grain and straw biomass following Jackson (1967).

Nutrient uptake (kg ha^{-1}) = [nutrient concentration (%) × yield (kg ha^{-1}]/100.

Results and Discussion

Crop Yields

Mustard yield was significantly influenced by the effects of the VC and fertilizer treatments at all sites (Table 5). Maximum grain yield was observed with the treatment T_6 (1.82, 1.26, and 1.49 t ha^{-1}) followed by the treatment T_8 (1.80, 1.23, and 1.46 t ha^{-1}) at all sites, but they were statistically identical with the T_4 and T_2 treatments. The treatment T_6 gave identical seed yield as treatments T_2, T_4, and T_8 at Rangpur and Mymensingh, while at Ishurdi the treatment T_6 gave a statistically similar yield as other treatments except the T_1 and T_2. The treatments T_4, T_6, and T_8 did not exert any significant differences in yield of mustard, which were also comparable to the treatment T_2 (100% chemical fertilizer). The lowest yield of mustard was observed with the treatment T_1 at all sites. Similar trends were observed on straw biomass of mustard at all sites.

Table 5 Effect of vermicompost with inorganic fertilizers on seed yield and straw biomass ($t\ ha^{-1}$) of mustard at different locations in 2017–2018 growing seasons.

Treatments	*Rangpur*		*Ishurdi*		*Mymensingh*	
	Seed Yield	*Straw Yield*	*Seed Yield*	*Straw Biomass*	*Seed Yield*	*Straw Yield*
T_1	0.58c	2.22b	0.29c	0.83b	0.26d	1.41b
T_2	1.72a	4.47a	1.22a	3.78a	1.43ab	3.23a
T_3	1.42b	4.36a	0.81b	3.27a	0.88c	3.03a
T_4	1.76a	4.51a	1.20a	3.73a	1.44ab	3.37a
T_5	1.53b	4.44a	1.07a	3.43a	1.03c	3.10a
T_6	1.82a	4.55a	1.26a	3.91a	1.49a	3.40a
T_7	1.54b	4.45a	1.13a	3.63a	1.34b	3.33a
T_8	1.80a	4.71a	1.23a	3.85a	1.46ab	3.32a
CV(%)	6.25	5.50	12.26	10.50	7.09	8.00

Notes: T_1: Native soil fertility, T_2: 100% chemical fertilizer (CF), T_3: 75% CF, T_4: 75% CF + Vermicompost (VC) @ 4 tha^{-1}, T_5: 85% CF, T_6: 85% CF + VC @ 4 tha^{-1}, T_7: 75% CF +VC @ 2 tha^{-1}, T_8: 85% CF+VC @ 2 tha^{-1}.

Chemical fertilization (75%) with 4-ton ha^{-1} VC or 85% CF with 2-ton ha^{-1} VC were equally effective to produce similar yields of mustard, which was comparable with the full dose chemical fertilizer (100%). Desai et al. (1999) reported that VC along with N fertilizer produced higher dry-matter and grain yield of wheat. Similarly, a positive response was obtained with the application of VC to other field crops, such as sunflower (Devi and Agarwal, 1999; Devi et al., 1998). Our results were in agreement with those findings. Therefore, 15–25% chemical fertilizer could be saved either with the application of 75% CF with 4-ton ha^{-1} VC or 85% CF with 2-ton ha^{-1} VC for mustard production in Bangladesh.

Grain yield of Boro rice was significantly influenced in response to the effects of experimental treatments (Table 6). Highest grain yield was observed with the treatments T_6 (6.64-ton ha^{-1}), T_4 (4.81-ton ha^{-1}), and T_8 (5.44 t ha^{-1}) at the Rangpur, Ishurdi, and Mymensingh sites, respectively. While the T_2, T_4, and T_6 and T_8 treatments produced statistically identical grain yields of Boro rice at all sites, the T_4, T_6, and T_8 treatments did not affect the grain yield, but were comparable to the treatment T_2 (100% CF). In contrast, the lowest yield of Boro rice was observed with the treatment T_1 at all sites. In the case of straw biomass, the highest yield was found in the treatment T_6 at the Rangpur and Ishurdi sites and the treatment T_5 at the Mymensingh site. The lowest straw biomass was observed with the treatment T_1 at all sites. The results indicated that 75% CF with 4 t ha^{-1} VC or 85% CF with 2 t ha^{-1} VC were equally effective to produce similar grain yields of Boro rice, which were comparable to the 100% CF. Therefore, 15–25% chemical fertilizer could be saved when applying 75% CF with 4 t ha^{-1} VC or 85% CF with 2 t ha^{-1} VC for Boro rice production. Our results were in agreement with the results of previous studies (Arya et al., 2007)

Both grain and straw biomass yields of Aman rice were significantly differed by the integrated treatments of VC and CF at all sites (Table 7). Maximum grain yield was observed with the treatment T_7 at all sites (6.26, 6.07, and 6.02 t ha^{-1} at Rangpur,

Table 6 Effect of vermicompost with inorganic fertilizers on grain yield and straw biomass ($t\ ha^{-1}$) of Boro rice at different locations during 2017–2018 growing seasons.

Trts	*Rangpur*		*Ishurdi*		*Mymensingh*	
	Grain Yield	*Straw Yield*	*Grain Yield*	*Straw Yield*	*Grain Yield*	*Straw Yield*
T_1	2.30e	2.52c	2.73c	3.04b	2.68e	2.42d
T_2	6.60a	6.13ab	4.72a	4.46a	5.26ab	4.95ab
T_3	5.28d	6.13ab	3.96b	4.23a	4.38d	4.84ab
T_4	6.50ab	6.07b	4.81a	4.36a	5.17abc	4.93ab
T_5	6.15bc	6.23ab	4.18ab	4.33a	4.70cd	5.04ab
T_6	6.64a	6.70a	4.80a	4.35a	5.38a	5.25a
T_7	6.00c	6.35ab	4.54ab	4.13a	4.80bcd	4.39c
T_8	6.53a	6.23ab	4.61ab	4.44a	5.44a	4.73bc
CV(%)	8.82	5.66	9.54	7.05	5.67	5.21

Notes: T_1: Native soil fertility, T_2:100% chemical fertilizer (CF), T_3: 75% CF, T_4: 75% CF + Vermicompost (VC) @ 4 tha^{-1}, T_5:85% CF, T_6: 85% CF + VC @ 4 tha^{-1}, T_7: 75% CF +VC @ 2 tha^{-1}, T_8: 85% CF+VC @ 2 tha^{-1}.

Table 7 Effect of vermicompost with inorganic fertilizers on grain yield and straw biomass ($t\ ha^{-1}$) of T. Aman rice at different locations during 2018 growing season.

Treatments	*Rangpur*		*Ishurdi*		*Mymensingh*	
	Grain Yield	*Straw Yield*	*Grain Yield*	*Straw Yield*	*Grain Yield*	*Straw Yield*
T_1	3.26d	4.03c	3.00d	4.05b	3.11c	3.92b
T_2	5.17a	5.72a	5.96a	6.18a	5.76a	5.54a
T_3	3.97c	4.76b	3.70c	5.67a	3.81c	5.66a
T_4	5.18a	5.80a	5.64a	6.02a	5.43a	5.87a
T_5	4.88b	5.44a	4.69b	6.05a	4.69b	5.37a
T_6	5.05a	5.53a	5.74a	6.35a	5.32ab	5.77a
T_7	5.26a	6.08a	6.07a	6.28a	6.02a	5.87a
T_8	5.13a	5.75a	5.94a	6.33a	5.81a	5.78a
CV(%)	7.92	6.53	9.0	9.45	9.69	9.69

Notes: T_1 : Native soil fertility, T_2 : 100% chemical fertilizer (CF), T_3 : 75% CF, T_4 : 75% CF + Vermicompost (VC) @ 4 tha^{-1}, T_5 : 85% CF, T_6 : 85% CF + VC @ 4 tha^{-1}, T_7 : 75% CF +VC @ 2 tha^{-1}, T_8 : 85% CF+VC @ 2 tha^{-1}.

Ishurdi, and Mymensingh, respectively) followed by the treatments T_2 (5.17 and 5.96 $t\ ha^{-1}$) at Rangpur and Ishurdi and T_8 (5.81 $t\ ha^{-1}$) at Mymensingh. However, the treatments T_2, T_4, T_7, and T_8 had statistically similar grain yield of Aman rice at all sites. The treatments T_4, T_6, T_7, and T_8 were comparable to the treatment T_2 (100% CF) for producing Aman rice grain yield. In contrast, the lowest grain yield of Aman was observed with the treatment T_1 (native soil fertility) at all sites. The highest straw biomass was found with the treatment T_7 at the Rangpur and Mymensingh sites whereas the treatment T_5 had the highest straw biomass at the Ishurdi site. The lowest straw biomass was observed with the treatment T_1 at all sites. Our results on Aman grain yield indicated that 75% CF with 2 $t\ ha^{-1}$ VC was equally effective to 100%

CF. The present findings are well supported by Bejbaruah et al. (2013). Therefore 15–25% chemical fertilizer (NPKS) could be saved with the application of 75% CF with 2 t ha^{-1} VC to soil.

Nutrients Uptake by Crops

Total N, P, and K uptakes by mustard were significantly affected by the fertility treatments at all sites (Table 8). The highest N uptake was observed in the T_8 treatment (85% CF + VC @ 2 t ha^{-1}) at both the Rangpur and Ishurdi sites, whereas the T_6 treatment (85% CF + VC @ 4 t ha^{-1}) showed maximum N uptake by mustard at the Mymensingh site. N uptake by mustard was identical with the treatments T_8, T_6, and T_4 at all sites. In contrast, the lowest N uptake was observed with the control treatment (T_1, native soil fertility), regardless of the sites. The highest P uptake of mustard was observed in the treatment T_6 (85% CF + VC @ 4 t ha^{-1}) at all sites followed by the treatment T_8. Likewise, the highest K uptake of mustard was obtained with the treatment T_6 at the Ishurdi and Mymensingh sites whereas the treatment T_8 showed maximum K uptake at the Rangpur site. Our results indicated that application of VC with CF significantly increased the nutrient uptake by mustard when compared to the CF alone. Total N, P, and K uptakes increases with the application of VC have been reported by other studies (Sreenivas et al., 2000; Jadhav et al., 1997; Narwal and Chaudhary, 2006)

Total N, P, and K uptake of Boro rice was significantly affected by treatments at all sites (Table 9). Maximum N uptake was observed in the treatment T_6 (85% CF+VC @ 4 t ha^{-1}) at the Rangpur and Mymensingh sites whereas the treatment T_4 (75% CF + VC @ 4 t ha^{-1}) showed maximum N uptake at Ishurdi. N uptake by Boro rice was statistically identical in the T_8, T_6, T_4, and T_2 treatments at all sites when compared to the lowest N uptake with the control treatment (native soil fertility) at all sites. The highest P uptake of Boro rice was observed in the treatment T_6 (85% CF + VC @ 4 t ha^{-1}) followed by the treatment T_8 (85% CF + VC @ 2 tha^{-1}) at all study sites. In contrast, the highest K uptake by Boro rice was recorded with the treatment T_6 at

Table 8 Total nitrogen, phosphorus, and potassium uptakes (kg ha^{-1}) by mustard as affected by integrated treatments of vermicompost and chemical fertilizers at different locations.

Treatment	*Rangpur*			*Ishurdi*			*Mymensingh*		
	N	*P*	*K*	*N*	*P*	*K*	*N*	*P*	*K*
T_1	28.8e	7.3e	34.8e	11.5d	3.0f	13.7f	14.8e	4.0f	22.0e
T_2	91.6b	25.2b	81.9b	65.1ab	18.5b	66.4b	72.6b	20.8c	63.0b
T_3	72.9d	19.1d	75.0d	44.4c	11.4e	53.4e	46.4d	13.0e	50.8d
T_4	93.8ab	25.6b	82.5b	52.8bc	18.1b	65.1b	73.9ab	21.5b	64.7a
T_5	82.2c	22.9c	77.2cd	57.5abc	16.1d	59.2d	55.1c	15.7d	55.7c
T_6	96.9a	27.3a	84.3ab	68.4a	19.6a	69.1a	76.4a	22.5a	66.4a
T_7	85.9c	24.0c	79.3c	63.2ab	17.6c	62.3c	70.9b	20.0c	62.5b
T_8	97.3a	26.5ab	85.1a	68.3a	18.6b	67.0ab	74.7ab	21.5b	63.9b
CV(%)	7.54	6.78	7.96	14.02	5.94	9.45	8.23	7.52	6.87

Table 9 Total nitrogen, phosphorus, and potassium uptakes (kg ha^{-1}) by Boro rice as affected by integrated treatments of vermicompost and chemical fertilizers at different locations.

Treatments	*Rangpur*			*Ishurdi*			*Mymensingh*		
	N	*P*	*K*	*N*	*P*	*K*	*N*	*P*	*K*
T_1	34.9d	7.1d	42.2d	42.2c	7.9c	48.7b	38.8d	8.8e	39.6d
T_2	105.3a	25.8ab	115.1ab	76.6a	18.1a	82.1a	84.3a	21.1ab	90.1ab
T_3	85.5c	18.8c	104.6c	63.4b	13.3b	71.6a	68.9c	15.4d	82.4c
T_4	104.0ab	25.5b	113.2b	77.7a	17.7a	80.4a	83.1a	19.8b	90.1ab
T_5	98.7b	23.5c	111.9bc	68.4ab	15.2b	76.6a	76.9b	18.3c	88.6b
T_6	108.5a	27.3a	122.4a	77.5a	18.1a	80.2a	86.2a	21.9a	95.0a
T_7	98.3b	25.1b	115.0ab	72.5ab	17.2ab	75.7a	75.4b	18.7c	79.5c
T_8	105.1a	26.4a	116.4ab	75.7a	17.8a	81.9a	85.8a	21.3ab	87.6b
SE	1.96	0.47	2.98	3.01	0.71	3.55	1.8	0.42	1.93
CV(%)	7.21	6.84	5.86	7.54	7.83	8.23	7.87	6.85	8.93

Table 10 Total nitrogen, phosphorus, and potassium uptakes (kg ha^{-1}) by mustard as affected by integrated treatments of vermicompost and chemical fertilizers at different locations.

Treatment	*Rangpur*			*Ishurdi*			*Mymensingh*		
	N	*P*	*K*	*N*	*P*	*K*	*N*	*P*	*K*
T_1	52.5e	11.5d	64.8c	50.1d	11.4d	65.3c	50.9d	10.8d	63.2b
T_2	85.7a	23.0a	101.5a	97.0a	26.5a	111.8a	91.8a	23.9a	100.5a
T_3	65.1d	15.9c	80.3c	65.1c	16.4c	93.3b	67.7c	15.7c	93.1a
T_4	83.8b	22.6a	101.7a	91.9a	24.8a	107.0a	89.3a	23.4a	103.5a
T_5	79.4c	20.7b	94.4b	79.5b	21.0b	102.5a	78.1b	19.6b	92.9a
T_6	83.4b	21.8ab	97.3b	94.0a	25.5a	112.2a	87.1a	22.4a	101.7a
T_7	88.3a	23.8a	107.1a	99.3a	27.0a	113.6a	97.5a	25.7a	107.4a
T_8	85.3a	22.9a	101.8a	97.4a	26.7a	113.8a	92.8a	23.8a	104.7a
SE	1.46	0.37	2.86	3.59	0.94	4.85	4.34	1.9	5.58
CV(%)	7.84	6.9	5.29	7.38	7.28	8.21	9.19	9.11	10.1

Rangpur and Mymensingh whereas the treatment T_2 showed maximum K uptake at Ishurdi, but the treatments T_2, T_4, T_6, and T_8 were statistically similar. The results of our study have shown that VC application with reduced rates of CF influenced the nutrient uptake of Boro rice, which was comparable to the 100% CF alone.

Total N, P, and K uptakes of Aman rice were significantly influenced by VC and CF treatments at all sites (Table 10). Maximum N, P, and K uptakes by Aman rice were observed in the treatment T_7 (85% CF + VC @ 2 tha^{-1}) at all sites except K uptake at Ishurdi. The treatment T_8 showed maximum K uptake at Ishurdi followed by the treatment T_7. The T_7 and T_8 treatments have shown statistically similar NPK uptakes to the treatment T_2 at all sites. As expected, the lowest NPK uptakes were observed with the control treatment (native soil fertility) at all sites. The results suggested that the VC application with reduced rates of CF consistently influenced the nutrient uptake of Aman rice, which was comparable to the sole application of 100% CF. It is reported that organic manuring along with chemical fertilization increased microbial activity (Rani and Srivastava, 1997), nutrient use efficiency (Narwal and Chaudhary, 2006), and improved availability of the nutrients to the crops, resulting higher nutrient uptake (Bejbaruah et al., 2013; Bhandari et al., 1992).

Changes of Soil Properties

Post-harvest soils analyzed for pH, organic carbon, total N, available P, exchangeable K, and available S have shown changes in soil properties (Tables 11–13). The changes in pH, organic carbon, and the nutrient contents were observed due to the integrated use of chemical fertilizers and VC. Soil pH increased at the Rangpur and Mymensingh sites where initial pH was slightly acidic to neutral in reaction. At the Ishurdi site, pH decreased where initial pH was slightly alkaline. Organic amendments enhanced the increasing of pH in acidic soil and decreasing pH in alkaline soil due to its chemical buffering capacity. An increase in organic carbon and N, P, K, and S contents in the VC-treated plots than the sole application of CF might be due to effects of nutrient-

Table 11 Changes in soil properties in response to integrated treatments of vermicompost and chemical fertilizers at Rangpur.

Treatments	*pH*	*OC (%)*	*N (%)*	*P (%)*	*K ($cmol_+ kg^{-1}$)*	*S ($mg\ kg^{-1}$)*
T_1	5.9	0.90	0.09	13.0	0.139	12.95
T_2	5.8	0.94	0.09	14.2	0.143	13.0
T_3	5.9	0.93	0.09	13.8	0.142	13.0
T_4	6.0	0.95	0.10	14.7	0.144	13.6
T_5	5.8	0.93	0.09	14.2	0.43	13.0
T_6	6.0	0.95	0.10	14.7	0.144	13.5
T_7	6.0	0.95	0.10	14.8	0.145	13.6
T_8	6.0	0.95	0.10	14.6	0.144	13.6
Initial	5.9	0.91	0.09	13	0.141	13

Table 12 Changes in soil properties in response to integrated treatments of vermicompost and chemical fertilizers at Ishurdi.

Treatments	*pH*	*OC (%)*	*N (%)*	*P (%)*	*K ($cmol\ kg^{-1}$)*	*S ($mg\ kg^{-1}$)*
T_1	7.1	0.83	0.09	13.8	0.170	12.9
T_2	7.1	0.88	0.10	14.1	0.172	13.30
T_3	7.1	0.85	0.10	14.3	0.171	13.0
T_4	7.0	0.91	0.11	14.7	0.173	13.8
T_5	7.1	0.87	0.10	14.4	0.171	13.2
T_6	7.0	0.91	0.11	14.7	0.173	13.7
T_7	7.0	0.91	0.11	14.7	0.173	13.7
T_8	7.0	0.91	0.11	14.7	0.173	13.8
Initial	7.1	0.84	0.10	14	0.171	13.0

Table 13 Changes in soil properties in response to integrated treatments of vermicompost and chemical fertilizers at Mymensingh.

Treatments	*pH*	*OC (%)*	*N (%)*	*P (%)*	*K (cmol kg^{-1})*	*S (mg kg^{-1})*
T_1	6.85	0.91	0.09	14.9	0.150	14.8
T_2	6.9	0.94	0.10	15.3	0.152	15.4
T_3	6.9	0.93	0.10	15.1	0.151	15.1
T_4	6.9	0.96	0.11	15.7	0.154	15.8
T_5	6.9	0.93	0.10	15.2	0.151	15.2
T_6	6.9	0.95	0.11	15.6	0.153	15.7
T_7	6.9	0.95	0.11	15.6	0.153	15.6
T_8	6.9	0.96	0.11	15.7	0.154	15.8
Initial	6.85	0.93	0.10	15	0.151	15.0

enriched organic matter amendments (Table 11–13). However, soil fertility did not vary consistently with the application of variable amounts of VC, which might be due to short-term application of VC (only one cropping cycle), but some positive changes of soil fertility were observed in the various doses of vermicompost.

Banik et al. (2004) reported that available N, P, and K contents were higher under fertilizer NPK as well as integrated application of chemical fertilizers and organic manure-treated plots, when compared to the control. Our results are well corroborated with the results of previous studies (Dekhane et al., 2014). Hence, the result suggested that soil amended with VC at 2–4 t ha^{-1} with reduced rates (15–25%) of chemical fertilizers (NPKS) is an appropriate option to sustain soil fertility.

Conclusions

Results indicated that 15–25% chemical fertilizer (NPKS) could be saved either with the application of 75% CF with 4 t ha^{-1} VC or 85% CF with 2 t ha^{-1} VC for mustard and Boro rice production, whereas 25% chemical fertilizers could be saved with the 75% chemical fertilization with 2 t ha^{-1} VC for Aman rice production.

Acknowledgements

The research work was supported by the grant of the World Bank and USAID which was provided through PIU-BARC-NATP-2, Dhaka, Bangladesh and the research works were part of CRG-802 subproject (PIU-BARC-NATP-2). The authors thank the PIU-BARC-NATP-2 for funding and support during the implementation of the project.

References

Alam, S.M.M., Bodruzzaman, M. and Sadat, M.A. (2016). Effect of different doses of vermicompost in combination with chemical fertilizers in a Potato-Wheat-Mungbean-T. Aman cropping pattern. *In*: M.A. Satter, Hossain, M.B. Boktiar S.M. Ahmed, S. (Eds.), *Proc. Res. Review and Planning Workshop of Soil Program of NARS Institutes, 2015.*, 76–79.

Ali, M.M., Shaheed, S.M. and Kubota D. (1997). Soil degradation during the period 1967–1995 in Bangladesh. II. Selected chemical characters. *Soil Sci. Plant Nutr., 43*: 879–90.

Arya, R.L. , Jay, G.V. and Kumar, L. (2007). Effect of integrated nutrient application in chickpea + mustard intercropping system in the semi-arid tropics of North India. *Commun. Soil Sci. Plant Analy., 38*: 229–40.

Banik, P.B. and Ranjita. (2004). Effect of vermicompost on rice (*Oryza sativa*) yield and soil-fertility status of rainfed humid sub-tropics. *Indian J. Agril. Sci., 74*: 488–91.

Bejbaruah, R., Sharma, R.C. and Banik, P. (2013). Split application of vermicompost to rice (*Oryza sativa* L.): its effect on productivity, yield components, and N dynamics. *Organic Agriculture, 3*(2): 123–28.

Bhandari, A.L., Sood, A., Sharma, K.N. and Rana, D.S. (1992). Integrated nutrient management in rice-wheat system. *J. Ind. Society Soil Sci., 40*: 742–47.

Black, C.A. (1965). *Methods of Soil Analysis*. Part I and II. American Soc. of Agron. Inc. Pub. Madison, Wisconsin, USA.

Cambardella, C.A., Richard, T.L. and Russell, A. (2003). Compost mineralization in soil as a function of composting process conditions. *Eur. J. Soil Biol., 39*: 117–27.

Dekhane, S.S., Patel, D.J., Jadhav, P.B., Kireeti, A., Patil, N.B., Harad, N.B. and Jadhav, K.P. (2014). Effect of organic and inorganic fertilizer on growth and yield of paddy cv gr 11. *International J. Infor. Res. and Rev., 1*(2): 026–028.

Desai, V.R, Sabale, R.N. and Raundal, P.V. (1999). Integrated nitrogen management in wheat-coriander cropping system. *Journal of Maharashtra Agricultural Universities, 24*: 273–75.

Devi, D. and Agarwal, S.K. (1998). Performance of sunflower hybrids as influenced by organic manure and fertilizer. *Journal of Oilseeds Research, 15*(2): 272–79.

Devi, D., Agarwal, S.K. and Dayal, D. (1998). Response of sunflower (*Helianthus annuus* L.) to organic manures and fertilizers. *Indian Journal of Agronomy, 43*(3): 469–73.

FAO-UNDP. (1988). *Land Resources Appraisal of Bangladesh for Agricultural Development: Agro-ecological Regions of Bangladesh*. Technical Report No. 2, FAO, Rome.

Goyal, S., Dhull, S. and Kapoor, K.K. (2005). Chemical and biochemical changes and compost maturity indices during vermicomposting of different organic wastes. *In*: Kapoor, K.K., Sharma, P.K., Dudeja, S.S., Kundu, B.S. (Eds.), *Management of Organic Wastes for Crop Production*, 111–16. CCS Haryana Agril. Univ. Hisar.

Gaur, A.C. (2005). Composting methods and biodynamic preparations for organic farming. *In*: Kapoor, K.K., Sharma, P.K., Dudeja, S.S., Kundu, B.S. (Eds.), *Management of Organic Wastes for Crop Production*, 25–39. CCS Haryana Agril. Univ. Hisar.

Hedge, D.M. and Dwivedi, B.S. (1992). Nutrient management in rice-wheat cropping system in India. *Fertilizer News*, 37: 27–41.

Jackson, M.L. (1967). *Soil Chemical Analysis*. Constable and Co. Ltd. London.

Jadhav, A.D., Talashilkar, S.C. and Pawar, A.G. (1997). Influence of the conjunctive use of FYM, vermicompost, and urea on growth and nutrient uptake in rice. *J. Maharashtra Agril. Univ., 22*(2): 249–50.

Khan, M.S., Shil, N.C. and Noor, S. (2008). Integrated nutrient management for sustainable yield of major vegetable crops in Bangladesh. *Bangladesh J. Agric. and Environ., 4* (Special Issue): 81–94.

Miah, M.A., Saha P.K., Islam, A., Hasan, M.N. and Nosov, V. (2008). Potassium fertilization in rice-rice and rice-wheat cropping system in Bangladesh. *Bangladesh. J. Agric. and Environ 4: 51-67, Special Issue.* 51–67.

Moral, R., Paredes, C., Bustamante, M.A., Marhuenda-Egea, R. and Bernal, M.P. (2009). Utilization of manure composts by high-value crops: Safety and environmental challenges. *Bioresource Technology, 100*: 5454–60.

Narwal, R.P. and Chaudhary, M. (2006). Effect of long-term application of FYM and fertilizer N on available P, K, and S content of soil. *Proc. 18th World Congress of Soil Sci., Philadelphia*, 9–15 July.

Olsen, S.R. and Sommers, L.E. (1982). Phosphorus.. *In*: Page, A.L. et al. (Eds.), *Methods of Soil Analysis: Part 2*, 403–30. Chemical and Microbiological Properties. *Agron. Mongr., Series 9*. (2nd Edn.), ASA and SSSA, Madison, WI.

Rani, R. and Srivastava, O.P. 1997. Vermicompost: A potential supplement to nitrogenous fertilizer in rice nutrition. *Int. Rice Res. Notes*, *22*(3): 30–31.

Sreenivas, C., Muralidhar, S. and Rao, M.S. (2000). Vermicompost, a viable component of IPNSS in nitrogen nutrition of ridge gourd. *Ann. Agri. Res.*, *21*(1): 108–13.

Williams, C.J. and Steinbergs, A. (1959). Soil sulphur fractions as chemical indicates of available sulphur in some Australian soils. *Aust. J. Agric. Res.*, *10*: 340–52.

Chapter 14

Vermicompost Impacts Sugarcane Yield and Soil Carbon Accumulation

Khandakar Mohiul Alam[1*], *Md. Aminul Haque*[2],
Gazi Md. Akram Hossain[2] and *Md. Amzad Hossain*[2*]

Introduction

Sugarcane is cash cum industrial crop that needs 12–14 months for its maturity. During this life span, sugarcane requires a high demand of nutrients for its growth and development. It produces heavy tonnage and removes considerable amount of nutrients from the soil. Soils, where sugarcane is grown, are losing their inherent capacity. As a result, sugarcanes are supplied inadequate supply of nutrients, which is reflected in poor yield and quality. Therefore, there is a need to add nutrients to the soil through fertilizers to get the desired yield and maintain soil fertility. Chemical or inorganic fertilizers are of paramount importance for enhancing sugarcane production to meet the sugar needs of the mounting population of the country. When chemical fertilizers are added in the soil, its continuous use degrades the soil health and quality, hence causing soil pollution. The adverse effect of these synthetic chemicals on human health and environment can only be reduced or eliminated by adopting new agricultural technological practices such as shifting from chemical intensive agriculture which includes the use of organic inputs such as vermicompost, manure, biofertilizers, biopesticides, slow-release fertilizers, and nano-fertilizers, etc. which would improve the application efficiency as well as use efficiency of the fertilizers (Chandini et al. 2019). Vermicompost is a carbon rich organic material which increases the organic carbon content of soil to initial value $4.5 gkg^{-1}$ when incorporated along with an NPK fertilizer (Kumar et al., 2006).

Plant nutrients have a pivotal role in sugarcane yield and sugar synthesis. Among all the nutrients required by a plant, nitrogen is of prime importance, being an integral

[1] Bangladesh Sugarcrop Research Institute, Jamalpur.
[2] Bangladesh; Bangladesh Sugarcrop Research Institute, Pabna, Bangladesh.
* Corresponding author: mohiulalam74@yahoo.com

part of chlorophyll, protoplasm, proteins, and several biochemical processes. Plant responses to phosphate fertilizer application are positive. Application of potassium fertilizers influenced cane tonnage and sucrose content significantly. Sulphur is the fourth important major plant nutrient as N, P, and K for the healthy growth and bumper yield of sugarcane. Higher sulphur content in the cane leaves is also closely associated with high sucrose, high recovery, and purity of juice (Alam et al., 2016).

Vermicompost is the microbial composting of organic wastes through earthworm activity to form organic fertilizer which contain higher level of organic matter, organic carbon, total, and available N, P, K, and micronutrients, microbial and enzyme activities (Parthasarathi et al., 2007). Vermicompost promotes humification, increased microbial activity, and enzyme production. It increases the aggregate stability of soil particles, resulting in better aeration (Perucci, 1990). Organic matter has a property of binding mineral particles such as calcium, magnesium, and potassium in the form of colloids of humus and clay, facilitating stable aggregates of soil particles for desired porosity to sustain plant growth (Haynes, 1986). Several workers observed that integration of vermicompost with inorganic fertilization tended to increase the yield of crop, viz. tomato, potato, rapeseed, groundnut, black gram, paddy, mulberry, and marigold (Manivannan et al., 2009).

Keeping in view this study was undertaken to find out the effect of vermicompost on sugarcane yield and to determine the carbon accumulation in soils from vermicompost application on Young Brahmaputra and Jamuna Floodplain.

Materials and Methods

Preparation of Vermicompost

Vermicomposting units were set up using vermitech pattern (Ismail, 2005). Cement charis were used and placed in a shaded elevated area as to facilitate the effective water drainage. Seven days dried cow dung and kitchen waste were used for composting. Twenty to twenty-five kilograms of mixture (8:1) were placed in a cement chari of dimension 100 cm × 50 cm, about 500 adult locally collected earthworms, *Eisenia foetida* were inoculated. The unit was then covered with 10 cm of agro waste. The entire unit was kept moist by sprinkling of water twice weekly and turned once weekly up to the 7th week. The unit should be maintained for 50 days, at 65–67% moisture, 30 ± 2°C temperature. and 65–70% RH. After 50 days fresh vermicompost was collected by hand from the chari and used for experiments (Fig. 1).

Experimental Site

An experiment was carried out in Bangladesh Sugarcane Research Institute (BSRI) Jamalpur substation farm soil at Jamalpur Sadar, Jamalpur (24°55′ N latitude and 89°56′ E longitude) under the Young Brahmaputra and Jamuna Floodplain soils of Bangladesh. It was in the Jamalpur series, Type-Predominantly Floodplain, agro-ecological zone (AEZ) 8. The climate of the location was tropical. The soil samples (0–15 cm) were collected from experimental plots, air-dried and passed through 2 mm sieve. The soil was a permeable silt loam in texture containing sand, silt and clay at the

Fig. 1 Different steps of vermicompost preparation.

rate of 50, 88, and 27%, respectively, with pH 6.5 and the following nutrient levels: Organic carbon, total N, available P, exchangeable K, and available S were 0.84%, 0.09%, 7.60 mg/kg, 0.18 cmol (+)/kg, and 10.10 mg/kg, respectively (Table 1).

Nutrient Compositions of Vermicompost

The well-prepared vermicompost was determined with pH 6.12 and the following nutrient composition levels: organic carbon, total N, available P, exchangeable K, and available S were 18.0%, 1.52%, 0.97 mg/kg , 0.90 cmol (+)/kg, and 0.69 mg/kg, respectively (Table 2).

Table 1 Initial soil properties.

Soil Analysis	*Texture*	*pH*	*SOC (%)*	*Total N (%)*	*Avail. (mg/kg)*	*Exch. K (cmol (+)/kg)*	*Avail. S (mg/kg)*
	Silt Loam	*6.5*	*0.84*	*0.09*	*7.6*	*0.18*	*10.10*
Fertility status	–	Slightly acidic	Low	Very low	Low	Medium	Low

Table 2 Vermicompost properties*

Elements	*pH*	*OC (%)*	*N (%)*	*P (mg/kg)*	*K (cmol(+)/kg)*	*S (mg/kg)*
	6.12	18.0	1.52	0.97	0.90	0.69

*Samples (soils, leaf and vermicompost) were analyzed from the laboratory of Soils and Nutrients Division, Bangladesh Sugarcrop Research Institute and Regional Laboratory of Soil Resource and Development Institute, Jamalpur.

Treatment Details and Fertilizer Application

The experiment consisted of eight treatments, viz., T_1 : Control (No fertilizer), T_2 : N_{165} P_{55} K_{120} S_{45} Mg_{20} $Zn_{2.5}$ (Recommended fertilizer doses as fertilizer recommendation guide, 2012), T_3 : 2.5 t/ha vermicompost + N_{165} P_{55} K_{120} S_{45} Mg_{20} $Zn_{2.5}$ (Recommended fertilizer doses as fertilizer recommendation guide, 2012), T_4 : 5 t/ha vermicompost + N_{165} P_{55} K_{120} S_{45} Mg_{20} $Zn_{2.5}$ (Recommended fertilizer doses as fertilizer recommendation guide, 2012), T_5 : 2.5 t/ha vermicompost + N_{124} P_{41} K_{90} S_{34} Mg_{15} $Zn_{1.9}$ (75% recommended fertilizer doses as fertilizer recommendation guide, 2012), T_6 : 5 t/ha Vermicompost + N_{124} P_{41} K_{90} S_{34} Mg_{15} $Zn_{1.9}$ (75% recommended fertilizer doses as fertilizer recommendation guide, 2012), T_7 : 2.5 t/ha vermicompost + N_{83} P_{28} K_{60} S_{23} Mg_{10} $Zn_{1.3}$ (50% recommended fertilizer doses as fertilizer recommendation guide, 2012), T_8 :5 t/ha vermicompost + N_{83} P_{28} K_{60} S_{23} Mg_{10} $Zn_{1.3}$ (50% recommended fertilizer doses as fertilizer recommendation guide, 2012) (Table 3). The experimental design was a randomized complete block design with three replications. Vermicompost and inorganic fertilizer were applied. Urea, triple super phosphate, gypsum, magnesium sulphate, zinc sulphate, and muriate of potash were used as sources of N, P, K, S, Mg, and Zn, respectively. Full amount of triple super phosphate, gypsum, magnesium sulphate, zinc sulphate, one-third muriate of potash and vermicompost were applied during final land preparation. Urea was applied after 30 days of planting. From the rest of the amount, one-third urea and one-third muriate of potash were applied after 150 days of plantation and rest of the amount of urea and potash were applied after 180 days of plantation.

Table 3 Vermicompost and inorganic fertilizer rates.

Treatments	*Nutrient Application Rate ($kg\ ha^{-1}$)*						*Vermicompost ($ton\ ha^{-1}$)*
	N	*P*	*K*	*S*	*Mg*	*Zn*	
T_1	0	0	0	0	0	0	0
T_2	165	55	120	45	20	2.5	0
T_3	165	55	120	45	20	2.5	2.5
T_4	165	55	120	45	20	2.5	5.0
T_5	124	41	90	34	15	15	2.5
T_6	124	41	90	34	15	15	5.0
T_7	83	28	60	23	10	10	2.5
T_8	83	28	60	23	10	10	5.0

* This rate is per fertilizer rate recommendation guide (2012).

Cultural Practices

Fifty days aged seedlings of sugarcane were planted on November 26, 2016. Each plot had an area of 6 m × 8 m in which 6 rows of cane were planted at row-to-row spacing of 1m; and 0.45 m was interplant spacing. Sugarcane variety BSRI Akh 42 was used as test crop in the experiment. Necessary intercultural operations like weeding, mulching, earthing-up, tying, and mechanical control of insect pests were done when required. Field crops were harvested in October 2017.

Analyses of Plant, Soil, and Vermicompost

Sugarcane leaf samples (3th–5th) from top of main cane stalks were collected from the sampled cane and separated into leaf blade and leaf sheath. Midribs were removed from the leaf blade and samples were dried at 65°C and milled for nutrient analysis. Cane Brix was estimated in randomly selected 10 canes from each replication by using a portable refractometer.

Soil textural class was determined by the hydrometer method. For soil and vermicompost, pH was measured in a 1:2.5 soil water suspension by a glass electrode pH meter. Total N was determined by the macro Kjeldahl procedure and organic carbon by the Walkley and Black method. Available soil P was extracted with 0.5 M $NaHCO_3$ and the amount was determined by Spectrophotometry. Exchangeable K (1N NH_4OAc extractable) was determined by flame photometry and available S by turbidimetric method (Black, 1965).

Statistical Analysis

The data obtained were statistically analyzed by using the computer package Statistix 10 program for Windows version. Significant differences among treatments were performed by using the Duncan Multiple Range Test (DMRT) at $p \leq 0.05$, unless otherwise mentioned.

Results and Discussion

Cane Yield and Brix

Cane yield and its components like tiller, millable cane stalk, and Brix are shown in Table 4. The perusal of data revealed that the application of different levels of vermicompost and inorganic fertilizers had significant effect of cane yield. The application of the recommended fertilizer dose with vermicompost @ 5 tha^{-1} (T_4) produced the highest cane yield (99.22 tha^{-1}) which was significantly superior over the rest of the fertilizer approaches. The lowest cane yield (60.69 tha^{-1}) was obtained from T_1 (control). The application of vermicompost with chemical fertilizer increased cane yield in T_3, T_4, T_5, T_6, T_7, and T_8 of 60.81%, 63.18%, 32.75%, 31.84%, 37.68%, and 46.98%, respectively, over control whereas the only chemical fertilizer T_2 increased the cane yield 51.70% over control. This could be attributed to application of the highest amount of fertilizer that contributed to higher yield attributing characters such as millable cane and number of tillers per clump resulting in a higher cane yield. Similar findings were also observed by Thakur et al. (2013). The application of inorganic fertilizer T_2 independently produced a similar yield (92.22 tha^{-1}), but the findings of Virdia and Patel (2010) who reported that organics released nutrients through decomposition and mineralization and that would have increased the availability of plant nutrient at a later stage and brought improvement of physical, chemical, and biological properties of the soil. As a result, the fertility status of soil might have escalated with the increasing of absorption of plant nutrients. Hence, more tillers were converted into millable cane which led to more cane yield. These results

Table 4 Effects of vermicompost and inorganic fertilization rates on sugarcane yield.

Treatment	*Tiller* ($\times 10^3$ ha^{-1})	*Millable Cane* ($\times 10^3$ ha^{-1})	*Yield* (ton ha^{-1})	*Brix* (%)
T_1	58.50d	49.00c	60.69d	20.50
T_2	83.33ab	61.77ab	92.22abc	21.00
T_3	91.66a	61.55ab	97.77ab	21.00
T_4	92.33a	63.88a	99.22a	20.00
T_5	80.77abc	53.78bc	80.66c	20.50
T_6	77.99bc	53.33bc	80.11c	21.00
T_7	68.11cd	48.33c	83.67c	21.00
T_8	79.00abc	54.77bc	89.34abc	21.00
$LSD_{(0.05)}$	6.28	4.17	7.93	ns

In a column, the figures having same letter do not differ significantly as per DMRT at 5% level.

are in agreement with the findings of Khan et al. (2005), Thakur et al. (2007), and Boikhtiar et al. (2015).

The number of millable cane stalk has significant variation at maturity stage due to application of different levels of vermicompost with inorganic fertilizer doses. The maximum millable cane was observed in T_4 (63.88×10^3 ha^{-1}) which was closely followed by T_3 (61.55×10^3 ha^{-1}). This might be due to the addition of vermicompost with inorganic fertilizer which reduced the plant mortality and produced a favorable condition for conserving tillers into millable cane. These results are in conformity with the findings of Sison et al. (2010), and Umesh et al. (2013).

The significant variation in number of tillers was found due to application of different levels of inorganic fertilizer doses with vermicompost. Among the different levels of fertilizer application, the highest number of tillers was observed in T_4 (92.33×10^3 ha^{-1}) which was identical to T_3 (91.66×10^3 ha^{-1}) having recommended fertilizer dose' 2012 along with vermicompost @ 2.5 tha^{-1}. The lowest number tiller was recorded in T_1 (58.50×10^3 ha^{-1}) receiving no fertilizer. Significant increase in tillers due to higher level of NPK fertilizer and compost was also observed by Thakur et al. (2013). The percent of Brix did not differ significantly due to application of different level of vermicompost and fertilizer. Similar observation was recorded by Alam et al. (2016).

Leaf Nutrient Contents

Application of vermicompost and inorganic fertilizer significantly influenced the nitrogen and potassium content of leaf tissues at tillering stage of sugarcane. It was observed (Table 5) that the treatment T_4 showed the highest concentration of nitrogen (1.41%) but the second highest concentration was in T_3 treatments (1.37%). There were significant differences among the treatments and all treatments showed higher concentration of N over control (1.04%). In respect of phosphorus, concentration varied significantly among the treatments and the highest phosphorus (0.028%) concentration showed in T_4. The potassium concentration in leaf varied significantly

Table 5 Sugarcane leaf nutrient content in response to vermicompost and fertilizer treatments.

Treatment	*Nutrient Concentration in Laf (%)*			
	N	*P*	*K*	*S*
T_1	1.04f	0.015e	1.00e	0.153f
T_2	1.31c	0.026b	1.05ab	0.205c
T_3	1.37b	0.026b	1.14b	0.208b
T_4	1.41a	0.028a	1.17a	0.212a
T_5	1.26d	0.023c	1.11c	0.188 d
T_6	1.30c	0.021d	1.11c	0.190d
T_7	1.18e	0.020d	1.07d	0.170 e
T_8	1.20e	0.020d	1.07d	0.172e
$LSD_{(0.05)}$	0.0117	6.009	0.0133	1.000

Means under each column separated by same letter do not differ significantly at 5% level.

among the treatments and the T_4 showed the highest concentration of potassium (1.17%). Sulphur concentration in leaves varied significantly among the treatments receiving different levels of vermicompost and inorganic fertilizers doses but they were higher than in control. Bokhtiar et al. (1998) reported that the N, P, and K content in sugarcane leaves varied inconsistently but showed a good indication of response to applied fertilizers.

Soil Fertility and Carbon Storage

The soils under investigation were low in total N, OC, available P and S, exchangeable K was medium, and soil pH was slightly acidic (Table 1). After application of vermicompost with chemical fertilizer, there was some indication of slight increase in OC, total N, available P and S, and exchangeable K, but a slight decrease in post-harvest soil pH (Table 6). Application of vermicompost with chemical fertilizer in post-harvest soils analyses indicated that soil pH varied from 6.3–6.4. Decreased soil pH observed in the soil treated with vermicompost might be due to the acidifying effects of organic acids produced during the course of decomposition of organic amendments. Brady (1988) reports that soil pH range between 6–7 seems to promote the availability of nutrients to the plants. The available nitrogen also increased due to application of different doses of vermicompost. This could be attributed to direct incorporation of organic matter that led to better root growth, more plant residue addition, and greater microbial activity which resulted in higher mineralization leading to higher N. This result corroborated with the findings of Thakur et al. (2012). The addition of vermicompost also had a positive effect on post-harvest soil P status over control. Increase in available P may be attributed to the decomposition of organic matter, besides organic matter forms a protective cover on sesquioxides, and this facilitates reduction in the phosphate fixing capacity of the soil (Tandon, 1987). Application of different levels of vermicompost with fertilizer increased K in the post-harvest soil. This may be attributed to the direct addition of K to the available pool of the soil. Beneficial effect of vermicompost or fertilizer may be due to the additional K supplied through them and it may be attributed to the reduction in

Table 6 Post-harvest soil properties as affected by vermicompost and fertilizer treatments.

Treatment	*Soil Properties*					
	pH	*OC (%)*	*N (%)*	*Avail. P (mg/kg)*	*Ex. K (cmol (+)/kg)*	*Avai. S (mg/kg)*
T_1	6.50a	0.55e	0.060e	6.08f	0.12b	8.50d
T_2	6.50a	0.72d	0.081d	7.80de	0.14b	11.10bcd
T_3	6.30c	0.91bc	0.088b	8.93b	0.12b	12.30abc
T_4	6.30c	0.99a	0.098a	9.27a	0.18a	13.30ab
T_5	6.30c	0.93abc	0.096a	7.90cd	0.13b	10.40cd
T_6	6.30c	0.97ab	0.092b	8.06c	0.14b	14.86a
T_7	6.40b	0.95ab	0.090b	7.62e	0.13b	10.30cd
T_8	6.30c	0.87c	0.085c	7.69de	0.13b	10.20cd
$LSD_{(0.05)}$	0.0378	0.0285	1.667	0.1035	0.0167	1.3333

Means under each column separated by same letter do not differ significantly at 5% level.

fixations of the released K due to interaction of organic matter with clay, beside the direct K addition to the available K pool of soil (Kumar and Prasad, 2008). The total nitrogen and available sulfur status were increased due to the application different doses of vermicompost with inorganic fertilizers.

Soil organic carbon content in the post-harvest soil was influenced by the application of different doses of vermicompost with inorganic fertilizer. Due to addition of vermicompost, the percent of soil organic carbon content ranged from 0.87–0.99% due to different levels of vermicompost application with inorganic fertilizers. The organic carbon accumulation increased to the tune of 58–80 per cent over control (T_1: 0.55%) with addition of vermicompost. This might be due to direct incorporation of organic matter as vermicompost which led to batter root growth and more plant residue in the soil. Singh et al. (2007) also observed substantial increase in soil organic carbon content due to addition of organic manure. They found 68.7% increase in organic carbon due to addition vermicompost @ 10 t ha^{-1}.

Conclusions

The results of this study suggest that the balance fertilization in soils, the application of vermicompost with inorganic fertilizes is extremely important for maximizing productivity of sugarcane and maintaining fertility status of the soils. Therefore, it appeared that the application of 5 ton vermicompost with inorganic fertilizers 165 kg N, 55 kg P, 120 kg K, 45 kg S, 20 kg Mg, 2.5 kg Zn per ha would be best combination at Young Brahmaputra and Jamuna Floodplain soils of Bangladesh for obtaining highest sugarcane yield as well as carbon accumulation in soil for maintained soil fertility.

References

Alam, K.M., Bokhtiar, S.M. and Paul, G.C. (2016). Nutrient requirement for sustainable sugarcane production. *The Planter, Kuala Lumpur*, *92*(1078): 17–24.

Black, C.A. (1965). Methods of soil analysis. *In*: Part 2: *Chemical and Microbiological Properties*. Agronomy Series No. 9, Madison Wisconsin, USA. American Society of Agronomy and American Society for Testing and Materials.

Bokhtiar, S.M., Roksana, S. and Moslehuddin, A.Z.M. (2015). Soil fertility and productivity of sugarcane influenced by enriched pressmud compost with chemical fertilizers. *SAARC Journal of Agriculture*, *13*(2): 183–97.

Brady, N.C. (1988). *The Nature and Properties of Soil*. Eurasia Publishing House Ltd., New Delhi.

Chandini, Randeep, K., Ravendra, K. and Om, P. (2019). The impact of chemical fertilizers on our environment and ecosystem. Chapter 5, 71–86. *In*: *Research Trends in Environmental Sciences*, 2nd Edn.

Haynes, R.J. (1986). The decomposition process mineralization, immobilization, humus formation and degradation.: Haynes, R.J. (Ed.), *Mineral Nitrogen in the Plant-Soil System*. Academic Press, New York.

Ismail, S.A. (2005). *The Earthworm Book*, 101 pp. Other India Press, Mapusa, Goa.

Khan, I.M., Khatri, A., Nizamani, G.H., Siddiqui, M.A., Raja, S. and Dahar, N.A. (2005). Effect of NPK fertilizer on the growth of sugarcane clone AEC 86–347 developed at NIA, Tendo Jam, Pakistan. *Pakistan Journal Biology*, *37* (2): 355–60.

Kumar, R., Pal, S. and Singh, N.B. (2006). Yield, uptake, and soil fertility in rice-wheat cropping system under integrated nutrient management. *Annals of Plant Soil Research*, *8*: 39–41.

Manivannan, S., Balamurugan, 1.M., Parthasarathi, 1.K., Gunasekaran, 1.G., Ranganathan, L.S. (2009). Effect of vermicompost on soil fertility and crop productivity: Beans (*Phaseolus vulgaris*). *Journal of Environmental Biology*, *30*(2): 275–81.

Parthasarathi, K., Ranganathan, L.S., Anandi, V. and Zeyer, J. (2007). Diversity of microflora in the gut and casts of tropical composting earthworms reared on different substrates. *Journal of Environ. Biology*, *28*: 87–97.

Perucci, P. (1990). Effect of the addition of municipal solid-waste compost on microbial biomass and enzyme activities in soil. *Biology and Fertility of Soils*, *10*: 221.

Sison, M.L.Q., Magbunua, J.P.V., Bugante, E.C., Torres, F.G., Nayve, F.R.P., Jr. Migo, V.P. and Fernandez, W.L. (2010). Agro-recycling of distillery effluent as liquid organic fertilizer for sugarcane. *Philippines Journal of Crop Science*, *35*(3): 1–8.

Tandon, H.L.S, (1987). *Phosphorus Research and Agricultural Production in India*, 160 pp. Fertilizer Development and Consultation Organization, 1987: Crops.

Thakur, S.K., Alam, M. and Umesh, U.N. (2007). Long-term effect of integrated nutrient management on productivity and sustainability of sugarcane of calciorthent. *Indian Journal of Sugarcane Technology*, *22*(1&2): 9–13.

Thakur, S.K., Jha, C.K., Alam, M. and Singh, V.P. (2012). Productivity, quality, and soil fertility of sugarcane (*saccharum* spp. complex hybrid) plant and ratoon grown under organic and conventional farming system. *Indian Journal of Agricultural Sciences*, *82*: 896–99.

Thakur, S.K., Kumari, G., Alam, M. and Singh, V.P. (2013). Improving soil fertility and yield and quality of sugarcane through integrated application of NPK and compost in calcareous soils in Bihar. *Indian Journal of Sugarcane Technology*, *27*(2): 12–15.

Umesh, U.N., Kumar, Vipin, Alam, M., Sinha, S.K. and Verma, K. (2013). Integrated effect of organic and in organic fertilizers on yield quality parameter and nutrient availability of sugarcane in calcareous soil. *Sugar Tech.*, *4*(15): 365–69.

Chapter 15

Integrated Nutrient Management on Sustainable Yield and Soil Fertility Under Maize–Mungbean–Aman Rice Cropping Systems in Bangladesh

M. Rokeya Khatun[1*], *Md. Shafiqul Islam*[2], *A.J.M. Sirajul Karim*[3], *A.R.M. Solaiman*[4] and *M.A. Haque*[5]

Introduction

Rice-maize cropping systems are distributed all over the South Asia but more particularly in Bangladesh. Production of hybrid maize increased due to increase in area as well as yield per hectare in all countries. Intensive cultivation of crops using high-yielding varieties removes more plant nutrients from the soils, severely depleting soil fertility. (Saleque et al., 2004; Yu et al., 2014). Use of imbalanced fertilizers causes nutrient loss from soil through mining, lessen soil fertility, and land productivity with the consequent decline in food production. A crop production system with high yield target cannot be sustainable unless nutrient inputs to soil are at least balanced against nutrient removal by crops (Rijpma and Jahiruddin, 2004). Available data indicate that the soil fertility in Bangladesh is in declining trend (Ali et al., 1997). Organic source of nutrients applied to preceding crop can benefit the succeeding crop (Hedge, 1998). The Integrated Nutrient Management (INM) is the maintenance of soil fertility for sustaining increased crop productivity through optimum use of organic and inorganic plant nutrients required for crop and appropriate to cropping system (Roy, 1986).

Mono crop-based fertilizer recommendations are preferred to the poor farmers. On the other hand, rich farmers use high doses of chemical fertilizers, specially urea

[1] Bangladesh Agricultural Research Institute, Gazipur, Bangladesh.
[2-4] Bangabandhu Sheikh Mujibur Rahman Agricultural University, Gazipur, Bangladesh.
[5] Bangladesh Agricultural Research Institute, Gazipur, Bangladesh
* Corresponding author: rokeyashimul@yahoo.com

which creates imbalance in soil nutrients. The nutrients added to the soil in the form of fertilizers are not being removed or utilized by the crops in one season. Some amounts are left over in the field. So proper fertilizer management is very important considering the residual effect of the nutrients. Pattern based crop cultivation including pulse crops is proved to be very effective on sustaining crop yield and maintaining soil health. Therefore, inclusion of a pulse crop in the cropping pattern would reduce the requirement of chemical fertilizers in the next crop maintaining a good soil health through biological nitrogen fixation and addition of organic matter to the soil (Singh et al., 2008; Stagnari et al., 2017; Urmi et al., 2022). Legume crops could play an important role in lowering the risk of climate change and meeting food security with sustainability principles.

The physical, chemical, and biological characteristics of soil are improved by the addition of organic manures to the soil. This can lead to the growth of larger and healthier root systems which help plants better handle stress (Piper and Pieter, 2010). The chemical properties of a soil can be improved by increasing available nutrients through organic manuring into the soil. The biological characteristics of a soil, such as biomass and biological activity, can also be improved through organic matter and through addition of pulse crop in a cropping system. These favorable changes in the soil's biology provide the short-term economic incentive by the use of pulse crops in maize-T. Aman cropping systems. Soil fertility monitoring under specific cropping system would be of great help in determining a better soil fertility management practice for sustained productivity at high level. For this reason, the present study has proposed to carry out with the objectives to (a) find out suitable nutrient combinations for maize-mungbean-T. Aman rice cropping pattern for sustainable yield, (b) improve soil fertility, (c) estimate nutrient uptake by the crops, and (d) increase sustainable yield of the crops.

Materials and Methods

The field experiment was conducted in the Grey Terrace Soil (AEZ-28) at Gazipur during 2011–2012 and 2012–2013. The initial soil samples collected from the depth of 0–15 cm were analyzed in the laboratory following standard methods. Initial values of some important chemical parameters of soil of the experimental field are presented in Table 1.

Table 1 Chemical properties of experimental soil (initial) at the Central Research Farm BARI, Gazipur during 2011–2012.

pH	*OM (%)*	*Ca*	*Mg*	*K*	*Total N %*	*P*	*S*	*Zn*	*B*
		meq/100 g soil				*mg g^{-1}*			
5.7	1.36	8.1	2.1	0.11	0.051	14	13	2.12	0.31
	Critical level	2.0	0.8	0.2	–	14	14	2.0	0.20

The experiment was laid out in randomized complete block design with four replications. Six different treatments, viz., T_1: Native fertility (control), T_2: Soil Test Based (STB) chemical fertilizers for moderate yield goal (MYG), T_3: MYG-STB chemical fertilizer + 5 tons cowdung (CD) ha^{-1}, T_4: STB chemical fertilizer for high yield goal (HYG), T_5: HYG-STB chemical fertilizer + 5 tons CD ha^{-1}, and T_6: HYG-STB chemical fertilizer + 3 tons poultry manure (PM) ha^{-1}. The fertilizer treatments used in the experiment were based on Fertilizer Recommendation Guide (BARC, 2005). Nutrient combinations of N-P-K-S-Mg-B for MYG were @ 200-30-65-20-10-1 kg ha^{-1} and for HYG were @ 240-40-80-25-15-1.2 kg ha^{-1} in maize. In T. Aman rice, N-P-K-S-Mg for MYG were @ 115-6-60-10-0 kg ha^{-1} and for HYG were @ 135-8-75-13-2 kg ha^{-1}. Chemical analysis data of poultry manure and cowdung used in the experimental field are presented in Table 2.

The unit plot size was 4 m × 5 m. The tested crop for maize was BARI hybrid maize-5. Maize seeds were sown on 15 November 2011 and 19 November 2012 in line-to-line distances 75 cm and seed to seed distances 20 cm. Two seeds were placed in each hole at a depth of 2.5 cm and thinned out to one when the seedlings were established properly. Seed rate of maize was @ 25 kg ha^{-1}. Seeds were treated with vitavex prior.

Table 2 Nutrient status of poultry manure and cowdung used in the experimental field.

Name of the Manure	*pH*	*OC*	*Ca*	*Mg*	*K*	*Total N*	*P*	*S*	*B*	*Zn*
		%								
Poultry manure PM)	8.1	18.7	13.0	2.80	0.87	2.13	2.55	0.90	0.011	0.11
Cow dung (CD)	7.5	12.4	1.50	0.44	0.53	1.15	1.32	0.38	0.013	0.15

Fertilizer N-P-K-S-Mg-B were supplied from urea, TSP, MP, gypsum, magnesium sulfate, and boric acid, respectively. All P-K-S-Mg-B and 1/3 of N were applied at the time of final land preparation. The remaining two-third of N were applied as top dress at 30 and 60 days after sowing of maize seed. Three irrigation and other intercultural operations were done as and when required. Harvesting of maize crop was done in April 2012 and 2013. From each plot central three undisturbed lines were harvested for grain and stover yields assessment. Maize grain yield was expressed at 14% moisture content.

A legume crop and the variety BARI mung-6 was cultivated after maize harvest. Only *Rhizobium* biofertilizer @ 1.5 kg ha^{-1} was applied in mungbean cultivation instead of urea. Mungbean seeds mixing properly by *Rhizobium* inoculum was sown on 12 May 2012 and 2 May 2013. Healthy seeds of mungbean @ 35 kg ha^{-1} were sown by hand as uniformly as possible in furrows. Line to line distance was 30 cm. In the year 2013, the cultivation of mungbean faced difficulties to establish the crop after maize due to occasional heavy rainfall resulting in low germination of seeds and consequently the yield was lower than the targeted yield. Thus, it was helpful for incorporation of brown biomass of mungbean into the soil. Mungbean crop was harvested on 12 July 2012 and 28 June 2013, respectively. After 2/3 picking of the pods the brown biomass was ploughed down into the soil before T. Aman rice transplantation in both the years. T. Aman rice crop and the variety BINAdhan-7 were cultivated after mungbean harvest. T. Aman seeds were sown by broadcasting on the prepared seedling nursery.

Twenty-five-days-old seedlings were transplanted in the experimental plots on 21 July 2012 and 30 July 2013. A distance of 25 cm from row to row and 15 cm

study, it may be mentioned that the study of Shahiduzzaman (1997) who reported that organic manures increased available B content in soil compared to the application of only inorganic fertilizer. The results stated above are in agreement with the findings of Hossain et al. (2008). These results are well accorded with that of Saha (2007).

Partial Budget Analysis

The results of economic analysis are shown in Tables 9a, and 9b. The analysis was done in order to find out the most profitable treatment based on costs and benefits of various treatments following the procedure of Perrin et al. (1979). In this analysis, variable costs have only been considered and fixed costs have been ignored. Variable costs included only fertilizer cost and variable opportunity means labour cost. Gross margin was calculated by subtracting the total variable cost from the gross return. Variable cost means fertilizer cost (the purchasing price of fertilizer and cowdung). Variable opportunity cost included the amount of money paid for carrying and broadcasting the fertilizers and cowdung. Gross margin was calculated as the total market value of grain yield of maize, mungbean, and T. Aman rice.

The results of partial budget analysis of the crops (maize, mungbean, T. Aman) taken together in the cropping pattern. Table 9a demonstrated that the highest gross margin (Tk. 236241) was obtained in T_6 which was followed by Tk. 224420, Tk. 224266, and Tk. 220939 noted by T_4, T_2, and T_5 treatments, respectively. The gross margin of Tk. 215892 was found in T_3 treatment. The farmers will preferably choose any of these treatments which can give a higher net benefit or gross margin with a lower variable cost, and which can sustain soil fertility.

Another attempt has been made to find out the marginal rate of return (MRR) against the undominated treatments (Table 9b). The MRR was calculated by dividing the marginal gross margin (MGM) by the marginal variable cost (MVC). For calculation of marginal benefit and marginal cost, the undominated treatments were arranged in the descending order of their gross margin. Normally the farmers would accept the technology when the MRR becomes at least 40% (Perrin et al. 1979). Under this criterion all the four undominated treatments except T_4 and T_2. Again, treatments T_3, T_5, and T_6 in this study have the acceptable range of MRRs. The highest MRR (68.9%) obtained in T_3 treatment followed by 51.0% in T_5 and 42.20% in T_6 treatments i.e., the MRR of the treatments was found to follow the sequence $T_3 > T_5 > T_6$. Thus, the farmers may be advised to go for treatments T_3, T_5, or T_6 which improved soil fertility and produced higher crop yield.

Conclusions

The two-year experiments on the maize-mungbean-T. Aman cropping pattern in Gazipur, Bangladesh, concluded that the T6 treatment (soil test-based chemical fertilizer for high yield goal (HYG) + 3 t ha^{-1} poultry manure) exhibited the highest yield and productivity, effectively increasing nutrient contents and uptake by crops. Treatment T5 (STB chemical fertilizer for HYG + 5 t ha^{-1} cow dung) yielded statistically similar results to T6. The best nutrient application combination for the pattern in Grey Terrace soil was $N_{240}P_{40}K_{80}S_{25}Mg_{15}B_{1.2}$ kg ha^{-1} with 3 t PM ha^{-1}

Table 9a Partial budget analysis of maize, mungbean and T. Aman rice in maize-mungbean T. Aman rice cropping sequence.

Treatments	*Average Grain Yield ($t\ ha^{-1}$)*			*Gross Return ($tk\ ha^{-1}$)*			*Total Gross Return ($tk\ ha^{-1}$)*	*Total Variable Cost ($tk\ ha^{-1}$)*	*Gross Margin ($tk\ ha^{-1}$)*
	Maize	*Mung.*	*Aman*	*Maize*	*Mung.*	*Aman*			
T1	2.54	0.62	4.47	40560	43691	80370	164621	0	164621
T2	6.77	0.89	5.53	108320	62125	99540	269985	45719	224266
T3	7.15	0.92	6.22	114400	64342	111870	290612	74719	215892
T4	7.13	0.91	5.69	114080	63525	102420	280025	55605	224420
T5	7.54	0.99	6.43	120640	69253	115650	305543	84605	220939
T6	8.01	1.04	6.57	128080	72596	118170	318846	83605	236241

Notes: Urea = Tk. 16 kg^{-1}, TSP = Tk. 25 kg^{-1}, MOP = Tk. 15 kg^{-1}, Gypsum = Tk. 15 kg^{-1}, $MgSO_4$ = Tk. 60 kg^{-1}, Boric Acid = Tk. 280 kg^{-1}, Cowdung = Tk. 2 kg^{-1}, Poultry manure = Tk. 4 kg^{-1}, Maize grain = Tk. 16 kg^{-1}, Mungbean = Tk. 70 kg^{-1}, Aman rice = Tk. 18 kg^{-1}.

Table 9b Marginal analysis of the undominated treatments.

Gross Margin	*Treatments*	*Total Variable Cost ($tk\ ha^{-1}$)*	*Change from Next Highest Benefit*		*MRR (%) (A/B)*
			MGM (A) ($tk\ ha^{-1}$)	*MVC (B) ($tk\ ha^{-1}$)*	
236241	T6	83605	11821	28000	0.422
224420	T4	55605	155	9885	0.016
224266	T2	45719	3327	–38885	–0.086
220939	T5	84605	5046	9885	0.510
215892	T3	74719	51272	74719	0.686
164621	T1	0	–	–	–

for maize, only *Rhizobium* inoculum biofertilizer @ 1.5 kg ha^{-1} for mungbean, and $N_{120}P_{25}K_{45}S_{15}Zn_4$ kg ha^{-1} with 3 t ha^{-1} PM for T. Aman (T6 treatment). Incorporating mungbean residues into the soil contributed to organic matter addition, nutrient recycling, and availability for subsequent crops. Mungbean residue incorporation reduced 14 kg N ha^{-1} for the subsequent T. Aman rice cultivation. Organic manuring increased total nitrogen, available phosphorus, sulfur, boron, exchangeable potassium, and magnesium contents in post-harvest soil compared to initial soil. The nutrient status in post-harvest soil was further improved by incorporating HYG-STB chemical fertilizers with poultry manure and cow dung. Economically, T6 treatment had the highest gross margin (Tk. 236,241 ha^{-1}), while T3 treatment had the highest marginal rate of return (68.6%). For poor farmers, T3 (STB chemical fertilizers for moderate yield goal with cow dung) might be a viable option due to its high MRR. However, wealthier farmers willing to invest more in fertilizers could opt for T6 or T5 to increase gross margin profitability while sustaining soil fertility and achieving higher crop yields. Ultimately, T6 and T5 are recommended for farmers considering soil health, increased crop yield, and their positive impact on climate change and food security in the country.

Acknowledgements

The authors are highly grateful to the authority of Bangladesh Agricultural Research Institute (BARI) for providing PhD scholarship and also heartfelt thanks to Dr. Md. Azizul Haque, chief scientific officer, soil science division, Bangladesh Agricultural Research Institute, Gazipur, Bangladesh for providing research field and laboratory facilities. I am thankful to my major Prof. Dr. Md. Shafiqul Islam, Bangabandhu Sheikh Mujibur Rahman Agricultural University, Gazipur, Bangladesh who encouraged me for completing my PhD thesis and also highly grateful to Prof. Dr. A.H.M. Mustafizur Rahman, University of Dhaka, Bangladesh for views and opinions expressed in this article.

References

Ali, M.M., S.M. Shaheed and D. Kubota. (1997). Soil degradation during the period of 1967–1995 in Bangladesh II. Selected chemical characters. *Soil Sci. Plant Nurt., 43*: 879–90.

Anonymous. (2006a). *Annual Report*, 34 pp. Soil Science Division, BARI, Joydebpur, Gazipur, Bangladesh.

Anonymous. (2006b). Improvement of soil fertility through integrated fertilizer management in maize-T. Aman rice cropping pattern. *Annual Research Report*, 20–40. Bangladesh Agricultural Research Institute, Gazipur-1701.

Anowar, M.M., M.Z. Ferdous and Mozidul Islam. (2012). Determination of nutrient management for potato-mungbean-T. Aman rice cropping pattern. *Bangladesh J. Prog. Sci. Tech., 10*(2): 173–76.

Baig, M.B., M.S. Ziaeldin and R.L. Mahler. (2004). Rehabilitation of problem soils through environmentally friendly technologies. *Arab Gulf J. Sci. Res., 22*(2): 51–59.

BARC (Bangladesh Agricultural Research Council). (2005). *Fertilizer Recommendation Guide 2005*, 25–128. BARC, Soils Pub. No. 14. 2007. Farmgate, Dhaka.

Hedge, D.M. (1998). Long-term sustainability of productivity rice (*Oryza sativa*)-wheat (*Triticum aestivum*) system in sub-humid ecosystem through integrated nutrient supply. *Ind. J. Agric. Sci., 68*: 144–48.

Hossain, M.A., M. Jahiruddin, M.R. Islam and M.H. Miah. (2008). The requirement of zinc for improvement of crop yield and mineral nutrition in the maize-mungbean-rice system. *Plant and Soil, 306*: 13–22.

Islam, M.S., Solaiman, A.R.M., Karim, A.J.M.S., Rahman K.M.M. and Haque, M.M. (2011). Effect of organic and inorganic phosphorus on the yield and nutrient use efficiency of T. Aman rice. *J. Sher-e-Bangla Agric. Univ.*, *5*(1): 1–11.

Katyal, V., Gangwar, K.S. and Singh, K.P. (2002). Changes in soil productivity and soil fertility under rice-wheat system as influenced by long-term integrated nutrient supply. *Indian J. Agron.*, *47*(4): 465–72.

Kumar, A., Thakur, K.S. and Sharma, S. (2005). Integrated nutrient management in maize (*Zea mays*)-Rabi season (*Brassica napa*) cropping system under rainfed condition. *Indian J. Agron.*, *50*(4): 274–77.

Mehla, D.S., Singh, J.P., Sekhon, K.S., Sihag, D. and Bhardwaj, K.K. (2008). Long-term effects of inorganic and organic inputs on yield and soil fertility in the rice-wheat cropping system in India. http:// www. Idd. go.th/18wcss/tech program/p12114.HTM.

Pathak, S.K., Singh, S.B. and Singh, S.N. (2002). Effect of integrated nutrient management on growth, yield and economics in maize (*Zea mays*)-wheat (*Triticum aestivum)* cropping system. *Indian J. Agron.*, *47*(3): 325–32.

Pathak, S.K., Singh, S.B., Jha, R.N. and Sharma, R.P. 2005. Effect of nutrient management on nutrient uptake and changes in soil fertility in maize (Zea *mays)*-wheat (*Triticum aestivum)* cropping system. *Indian J. Agron.*, *50*(2): 269–73.

Perrin, R.K., WinKelman, D.L., Moscardi, E.R. and Anderson, J.R. (1979). *Economic Training Manual.* Information Bull. No. 27. CIMMYT, Mexico.

Piper, C.V. and Pieter, A.J. (2010). *USDA Farmers Bulletin*, 1250–95. Green Manuring (Ed.).

Rahman, M.S., Islam, M.R., Rahman, M.M. and Hossain, M.I. (2009). Effect of cowdung, poultry manure, and urea-N on the yield and nutrient uptake of BRRI dhan 29. *Bangladesh Res. Publications J.*, *2*: 552–58.

Rahman M.H., Islam, M.R., Jahiruddin, M. Haque, M. (2012). Management of organic manure and inorganic fertilizer in the maize-mungbean/dhaincha-T. Aman rice cropping pattern for increased crop production. *Bangladesh J. Agric. Res.*, *37*(2): 225–34.

Rahman M.H., Islam, M.R., Jahiruddin, M., Rafii, M.Y., Hanafi, M.M. and Malek, M.A. (2013). Integrated nutrient management in maize-legume-rice cropping pattern and its impact on soil fertility. *Journal of Food, Agriculture, and Environment*,*11*(1): 648–52.

Rijpma J. and Jahiruddin, M. (2004). *Strategy and Plan for Use of Soil Nutrient Balance in Bangladesh.* Final Report of Short-term. Assignment. SFFP/DANIDA.

Roy, R.N. (1986). FAO, fertilizer programme with special reference to integrated plant nutrition system. *Fertilizer in Agricultural Development*, 149–69. Krishak Bharati Cooperative Limited, New Delhi, India.

Saha, P.K. (2007). Long-term integrated nutrient management for rice-based cropping pattern: Effect on growth, yield, nutrient uptake, nutrient balance sheet, and soil fertility. *Commun. Soil Sci. Plant Annal.*, *24*: 1679–89.

Saha, M. and Mondal, S.S. (2006). Influence of integrated plant nutrient supply on growth, productivity and quality of baby corn (*Zea mays*) in Indo-Gangetic plains. *Indian J. Agron.*, *51*(3): 202–205.

Saha P.K., Rahman, F., Akter, M., Islam, R., Hossain, A.T.M.S. and Ali, M.G. (2016). Integrated Nutrient Management for Potato-Maize-T. Aman Rice Cropping Pattern. *Bangladesh Rice J.*, *20* (1): 51–58.

Saleque, M.A., Abedin, M.J., Bhuiyan, N.I., Saman, S.K. and Panaullah, G.M. (2004). Long-Term Effects of Inorganic and Organic Fertilizer Sources on Yield and Nutrient Accumulation of Low Land Rice. *Field Crops Res.*, *86*: 53– 65.

Satyanarayana, V., Prasad, P.V.V., Murthy, V.R.K. and Boote, K.J. 2002. Influence of integrated use of farmyard manure and inorganic fertilizers on yield and yield components of irrigated low land rice. *J. Plant Nutr.*, *25*(110): 2081–90.

Sengar, S.S., Wade, L.J., Bagnel, S.S., Singh, R.K. and Singh, G. (2000). Effect of nutrient management on rice *(Oryza sativa)* in rainfed low land of southwest Madhya Pradesh. *Indian J. Agron.*, *45*(2): 315–22.

Shahiduzzaman, M. (1997). *Effect of Inorganic and Organic Sources of Nitrogen and Potassium on BGA Population and on Growth and Yield of BR11 Rice.* MS thesis. Dept. of Soil. Science. (Janu.–June Sem.): 51.

Singh, K.K., Srinivasarao, Ch. and Ali, M. (2008). Phosphorous and Mung Bean Residue Incorporation Improve Soil Fertility and Crop Productivity in Sorghum and Mungbean-Lentil Cropping System. *J. Plant Nutri.*, 31: 459–471.

Stagnari, F., Maggio, A., Galieni, A. and Pisante, M. (2017). Multiple benefits of legumes for agriculture sustainability: an overview. *Chem. Biol. Technol. Agric.*, 4: 2.

Steel, R.G.D. and Torrie, J.H. (1980). *Principles and Procedures of Statistics. A Biometrical Approach*, 2nd Edn. McGraw-Hill Book Company, New York.

Urmi, T.A., Rahman, M., Islam, M., Islam, A., Jahan, N.A., Mia, A.B., Akhter, S., Siddiqui, M.H. and Kalaji, H.M. (2022). Integrated Nutrient Management for Rice Yield, Soil Fertility, and Carbon Sequestration. *Plants*, *2022* (11): 138.

Vanja, M. and Raju, A.S. (2002). Integrated nutrient management performance in rice crop. *Abbakas Agric. Res.*, *2391*: 177–82.

Yu, Y., L. Xue, Yang (2014). Winter legumes in rice crop rotations reduces nitrogen loss, and improves rice yield and soil nitrogen supply. *Agron. Sustain. Dev.*, *34*: 633–40.

Chapter 16

Addressing Food Security in Bangladesh Associated with Climate Change

Rakhi R. Sarker[1], *M.M.R. Jahangir*[2*] and *Rafiq Islam*[3]

Introduction

As one of the most climate-vulnerable countries in South Asia, Bangladesh and its fragile ecosystems (Haors, Chars, Coasts, Hills, and Barind, which cover 9.24 M ha area or 64% of its total area), faces great threats to its agriculture and food security associated with climate change.

Bangladesh had one of the highest rates of acute food insecurity in 2020–2022 among more than 50 countries (FAO, 2022). The population of Bangladesh was 169.4 million in 2021 (BBS, 2021), and increasing by approximately 2 million every year. At this rate, the population of Bangladesh would be around 200 million people by 2030, meaning rice production would need to double in order to meet food security demands. The expected rate of growth of the world's population will necessitate a tremendous increase agricultural production (Timmusk, 2017). Bangladesh is losing its tillable lands and productive capacity due to soil organic matter (SOM) depletion. Approximately 25 million people in Bangladesh are food unsecured, with approximately 11 million of them experiencing acute hunger as a result of the pandemic and other circumstances (WFP, 2021). Annually 69,000 ha of land (0.73%) is lost from agricultural production due to urbanization, human settlement, infrastructure, and riverbank erosion. About 108 million people living in 48 countries were experiencing acute food insecurity in 2016 (FAO, 2018).

Food security depends on the availability of food, access to food, and utilization of food. Climate change affects agricultural production, food quality, and increases food insecurity due to various climate hazards – floods, droughts, soil and water salinity,

1 Bangladesh Institute of Nuclear Agriculture, Mymensingh, Bangladesh.
2 Bangladesh Agricultural University, Mymensingh, Bangladesh.
3 The Ohio State University, Columbus, USA.
* Corresponding author: mmrjahangir@bau.edu.bd

cyclones, etc. in the country. The food security situation is worsening due to the global food crisis, high national inflation rates, increasing food prices for several key commodities, and high fuel and chemical fertilizer prices (WFO, 2023). Agriculture and climate change influence one another through the emission of greenhouse gasses (GHGs). The national goal of water and food security, poverty reduction will be threatened due to climate change extremities. As soil and water are sources of life, we have to promote sustainable technologies to conserve soil resources to build a food-secure world. Climate resilient and sustainable agriculture management should be adopted in order to achieve universal access to food security by 2030, as outlined in Sustainable Development Goals 2 (zero hunger) and 6 (water and sanitation), which are crucial for the nation's food and water security, and for present and future generations.

Food Security Challenges in Bangladesh

Bangladesh is one of the most vulnerable countries to climate change, and so far has experienced 191 climate-related extreme events between 1999 and 2018, according to the Global Climate Risk Index 2020 (Eckstein et al., 2020). It is the most densely populated country in the world and this demographic burden places tremendous economic, social, and environmental strains on the country and affects its ability to provide food and services for its citizens (Fig. 1).

The country has achieved self-sufficiency in rice production since 2000, and has become surplus from 2005 by 370,000 Mt to 2021 by 7,500,000 Mt. Rice production grew from 12 Mt in 1973 to 37 Mt in 2021 (MoEFCC, 2018). According to estimates by the Bangladesh Agricultural Research Council (BARC), Bangladesh will have a surplus on rice, potato, vegetables, and fruits, but a food deficit on wheat, maize, pulses, and edible oils. About 12 Mt rice in the late 1970s has increased to 37,600,000 Mt in 2021 due to the adoption of high yielding varieties and good management practices (BBS, 2022).

Over the years, substantial economic losses have occurred in the agriculture sector due to climate change effects (Aryal et al., 2020). The average global temperature is growing continuously and is expected to rise by 2°C by 2100. Frequency of severe climatic events such as floods, droughts, salinity, heat waves, and storms are most likely to be observed more frequently in southern and eastern Asian countries (Malhi et al., 2021). Rising temperatures are projected to result in an annual gross domestic product (GDP) loss of around 2% by 2050 and 9% by 2100 without adaptation of climate change management policies (MoEFCC, 2018). During 2016–2021, 1.3% of annual GDP loss occurred due to climate change-induced disasters (BBS, 2017, 2021).

Asian agriculture is currently facing serious challenges due to climate change, although there are low emissions of GHGs in the region (Rahman et al., 2022). South Asia consists of 262 M malnourished inhabitants, which has made it the most food insecure region across the world. Around 25 million people in Bangladesh are food insecure, of which about 11 M of them are experiencing severe hunger as a result of the pandemic (WFP, 2021).

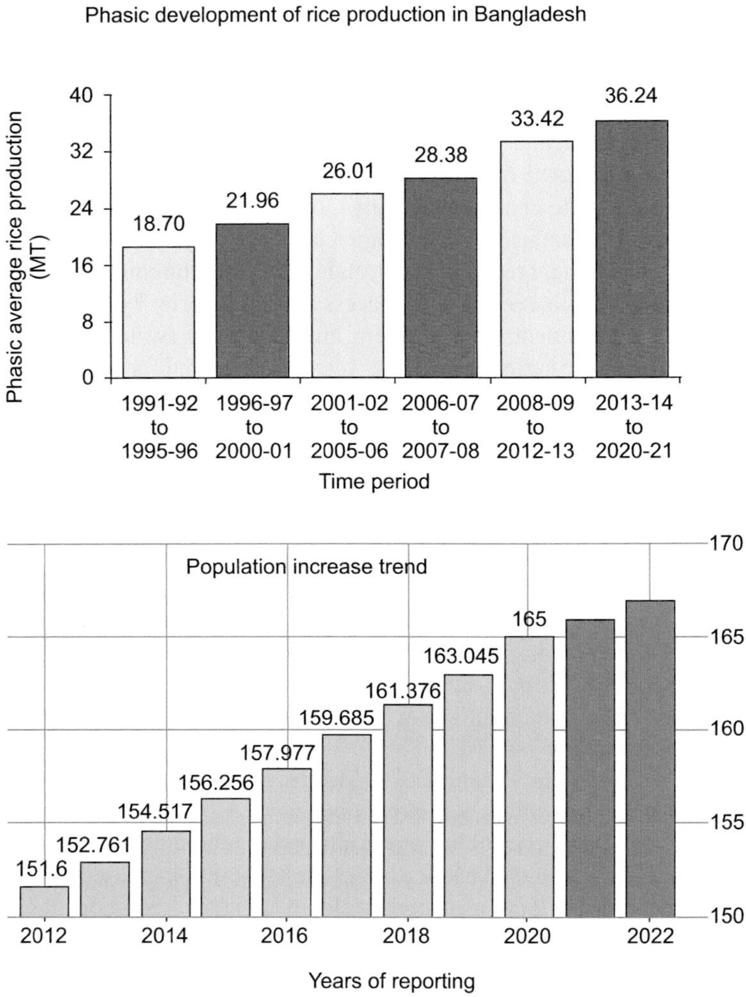

Fig. 1 Comparative display of rice production and population increase (*Source:* BBS, 2021; *Bangladesh Agriculture Under Climate Change: Future Perspectives* by M. Jahiruddin).

Effect of Agriculture on Climate Change

Agriculture influences climate change through the effects of emissions of GHGs, which are significant contributors to anthropogenic global warming. The global food system is responsible for 21–37% of annual emissions of GHGs (Mbow et al., 2019). While CH_4 and N_2O emissions from agriculture are less, they could still play a significant role in climate change mitigation due to their increased global warming potentials (Lynch et al., 2021). Plants consume a large amount of CO_2 through the process of photosynthesis, but most of the plant products are converted back to CO_2 during decomposition. Considering the total pool cycling in the system, the net emissions of CO_2 are small in agriculture.

South Asia has the largest share of CH_4 emissions. Major sources of CH_4 emissions include the livestock sector and wetland rice production, while the application of manures and fertilizers comprise the prime source of N_2O emissions. From rice cultivation only, South Asia emits almost 4,900 Gg CH_4 per year, which is the third-highest level of CH_4 emissions from rice cultivation globally (IRRI, 2018). The annual GHG emissions from agricultural production, from 2000–2010, were estimated at 5.0–5.8 Gt CO_2e yr^{-1} (IPCC, 2014).

Agriculture contributes to 39% of the country's GHG emissions, from which rice cultivation accounts for 39%, synthetic fertilizer 7%, crop residue 5%, cultivation of organic soil 4%, and burning crop residue 1% emissions. Cropland and livestock production contributed equally to agricultural GHGs in 2013 (Shelley et al., 2016). Compared to other sectors whose emissions increased by almost 500% between 1990–2013, agricultural emissions in Bangladesh have remained more stable, increasing by about 30% during that same time period (WRI, 2013). Nevertheless, the agriculture and livestock sector remain the main source of emissions contributing to climate change effects.

Conventional puddle transplanted rice is a main source of CH_4 emissions. The emissions of N_2O from agriculture have increased over time due to the effects of excessive chemical fertilization in South Asia. The use of groundwater irrigation could increase energy consumption from the agriculture sector and hence substantially increase GHG emissions (Aryal et al., 2020).

Effect of Climate Change on Agriculture

Agricultural growth and development are key to food security in Bangladesh. Global climate change causes significant losses to agriculture and crop production by influencing climate-induced disasters, coastal tropical cyclones, monsoon flooding, flash floods, droughts, sea level rise, salinity, urban floods, etc. Therefore, agriculture production is directly dependent on climate change. The current rate of annual loss to GDP is about 1.3% because of climate-induced disasters, which may rise to 2% by 2050 and to over 9% by 2100 under extreme scenarios (MoEFCC, 2018). Climate change may cause a substantial loss in agricultural crop production due to poor soil quality, salinization, excessive rainfall or drought, water quality degradation, and pest and weed infestations (Gupta et al., 2020). The Bangladesh Delta Plan (GED, 2018) has shown the possible effects of climate change on overall rice production, expecting a decreasing rate of 17% by 2050, which would impede food and nutrition security.

Salinization

Soil and water salinity are major environmental concerns throughout the world and have adverse effects on food security. Bangladesh is highly susceptible to secondary salinization of soil-water ecosystems due to its geographic location and nature of the delta. The coastal region of Bangladesh covers about 20% of the country and more than 30% of the cultivable land (SRDI, 2010). Coastal and offshore land area is estimated at 2.86 M ha of which about 1.06 M ha are affected by different degrees of soil salinity (MoEFCC, 2022). Salinity intrusion in the southwest region is responsible for decreased agricultural production in the coastal zone. Climate change-induced sea level rise has significantly increased river water salinity during the dry season (Fig. 2). From 1973–2009, land affected by salinity in Bangladesh increased by about 26.7%. Around 7 and 13 M people will be at risk in the coastal region by 2025 and 2050, respectively (World Bank, 2021). It is expected that around 39% of salinity will increase across the coastal belt in Bangladesh by 2050. High levels of soil and water salinity are expected to have adverse effects on crops, fish, and livestock across the coastal belt in Bangladesh. It is reported that salinity is responsible for a 20% yield reduction of coastal region agriculture (Alam et al., 2016).

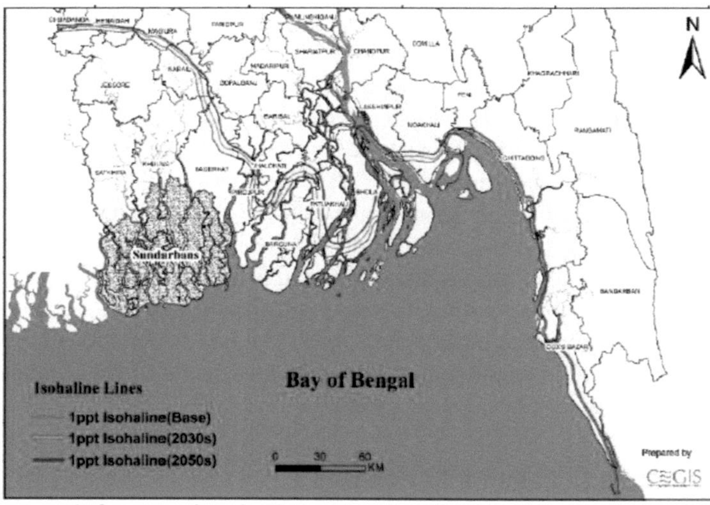

Fig. 2 Surface water salinity distribution in coastal Bangladesh due to climate change (*Source:* CEGIS Bay of Bengal Model, 2021).

Acidification

Soil acidity is the main concern in the Bengal Delta of Bangladesh and India for crop production due to climate change. Acidic soils' distribution in Bangladesh is about 36% of the land coverage, while it is 57% of the world's land coverage (Islam et al., 2021). These soils are low in organic matter and nutrients and need to be ameliorated to regenerate and/or maintain soil health for increased crop production to ensure food security. Soil acidity affects plant growth by influencing soil microbial diversity and efficiency, nutrient leaching, and heavy metals toxicity.

Currently, acid soils are a concern in Bangladesh because of their negative impact on soil fertility and agricultural productivity. Acid soils, especially acid sulfate soils, acid basin clays, and terraced and eroded hill soils are all moderately to extremely acidic in nature. The primary causes of soil acidity are the loss of basic cations, increased solubility of iron and aluminum, and the continuous and excessive usage of urea fertilizers. Continuous and unbalanced use of chemical fertilizers in acid soils causes acute nutrient deficiencies and impairs soil physicochemical characteristics by increasing acidity, resulting in rapid degradation of soil health, crop productivity, ecological stability, and environmental sustainability (Rahman et al., 2022).

The major problems of acid soils are Fe, Al, and Mn toxicity; Mo, Ca, and Mg deficiency; legume nodule failures; and increased plant diseases. Soil acidity reduces the availability of plant nutrients like P and Mo and increases the availability of heavy metals, particularly Fe, Al, and Mn. According to the Bangladesh Bureau of Statistics (BBS), the current rate of agricultural land use intensification was 194% by 2020. About 30% of the land is managed for crop production because of acid soils (Ahmed et al., 2018). The soil acidity is increasing over time, with a 13% increase in very highly acid and strongly acid soils in Bangladesh between 1988–2010 (SRDI, 2010). The majority of Bangladesh's acid sulfate soils are currently fallow, with the exception of a few that are being used for single or double cropping of flooded rice, as well as shrimp and salt production, depending on the topography of the land, irrigation facilities, drainage, and acidity of the soils. Khan et al. (2014) reported that in Bangladesh's coastal zone, about 700,000 ha of acid sulfate soils have been identified as problematic soils.

Desertification

The average temperature in Bangladesh has been rising sharply in the last three decades compared to the previous three decades. Because of extreme and prolonged drought and high temperatures, desertification symptoms have been observed in the northwestern region of Bangladesh. Drought conditions become extremely severe in April and May with less rainfall, extreme summer heat of more than 40°C, and low soil fertility, leading to land degradation and desertification.

The dry-regions of Bangladesh located along the western border are most susceptible to meteorological droughts in pre- and post-monsoon periods. Mean annual rainfall in the dry-zone is around 1,250–1,750 mm from May–June to September–October (Ahmed and Suphachalasai, 2014). Due to the combined effects of low soil moisture-holding capacity, there were an increasing number of dry days in those regions. There were 19 drought events in Bangladesh between 1960–1991. Bangladesh, over time, experienced severe droughts in 1951, 1957, 1961, 1972, 1976, 1979, 1989, and 1997.

The groundwater table has decreased by at least 10 m in the Barind tract of the northwest region of Bangladesh during the past 14 yr. (Ali et al., 2021). The contribution of groundwater has increased from 41% in 1982–1983 to 79% in 2016–2017 (BBS, 2017). On average, 2.32 (Kharif seasons) and 1.2 M ha (Rabi season) of agricultural lands per yr. are affected during a typical drought event (CEGIS, 2013). According to BDP (2100) studies, Aus rice production would decline by 27% under a moderate climate change scenario while wheat production would fall to 61%. Under a severe climate change scenario, dry-season Boro rice yield might decrease by 55–62%. In the coming years, soil moisture stress might force farmers to reduce Boro cultivation.

Temperature and Drought

Agricultural productivity in Asia is facing alarming challenges due to climate change. The global mean temperature is expected to increase by 1.5°C between 2030–2050, if it continues to increase at the current rate predicted by various climatic models (IPCC,

2019). At the same time, there will be a 10 cm rise in sea level by 2100 across the globe, including Bangladesh, due to climatic abnormalities (IPCC, 2019).

Global warming will pose a severe threat to the world's food security, but if it is limited to 1.5°C, 76% of the developing countries' vulnerability will be reduced compared to the same regions at 2°C (Malhi et al., 2021). The average temperature in Bangladesh has been rising sharply during the last three decades compared to previous three decades (Fig. 3). In recent decades, the mean temperature has been rising very rapidly. From 1991 to 2000, the mean temperature increased by 0.39°C, which further increased by 0.53°C during 2001–2010 and 1.06°C during 2011–2019 (CEGIS, 2021).

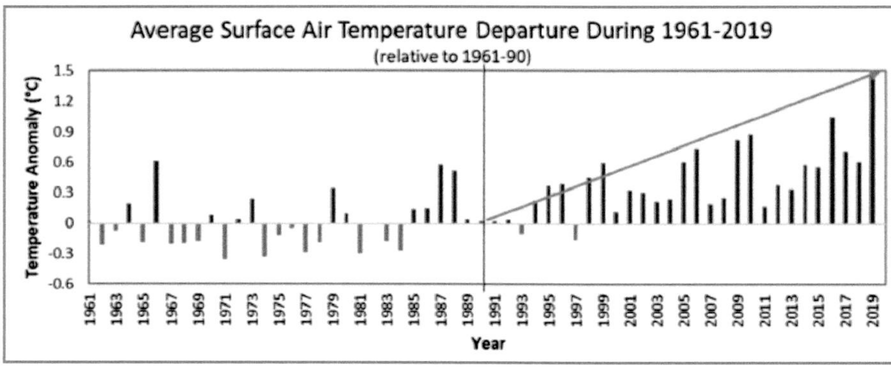

Fig. 3 Departure of average annual temperature in Bangladesh relative to climate normal of 1960–1990
(*Source:* CEGIS analysis based on BMD data).

The rice-wheat rotation, a major cropping system that fills half of the food demand in Asia, is under threat due to climate change. The protein content and grain yield of wheat have reduced because of the adverse impacts of increasing temperature (Asseng et al., 2019). Drought and high temperatures are key abiotic stresses impacting cereal yields. Similarly, if the temperature rises about 30°C during floret development, it can cause sterility in cereals. It was reported that wheat yield decreased by 10%, maize decreased by 8.3%, and about 3–4% reduction in wheat yield takes place with every 1°C rise in temperature (Brown and Plan, 2008).

Flooding

In 2020, super cyclone 'Amphan' and flash floods in the northern and Haor regions of Bangladesh caused substantial damage to the local population and economy. The northeastern areas of Bangladesh are more prone to flash flooding than other areas of the country. Flash floods inundate crops near the harvesting time, damage infrastructure, and often cause loss of lives and properties (CEGIS, 2021).

Flash floods are caused by heavy rainfall or upstream flooding in a short period of time. Flash floods are most common from April–July and from September–October. Boro rice is the only major crop in the northeastern areas of Bangladesh. Almost 80% of the area is covered by dry-season rice from January–May, producing 18% of Bangladesh's total rice production (BHWDB, 2012). Flash floods suddenly inundate crops near harvesting time, damage infrastructure, and often cause loss of lives and properties. The flash flood events of 2017 and 2022 were the most devastating, disturbing roads and embankments and damaging pre-mature dry season crops, which posed a threat to food security (UN RC Bangladesh, 2022). The Eastern hill regions are also extremely prone to flash floods. From 1985–2015, 12 flash flood events occurred in the region. These events caused substantial damage to the local population and economy. Future climate change will increase rainfall in the pre-monsoon and monsoon seasons in the northeastern areas, resulting in more chances of flash floods.

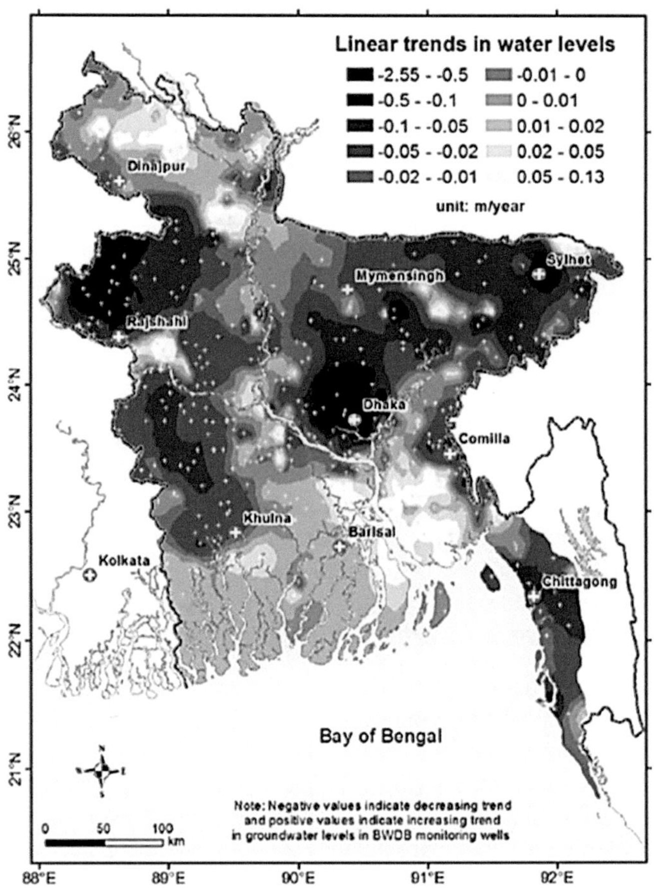

Fig. 4 Long-term (1985–2005) trends in shallow groundwater levels throughout Bangladesh (Shamsudduha et al., 2009). Groundwater levels are declining over dark areas but stable to rising over light-gray to white areas (https://www.researchgate.net/profile/Mohammad).

Ground Water Level Decline

Currently, two-thirds of the world's population is experiencing water scarcity at least one month out of the year (Mekonnen and Hoekstra, 2016). Though water scarcity is a global problem, this is the most crucial challenge for the northwest region of Bangladesh due to climate change. Climate change has affected the groundwater recharge, water cycle, soil moisture availability, aquatic species, and deteriorated water quality for irrigating crops (Gupta et al., 2020). In Bangladesh, groundwater withdrawal has increased beyond its capacity to recharge since the 1980s. Currently, it supplies 79% of the water for irrigation, livestock, and household and industrial usages (BBS, 2017). Bangladesh ranks No. 6 in the world with the largest estimated annual groundwater extraction, as per the UN World Water Development Report of 2022. The groundwater situation in the northwestern region of Bangladesh is considered more crucial than other regions of the country due to drought, less and uneven rainfall distribution, agricultural intensification, and accelerated urbanization (Fig. 4).

It is estimated that 1.2 billion people do not have clean water to drink, and 2.6 billion people do not have water for their basic sanitation needs (Kurniawan et al., 2012). Agriculture accounts for 72% of global freshwater withdrawals, but like all natural resources, freshwater is not infinite. Groundwater has become the main source of irrigation in South Asia for rice cultivation, leading Bangladesh to be the third largest rice-producing country globally. The groundwater table in few regions of the Barind tract has decreased by at least 10 m during the past 14 years (Ali et al., 2021). The World Health Organization reported that almost 1.7 M people have died due to water pollution and 4 billion cases of different health issues are reported annually due to water-borne diseases (Briggs et al., 2016).

The depth to the groundwater table could be almost 1.5 times deeper by 2050 and twice as much by 2075 (Ali et al., 2023). Over extraction of groundwater for irrigation, less rainfall, fewer sources of surface water, higher temperatures, and heavy clay soil that limit the natural recharge are the major causes of continuous depletion of groundwater table in the Barind Tract (Hossain et al., 2021). Excess volumes of groundwater are being extracted for irrigation in this region. Less rainfall, dry climate, clayey red soil, and relatively high temperature are exceptionally significant compared to other regions of the country (BMDA, 2018).

Land Degradation Due to Riverbank and Hill Erosion

Land degradation occurs due to nutrient and organic matter loss, soil acidification, drought and lowering of the groundwater table, soil sealing and pollution, water logging and soil salinization, ecosystem vulnerability, flash floods, riverbank erosion, deforestation, and other causes (Fig. 5). The frequency and severity of riverbank erosion is alarming in Bangladesh. Currently, average erosion in Bangladesh is 3,000 ha per year, displacing approximately 25,000 people annually. Due to climate change, erosion might escalate further in the near-term as high-intensity rainfall and peak flow in the major rivers will increase and carry more sediment from upstream catchments. Riverbank erosion is the most common natural disaster in Bangladesh and causes a yearly loss of 8,700 ha of homestead lands, displacing almost 200,000 people (Rahman and Rahman, 2015). It is responsible for food shortages in households, and 73.3% of the people lost their income sources due to the disaster.

Land degradation is a serious concern in which there is a decline in land quality due to human-induced activities. Deforestation, one of the causes associated with land degradation, leads to flooding and soil erosion. Deforestation also leads to

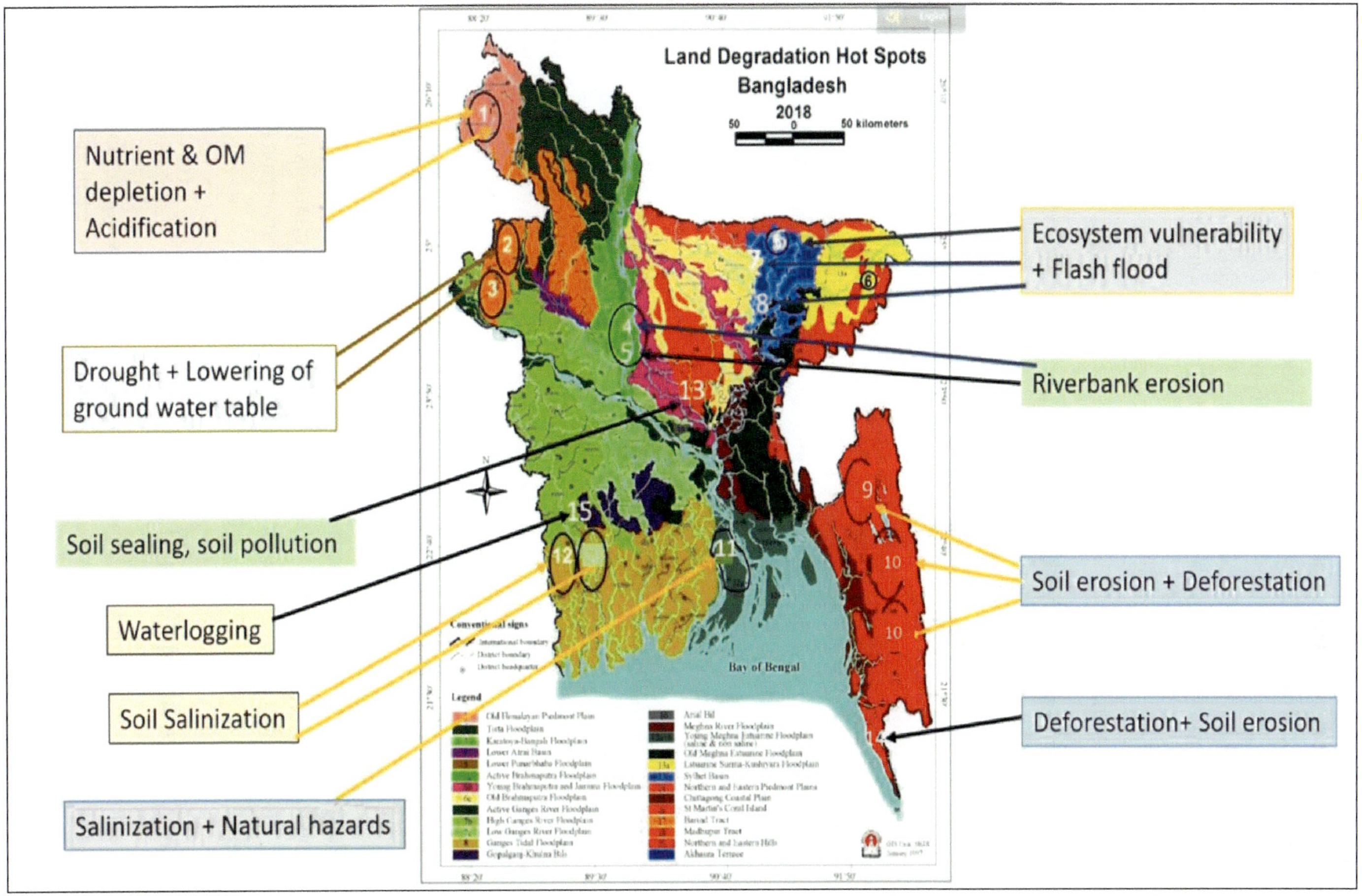

Fig. 5 Land degradation hot spots in Bangladesh (Source: Under the DS-SLM project, SRDI conducted the baseline study of land degradation processes from the year 1985 to 2000 (https://encrypted-tbn0.gstatic.com/images?q=tbn:ANd9GcQRd7_cHa6geCH8Qzt8ij_wIpfIzVHQjrGCSUxmvri9HQ&s).

soil erosion by wind and desertification of land due to decreases in annual and seasonal precipitation (Gupta, 2019). According to FAO estimates, about 52% of agricultural land worldwide, amounting to more than 2 billion ha, is moderately or severely degraded, adversely affecting 1.5 billion people with a disproportionately adverse impact on women, children, and the rural poor (FAO, Land and Water, 2018). Another form of land degradation is due to the deposition of silt or infertile soil on agricultural land by breach of embankment during floods. This phenomenon is the result of deforestation of the hills and slash-and-burn cultivation practice in the upper catchment areas.

Soil Nutrient Depletion

Management of soil nutrients is a key factor for the productivity of agricultural lands. Nutrient and irrigation management were the main sources of differences in potential yield gaps (Mueller et al., 2012). This dimension of food security is directly related to agricultural management and availability of land to support crop production. The extensive production of cereals and conventional soil management practices are increasing soil erosion; depleting soil organic matter, soil fertility, water resources, and increasing salinity, micronutrient deficiencies, the degrading soil structure, and global warming are serious sustainability issues in South Asia (Nawaz et al., 2021). The nutrient deficiency of the country's arable soil is acute. It is estimated that the overall nitrogen balance of Bangladesh soils is negative, phosphorus balance is close to zero, and potassium balance is negative. The same situation exists for sulfur, zinc, boron, and other organic matter, as well as pH status (SRDI, 2010).

Human Nutrition Depletion and Deficiency

Micronutrient deficiencies in people are most common where grains have low nutritional contents in South Asian countries. Micronutrient deficiency is one of the leading contributors to global disease burden, affecting approximately 50% of the world's population. Low levels of soil Zn, Cu, and Mn, in particular, have been linked to increased child mortality (Berkhout et al., 2019). A sharp increase in water-borne diseases, births of children with disabilities, skin diseases, and diarrhea occur in coastal areas and during floods which results in a shortage of foods. Pre-eclampsia and gestational hypertension rise during the dry season when salinity of water have increased (Khan et al., 2014). Heatstroke and vector-borne diseases will spread with sudden temperature rise as well. The number of internal climate migrants may reach 19.9 M by 2050, comprising half of those in the entire South Asian region (Clement et al., 2021).

Climate Change Adaptation and Mitigation

Use of abiotic and biotic stress tolerance crops and varieties, growing high water-use efficient crops, adaptation of climate-smart agriculture, shifting of planting time, conservation tillage, and water-saving practices need to be adopted. Mitigating GHG emissions is possible either by CO_2 sequestration via plant photosynthesis or the reduction of CH_4 and N_2O emissions from wetland rice cultivation. Agriculture offers nature-based options to remove CO_2 from the atmosphere, through carbon sequestration in biomass and soils. Direct (agriculture) emission reductions would represent 29% of this total, soil carbon sequestration 9%, and reduction of deforestation and other land use change emissions 62% (OECD, 2022). According to the recent report of the Intergovernmental Panel on Climate Change (IPCC), the current level of emissions of GHG needs to be reduced to limit the temperature rise to below 2°C (UNFCCC, 2015).

Soil Health Improvement

Maintaining and/or improving soil health is a precondition for sustainable agriculture and food security. Soil microbes are considered as one of the management-induced sensitive and early soil quality indicators to support and provide agroecosystem services (Sun et al., 2016). In conservation agricultural (CA) systems, reduced or no-till tillage along with cropping diversity with cover crops and residue management play key roles in determining soil biodiversity, especially fungal dominance, as residues serve as a steady source of food, energy, and nutrients for microbes, and retention of these crop residues on the soil surface increases their abundance by providing suitable habitats for their growth and survival (Islam et al., 2021). Climate change affects the microbial population and their enzymatic activities in soil (Malhi et al., 2021). Due to excessive use of chemical fertilizers, beneficial microorganisms in the soil are rapidly declining. With regard to the rejuvenation of soil quality, conservation agriculture-based cropping with legumes improved soil biodiversity and increased soil health. Reduced tillage with crop residue management increased soil microbial populations with higher fungal-to-bacterial ratios (Sarker et al., 2023a). Tillage disturbance is responsible for the largest impact on microbial ecology compared to fertilizer management practices to influence climate change (Srour et al., 2020).

In CA-based management, crop residue management provides a greater amount of labile carbon and nutrients under an undisturbed environment favorable for microbial growth and higher C-use efficiency (Islam et al., 2021). Soil microbial communities and their species composition were influenced by reduced tillage combined with residue retention by improving soil moisture availability, diversity in C and nutrients pools, and altering the soil temperature and aeration (Amoakwah et al., 2021). Diversification of cropping systems and the adoption of conservation agriculture may enhance the sequestration of soil carbon and increase biodiversity (Nawaz et al., 2021). Recent studies have shown that conservation tillage protects life cycles of arthropods, increasing their diversity and playing a major role in shaping microbial communities (Kraut-Cohen et al., 2020). It protected and stimulated the soil biological functions and significantly increased microbial abundances, and helped bind and retain nutrients in soil.

Conservation Agriculture

Conservation agriculture is an eco-friendly farming system that enriches soil health and decreases the cost of food production (Chatterjee et al., 2022). Globally, CA is an important management strategy to address diverse challenges of sustainable food production and climate change adaptation and mitigation. With CA, it becomes feasible to make a substantial contribution to food security and implement a primary approach for mitigating climate change through soil carbon sequestration (Yimam et al., 2022). While CA practices increase the N-use efficiency and N fertilizer management, it inevitably led to decreased emissions of N_2O (Sapkota et al., 2014). Due to crop diversification and minimum tillage practices, CA increases SOC stocks, hence climate change mitigation occurs through the transfer of C from atmosphere to the land via plant photosynthesis. It is reported that each 1 kg of N fertilizer is responsible for an emission of 11.7 kg CO_2e, equivalent to a change of 3.2 kg SOC (Smith et al., 2012). Crop residue management is the other major component of CA, and thus, it reduces the burning of crop residues and minimizes air pollution. Cover crops and fallow rotation suits warm and moist climates the best, annually mitigating 0.98-ton CO_2e ha^{-1} (Aryal et al., 2020). With CA, there is a potential to mitigate GHG emission, either by the sequestration of CO_2 or the reduction in CH_4 and N_2O emissions.

Tillage and crop residue management: Intensive tillage disturbance on SOM decomposition results in SOC loss, thereby increasing greenhouse gas emissions. The CA system with minimum tillage reduced CO_2 emissions. Annual plowing degraded and dispersed soil macro-aggregates, which reduced the labile C pool (Jahangir et al., 2021). Undisturbed soils in a CA system may have harbored a diversified microbial community to increase SOC sequestration. Crop residue retention and the addition of bio-inoculants help to maintain soil C and N balance and enrich the labile C pool in rice-legume based cropping systems (Thakuria et al., 2009).

Soil can be a source or sink of tropospheric CO_2, depending on land use and agricultural practices. SOC is not only a vital source of C for soil functional processes, but also a sink for SOC sequestration. While plowing can decrease SOC content, lead to soil quality deterioration, and affect soil productivity, by contrast, the adaptation of a CA can decrease the adverse effects on crop production via climate change mitigation and adaptation (Islam et al., 2022). Reduced tillage with residue management is beneficial for sustainable agriculture by increasing labile C and SOC stocks under rice-based cropping systems in the South Asia. Higher C management index in a CA system indicates higher SOC sequestration potential (Sarker et al., 2023b).

Crop diversification: The crop diversification in South Asia was influenced by per capita gross domestic product, minimum temperature, pesticide consumption, food crop yield index, and non-food crop yield index. It is essential to achieve higher farm income, create jobs, protect soil and water resources, consumer preferences for high-value and nutrient-enriched food and feed, rural livelihood, sustainable use of natural resources, and poverty alleviation (Singh et al., 2022). To achieve sustainable food production, the cultivation of cereals, citrus, fiber, fruits, roots, and vegetables increased significantly in Bangladesh from 1991–2020; however, the cultivation of oilseeds, pulses, and sugar crops decreased significantly.

A significantly higher SOC content observed in the diversified legume-based cropping system compared to the rice-based systems was due to the effects of N recycling from leguminous crop residues. The contribution of crop diversification to agricultural growth has been declining over time; however, this can be used as a warning sign for countries to frame appropriate agriculture policies for making agriculture more diversified in the coming decades for sustaining food and nutritional security.

Adaptation and Mitigation Using Organic Amendments

While Bangladesh has a large number of problem soils, diverse organic amendments (OA) have been shown to enhance soil health for higher crop output via improving soil quality properties and processes. Resource conservation technologies comprising soil and crop management practices (biochar, organic fertilizer, poultry manure, vermicompost, etc.) can improve soil resilience and increase its productive capacity through sustainable intensification and diversification in the nutrient poor and problem soils.

The Integrated Plant Nutrient System (IPNS), which combines chemical and organic fertilizer, is a well-accepted technology for sustaining soil health and supplying crops with their nutrient demands (Jahangir et al., 2021). Frequent chemical fertilization causes serious soil-plant-water issues, such as food contamination, nutrient shortages, and physiochemical property degradation by increasing soil acidity, all of which lead to the accelerate degradation of soil health and productivity, and both stability and sustainability. The degraded soils can be regenerated via OA inputs to increase the SOM contents and nutrients. Biochar, as a soil conditioner, can be utilized to sequester SOC, as it is a stable form of waste-recycled carbon that remains in soil for a long time. The influence of biochar addition on crop productivity is visible in sandy and degraded soils as compared to highly-fertility soils (El-Naggar et al., 2019).

Biochar with vermicompost often boosted microbial biomass and soil enzymatic activities. It is reported that biochar with vermicompost increased organic manure's nutritional status, CEC, microbial biomass and associated activities, and enzymatic activities with SOC accumulation (Kumar et al., 2020). The poultry manure biochar (PMB) and compost-based IPNS had the highest impact on the carbon lability index of MBC in acidic soil and the carbon management index of MBC in diverse soils, indicating increased SOC stocks. The PMB and compost-based IPNS can be recommended as the best management practices in degraded acidic and estuarine alluvial soils (Char land), especially in terms of soil health improvement or restoration (Rahman et al., 2022).

Water Management

Conventionally puddled transplanted rice growing practice is a major source of methane and nitrous oxide emissions. Likewise, conventional flood irrigation causes greater methane emissions than the intermittently irrigated and drained soils. The IPCC (2006) Guidelines for National GHG Inventories claims a 48% decrease in CH_4 emissions from soils under alternate wetting and drying practices, whereas on-farm experiments on alternate wetting-drying shows 20–70% reduction in CH_4 emissions (Richards and Sander, 2014). Besides reduced CH_4 emissions, alternate wetting-drying also supports rice production possible when water resources are scarce. Rice production with alternate wetting-drying of water management saves 25–30% of water compared to the conventional flooding system (Richards and Sander, 2014). Gathala et al. (2013) show that adopting alternate wetting-drying of soil can save

30% water used for irrigation in puddled transplanted rice, thereby reducing GHG emissions due to reduced energy used for pumping irrigation. Alternate wetting-drying also decreases irrigation input in rice fields by up to 38% without any declines in rice yield (Lampayan et al., 2015). Therefore, alternate wetting-drying serves the dual purpose of mitigating GHG emissions by reducing CH_4 and adapting to water stress (Aryal et al., 2020).

Adaptation Technology, Local Practices, and Stress Tolerant Crop Varieties

Changes in management practices such as tillage, irrigation, and fertilization can significantly reduce GHG emissions from agricultural lands. Furthermore, modification of traditional agroforestry systems and improved livestock management in the livestock-crop mixed-system also reduce GHG emissions without reducing agricultural production (Aryal et al., 2020). Promotion of climate resilient crops (drought and salinity tolerant, lowland and short duration cultivars) and adoption of climate-smart technologies and sustainable soil, water, and nutrient management practices can significantly reduce GHG emissions from croplands. Alternate wetting-drying, construction of a mini-pond in the crop field, sustainable groundwater and irrigation management, and growing of high water-use efficient crops in drought-prone areas, converting to natural rainfall-matching cropping pattern (Aus rice), and changing the monoculture of rice-based cropping patterns need to be adopted for achieving food security.

The utilization of stress-tolerant crop varieties is anticipated to be well-suited for cultivation in the delicate ecosystems of Bangladesh. Noteworthy varieties include those resilient to salinity (BRRI dhan-53, BRRI dhan-54, and BRRI dhan-73 for Aman; BRRI dhan-47, BRRI dhan-67, BRRI dhan-97, and BRRI dhan-99 for Boro; Binadhan-8, Binadhan-10, and Binadhan-23); submergence-tolerant strains (BRRI dhan51, BRRI dhan52, BRRI dhan97 for Aman; BINAdhan-11, BINAdhan-12, and BINAdhan-23 for Aman); and drought-tolerant varieties, such as rice (BRRI dhan-56, BRRI dhan-57, BRRI dhan-66, and BRRI dhan-71 for Aman; Binadhan-19 and Binadhan-21 for Aus), wheat (BARI Gom-33, BWMRI Gom-1, 2), and chickpea (BARI Chola-5, BinaChola-6).

To address the specific challenges posed by climate change, location-specific technologies must be embraced. Strategies include adopting alternate wetting-drying practices for the Barind tract and northwest drought-prone areas, constructing mini-ponds for crop fields in the north and northwest, cultivating less water-dependent crops in districts like Rajshahi, Rangpur, and Dinajpur in the north and northwest, employing the Sarjan method for cultivation in southern regions and salinity-prone areas, implementing zero or minimum tillage (Conservation Agriculture-CA) in flood-prone areas, and utilizing floating gardens in flooded or waterlogged areas of Bangladesh.

Rainwater harvesting along with water-saving technologies and integrated water resources management need to be prioritized (MoEFCC, 2022). Mulch can be used in drought- and saline-prone areas. Artificial recharge to the aquifers need special attention in areas where groundwater levels drop below the critical limit of water pumps. Alternate technology for pumping groundwater needs to be made available to meet household water requirements (Mojid et al., 2019). Mulching can be used to reduce evaporation, thereby retaining soil moisture in drought-prone areas. In saline soil, mulching restricts the capillary rise of saline water by reducing evaporation from surface soil (MoEFCC, 2022). Floating agriculture (locally known as vasoman/dhap chash) is practiced in the low-lying areas of southern region (Pirojpur, Gopalganj, and Barisal districts) to cope with the flooded/submerged conditions in Bangladesh.

Agroforestry

Agroforestry contributes to climate-change mitigation by storing carbon above ground in the form of biomass and below ground in the form of carbon in SOM. Soil organic carbon can be enhanced by planting trees and grasses in degraded lands (Kar et al., 2009). The mitigation strategies used include afforestation and reforestation programs. Coastal afforestation programs have been undertaken to stabilize the coastline and create greenbelts by engaging community people via food for program activities.

Tree coverage once was only 7%, and has since been increased to 22%. The national strategy to restore all degraded forests and bring all newly accreted coastal lands under afforestation can increase the tree cover of Bangladesh from its current 22% to 25% by 2030. Increasing national forest and tree coverage can be a holistic strategy for adaptation, reducing emissions, and enhancing forest carbon stocks. Bangladesh completed its first National Forest & Tree Inventory in 2019 and assessed a carbon sequestration capacity of 1275.5 M tons (MoEFCC, 2022).

Mechanization

Agricultural mechanization plays a vital role in improving farm production to feed the growing population. To address the looming global threat of food scarcity, there is a forecast to increase grain production by 40% by 2020. Achieving this target primarily hinges on enhancing crop yields and minimizing losses through judicious mechanization. Climate-smart farming technologies can contribute to meet the increasing food demand and sustainable farm production, which helps for achieving Sustainable Development Goals (SDGs) in the country. Agricultural production can be increased by promoting climate-smart farming technologies, which will lead to sustainable mechanized farming to achieve SDG (MOA, 2016). Currently, agriculture in Bangladesh is facing a labor shortage problem due to rapid urbanization, which causes migration of labor in urban areas for seeking better employment opportunities. At the same time, daily agricultural wages are getting higher. During the peak season, agricultural daily wages increased by 3.5 times in the last 10 years (MoEFCC, 2022).

Flash floods often inundated crops nearing harvesting time in the Haor region of Bangladesh, damaging pre-mature Boro rice, which posed a threat to the overall food security of the country. Recently, combined harvesters were used by the farmers for paddy harvesting on custom hiring due to labor shortage and timely harvest (Mottaleb et al., 2016). Rice transplanters, paddy threshers, maize shellers, and reapers are becoming popularized with farmers for reducing operational costs and minimizing crop loss and harvesting time. The demand for agricultural machinery is increasing for tillage, irrigation, and postharvest operations. It has been well-accepted that mechanization creates an employment opportunity in rural areas by diversifying the business opportunities, especially development of skilled manpower, marketing, and supply stores at the village level. It helps in poverty reduction and better livelihood of the rural poor people.

Global Interests

Bangladesh is a country of agrarian economy, and agriculture is the single largest production sector of the economy, as it comprises about 12.7% of the country's GDP (BBS 2022). In terms of production status in global ranking, Bangladesh is No. 2 in jackfruit; No. 3 in rice, vegetables, and onion; No. 7 in potato and mango; No. 8 in guava; and No. 3 in aquaculture fish production. However, the current rate of annual loss to GDP of approximately 1.3% due to climate-induced disasters may rise to 2% by 2050 and over 9% by 2100 under extreme scenarios (MoEFCC, 2022). Bangladesh has become a lower middle-income country and is set to upgrade from least development country status by 2026, which is putting extra pressure on available funds from domestic public and private sources.

The Bangladesh Delta Plan indicates the possible impact on overall rice production, predicting a declining rate of 17% by 2050, which would impede food and nutrition security (GED, 2018). Asia contributes to more than 10% of GDP in most regional economies and provides jobs for over one-third of the working population in the region, and a large majority of them are dependent on agriculture. Consequently, agricultural development has important implications for the reduction of poverty in Southeast Asia.

National Policies for Food Security in Bangladesh

Multiple policy frameworks have been established to address Food Security and Agricultural Development in Bangladesh, notably within the Seventh Five Year Plan (2016–2020). In pursuit of SDGs, the Bangladesh Government crafted an SDG Action Plan in 2017, aiming to realign national development efforts. Other key initiatives include the National Adaptation Program of Action (NAPA, 2005), the Bangladesh Climate Change Strategy and Action Plan (BCCSAP, 2009), the Bangladesh Delta Plan 2100, that integrates effective climate change measures. Notably, Bangladesh has recently devised a strategic Delta Plan 2100.

Conclusions

Global climate change, agriculture, soil productivity, and food security are interconnected concepts. Among them, food security is becoming more challenging due to climate-induced disasters in the South Asian region. Bangladesh climate change is considered one of the most serious threats to the national economy and sustainable development due to overexploitation of natural resources, climate change effects, and increasing demand for population boom. Climatic change is a critical challenge for South Asia, especially Bangladesh, for increasing food demand for an increasing population. Climate vulnerability has an adverse effect on soil-plant-water ecosystems. Sustainable agricultural practices need to manage environmental footprints and are directly concerned with attaining food security and promote sustainable agriculture, which is one of the goals of SDGs. Ecologically novel and holistic approaches are needed for sustainable land management and economic crop production to address food security for an increasing population. Development and adoption of climate-smart technologies in different agricultural production systems should be adopted to achieve food and nutrition security. Policy formulation and implementation of crop adaptation and mitigation measures are important to cope with the climate change effects and ensure food and nutrition security.

References

Alam, M.Z., Carpenter-Boggs, L., Rahman, A., et al. (2016). Water quality and resident perceptions of declining ecosystem services at Shitalakka wetland in Narayanganj city. *Sustain Water Quality Ecol.*, *9–10*, 53–56.

Ali, M.H., Zaman, M.H., Islam, M.A., Biswas, P., Karim, N.N. and Kader, M.A. (2021). Recent Trend of Precipitation and Crop Planning in Rajshahi Region of Bangladesh. *Asian J. Advance Agricul. Res.*, *16*: 28–39.

Ali, M.H., Zaman, M.H., Biswas, P., Islam, M.A., Islam, M.T. and Kader, M.A. (2023). Water-Table Dynamics, Its Trend and Sustainability Considerations at Two Upazila of North-Western Bangladesh. *Global Res. Environ. Sustain.*, *1*: 1–9.

Ahmed, M. and Suphachalasai, S. (2014). *Assessing the Costs of Climate Change and Adaptation in South Asia*. © Asian Development Bank.

Ahmmed, et al. (2018). *Fertilizer Recommendation Guide-2018*. Bangladesh Agricultural Research Council (BARC), Farmgate, Dhaka-1215.

Amoakwah, E., Arthur, E., Frimpong, K.A. and Islam, K.R. (2021). Biochar amendment influences tropical soil carbon and nitrogen lability. *J. Soil Sci. Plant Nutri.*, *21*: 3567–79.

Aryal, J.P., Rahut, D.B., Sapkota, T.B., Khurana, R. and Arun Khatri-Chhetri. (2020). Climate change mitigation options among farmers in South Asia. *Environ. Develop. Sustain.*, *22*: 3267–89.

Asseng, S., Martre, P., Maiorano, A., Rotter, R.P., O'leary, G.J., Fitzgerald, G.J., et al. (2019). Climate change impact and adaptation for wheat protein. *Glob. Change Biol.*, *64*: 155–73.

BBS (Bangladesh Bureau of Statistics). (2017). *Yearbook 16, Series 28*. Statistics and Informatics Division (SID), Ministry of Planning, Government of the People's Republic of Bangladesh, Dhaka, Bangladesh.

BBS. (2021). *Statistical Yearbook Bangladesh*. Statistics and Informatics Division (SID), Ministry of Planning, Dhaka, Bangladesh.

BBS. (2022). *Statistical Yearbook Bangladesh*. Statistics and Informatics Division (SID), Ministry of Planning, Dhaka, Bangladesh.

BDP (Bangladesh Delta Plan) 2100. Volume 1: Strategy. (2022). *General Economics Division. Bangladesh Planning Commission Ministry of Planning. Government of the People's Republic of Bangladesh.* http://www.plancomm.gov.bd/

Berkhout, E.D., Malan, M. and Kram, T. (2019). Better soils for healthier lives? An econometric assessment of the link between soil nutrients and malnutrition in Sub-Saharan Africa. *PLOS One*, *14* (2–3). doi:10.1371/ journal.pone.0210642.

BHWDB. (2012). *Master Plan of Haor Area: Summary Report*. Volume I. Ministry of Water Resources, Government of the People's Republic of Bangladesh, Dhaka, Bangladesh.

BMDA. (2018). *Development Activities of BMDA*. Rajshahi, Bangladesh.

Briggs, A.M., Cross, M.J., Hoy, D.G., Blyth, F.H., Woolf, A.D., March, L., Musculoskel, et al. (2016). Health Conditions Represent a Global Threat to Healthy Aging: A Report for the 2015 World Health Organization World Report on Ageing and Health. *Gerontol.*, *56*: 243–55.

Brown, L.R. and Plan, B. (2008). *Mobilizing to Save Civilization* (Substantially Revised). W.W. Norton & Company: New York, NY, USA.

CEGIS. (2021). *CEGIS Technical Note: Impacts of Projected Sea Level Rise on Coastal Embankments*. CEGIS, Dhaka, Bangladesh.

CEGIS. (2013). *Assessing the Economic Impact of Climate Change On Agriculture, Water Resources, and Food Security and Adaptation Measures Using 5 Seasonal And Medium Range Of Forecasts*. Dhaka.

Chatterjee, R., Islam, R., Acharya, S.K. and Biswas, A. (2022). Conservation agriculture: Analysis and prioritization of socio-ecological factors operating at farm levels in Ohio, USA. *Environmental Science and Policy*, *138*: 1–10.

Clement, V., Rigaud, K.K., Sherbinin, A., Jones, B., Adamo, S., Schewe, J., Sadiq, N. and Shabahat, E. (2021). *Groundswell Part 2: Acting on Internal Climate Migration*. World Bank, Washington, DC.

Eckstein, D., Künzel, V., Schäfer, L. and Winges, M. (2020). Global climate risk index 2020 who suffers most from extreme weather events? https://germanwatch.org/sites/germanwatch.org/files/20-2-01eGlobalClimateRiskIndex2020_14.pdf.

El-Naggar, A., Lee, S.S., Rinklebe, J., Farooq, M., Song, H., Sarmah, A.K., Zimmerman, A.R., Ahmad, M., Shaheen, S.M., Ok, Y.S. (2019). Biochar application to low fertility soils: A review of current status, and future prospects. *Geoderma*, *337*: 536–54.

FAO. (2018). Food and Agriculture Organization Land and Water. http://www.fao.org/land-water/land/httpwwwfaoorgsoils-portalen/en/

FAO. (2022). The U.N. Food and Agriculture Organization (FAO), in the 2022 Global Report on Food Crises.

Gathala, M.K., Kumar, V., Sharma, P.C., Saharawat, Y.S., Jat, H.S., Singh, M., et al. (2013). Optimizing intensive cereal-based cropping systems addressing current and future drivers of agricultural change in the northwestern Indo-Gangetic Plains of India. *Agricul. Ecosys. Environ.*, *177*: 85–97.

GED. (2018). *Bangladesh Delta Plan 2100*. Dhaka: General Economics Division (GED), Planning Commission, Government of the People's Republic of Bangladesh.

Gupta, G.S. (2019). Land Degradation and Challenges of Food Security. *Review of European Studies*, *11*: 1.

Gupta, A.K., Yadav, D., Gupta, P., Gupta, V., Ranjan, S. and Badhai, S. (2020). *Effects of Climate Change on Agriculture*. http://www.foodagrispectrum.org/

Hossain, M.I., Bari, M.N., Miah, S.U., et al. (2021). Application of modified managed aquifer recharge model for groundwater management in drought-prone water-stressed Barind Tract, Bangladesh. *Environ. Challen.*, *4*(2021): 100173.

IPCC. (2006). 2006 IPCC Guidelines for National Greenhouse Gas Inventories Programme, Eggleston, H.S., Buendia, L., Miwa, K., Ngara, T. and Tanabe, K. (eds.) IGES, Japan. http://www.ipcc-nggip.iges.or.jp/public/2006gl/

IPCC. (2014). Climate change: Mitigation of climate change. *In*: O. Edenhofer, R. Pichs-Madruga, Y. Sokona, E. Farahani, S. Kadner, K. Seyboth, A. Adler, I. Baum, S. Brunner, P. Eickemeier, B. Kriemann, J. Savolainen, S. Schlömer, C. von Stechow, T. Zwickel, J.C. Minx (Eds.), Contribution of Working Group III to the *Fifth Assessment Report of the Intergovernmental Panel on Climate Change*. Cambridge, New York, NY: Cambridge University Press, Intergovernmental Panel on Climate Change (IPCC).

IPCC. (2019). *Global Warming of 1.5°C. Summary for Policy Makers*. Switzerland: World Meteorological Organization, United Nations Environment Program, and Intergovernmental Panel on Climate Change. Bern.

IRRI. (2018). *GHG Mitigation on Rice*. International Rice Research Institute. http://ghgmitigation.irri.org/our-work/bangladesh.

Islam, K.R., Roth, G., Rahman, M.A., Didenko, N.O. and Reeder, R.C. (2021). Cover crop complements flue gas desulfurized gypsum to improve no-till soil quality. *Comm. Soil Sci. Plant Anal.*, *52*: 926–47.

Islam, M.A., Bell, R.W., Johansen, C., Jahiruddin, M., Haque, M.E. and Vance, W. (2022). Conservation agriculture effects on yield and profitability of rice-based systems in the Eastern Indo-Gangetic Plain. *Experi. Agricul.*, *58*: e33.

Jahangir, M.M.R., Jahiruddin, M., Akter, H., Pervin, R. and Islam, K.R. (2021). Cropping diversity with rice influences soil aggregate formation and nutrient storage under different tillage systems. *J. Plant Nutri. Soil Sci.*, *184*: 150–61.

Kar, A., Garg, B., Singh, M. and Kathju, S. (2009). *Trends in Arid Zone Research in India*. Jodhpur, Rajasthan: Central Arid Zone Research Institute.

Khan, A.E., Scheelbeek, P.F., Shilpi, A.B., Chan, Q., Mojumder, S.K., Rahman. A., Haines, A. and Vineis, P. (2014). Salinity in drinking water and the risk of (pre)eclampsia and gestational hypertension in coastal Bangladesh: A case-control study. *PlOS One*, *9*: e108715.

Kraut-Cohen, J., Zolti, A., Shaltiel-Harpaz, L., Argaman, E., Rabinovich, R., Green, S.J. and Minz, D. (2020). Effects of tillage practices on soil microbiome and agricultural parameters. *Sci. Total Environ.*, *705:* 135791.

Kumar, R., Naresh, R.K., Shipra, Y. and Chandra, M.S. (2020). Influence of balanced fertilization on productivity, nutrient use efficiency, and profitability of rice in inceptisol: A review. *Inter. J. Current Microbiol. Applied Sci.*, *9*: 568–90.

Kurniawan, T.A., Sillanpaa, M.E.T. and Sillanpaa, M. (2012). Nano adsorbents for remediation of aquatic environment: Local and practical solutions for global water pollution problems. *Crit. Rev. Environ. Sci. Technol.*, *42*: 1233–95.

Lampayan, R.M., Rejesus, R.M., Singleton, G.R. and Bouman, B.A.M. (2015). Adoption and economics of alternate wetting and drying water management for irrigated lowland rice. *Field Crops Res.*, *170*: 95–108.

Lynch, J., Cain, M., Frame, D. and Pierrehumbert, R. (2021). Agriculture's Contribution to Climate Change and Role in Mitigation is Distinct from Predominantly Fossil CO2-Emitting Sectors. *Front. Sustain. Food Syst.*, *4*: 518039. doi:10.3389/fsufs.2020.518039.

Malhi, G.S., Kaur, M. and Kaushik, P. (2021). Impact of Climate Change on Agriculture and Its Mitigation Strategies: A Review. *Sustain.*, *13*: 1318.

Mbow, C., Rosenzweig, C., Barioni, L.G., Benton, T.G., Herrero, M., Krishnapillai, M., et al. (2019). Food security. *In*: P.R. Shukla, J. Skea, E. Calvo Buendia, V. Masson-Delmotte, H.-O. Pörtner, and D.C. Roberts (Eds.), *Climate Change and Land: An IPCC Special Report on Climate Change, Desertification, Land Degradation, Sustainable Land Management, Food Security, and Greenhouse Gas Fluxes in Terrestrial Ecosystems*.

Mekonnen, M.M. and Hoekstra, A.Y. (2016). Four billion people facing severe water scarcity. *Sci. Adv.*, *2*(2): e1500323.

MOA. (2016). *Agricultural Mechanization Road Map 2021, 2031, 2041*. Ministry of Agriculture, People's Republic of Bangladesh, Dhaka.

MoEFCC. (2018). *Nationwide Climate Vulnerability Assessment in Bangladesh*. Dhaka: Ministry of Environment, Forest and Climate Change, Gov. People's Republic of Bangladesh.

MoEFCC. (2022). *Knowledge Management Plan for Climate Change Adaptation*. Dhaka: Ministry of Environment, Forest, and Climate Change, Gov. People's Republic of Bangladesh.

Mojid, M.A., Parvez, M.F., Mainuddin, M. and Hodgson, G. (2019). Water Table Trend: A sustainability status of groundwater development in northwest Bangladesh. *Water*, *11*: 1182.

Mottaleb, K.A., Krupnik, T.J. and Erenstein, O. (2016). Factors associated with small-scale agricultural machinery adoption in Bangladesh: Census findings. *J. Rural Stu.*, *46*: 155–68.

Mueller, N.D., Gerber, J.S., Johnston, M., Ray, D.K., Ramankutty, N. and Foley, J.A. (2012). Closing yield gaps through nutrient and water management. *Nature*, *490*(7419): 254–57.

NAPA (Formulation of the Bangladesh Programme of Action for Adaptation to Climate Change), (2005). *Document Sectoral Consultant Report (Fisheries)*. Bangladesh Agriculture Research Council, Dhaka, Bangladesh

Nawaz, A., Farooq, M., Ul-Allah, S., et al. (2021). Sustainable Soil Management for Food Security in South Asia. *J. Soil Sci. Plant Nutr.*, *21*: 258–75.

OECD. (2022). *Enhancing Climate Change Mitigation through Agriculture*. OECD Publishing, Paris. http://www.oecd.org/agriculture/ministerial

Rahman, M.A. and Rahman, S. (2015). Natural and Traditional Defense Mechanisms to Reduce Climate Risks in Coastal Zones of Bangladesh. *Weather Clim. Extreme*, *7*: 84–95.

Rahman, M.M., Islam, M.R., Uddin, S., Rahman, M.M., Gaber, A., Abdelhadi, A.A. and Jahangir, M.M.R. (2022). Biochar and Compost-Based Integrated Nutrient Management: Potential for Carbon and Microbial Enrichment in Degraded Acidic and Charland Soils. *Front. Environ. Sci.*, *9*.

Richards, M. and Sander, B.O. (2014). Alternate wetting and drying in irrigated rice: Implementation guidance for policymakers and investors. *In: Practice Brief on Climate-Smart Agriculture*. Copenhagen, Denmark: CGIAR Research Program on Climate Change, Agriculture and Food Security (CCAFS).

Richards, M. and Sander, B.O. (2014). Alternate wetting and drying in irrigated rice: Implementation guidance for policymakers and investors. *In*: *Practice Brief on Climate-Smart Agriculture*. Copenhagen, Denmark: CGIAR Research Program on Climate Change, Agriculture, and Food Security (CCAFS).

Sapkota, T.B., Majumdar, K., Jat, M.K., Kumar, A., Bishnoi, D.K., McDonald, A.J. and Pampolino, M. (2014). Precision nutrient management in conservation agriculture-based wheat production of Northwest India: Profitability, nutrient use efficiency, and environmental footprint. *Field Crop Res.*, *155*: 233–44.

Sarker, R.R., Rashid, M.H., Islam, M.A., Jahiruddin, M., Islam, K.R. and Jahangir, M.M.R. (2023a). Tillage and Residue Management Impact on Microbial and Nematode Abundance Under Diverse Rice-based Cropping Systems in Calcareous and Non-calcareous Floodplain Soils. *J. Soil Sci. Plant Nutri.* https://doi.org/10.1007/s42729-023-01168-9.

Sarker, R.R., Rashid, M.H., Islam, M.A., Jahiruddin, M., Islam, K.R. and Jahangir, M.M.R. (2023b). Conservation agriculture's impact on total and labile organic carbon pools in calcareous and non-calcareous floodplain soils under a sub-tropical rice-based system. *PLOS One*, *18*(11): e0293257.

Shamsudduha, M., Chandler, R.E., Taylor, R.G. and Ahmed, K.M. (2009). Recent trends in groundwater levels in a highly seasonal hydrological system: The Ganges-Brahmaputra-Meghna Delta. *Hydrol. Earth Syst. Sci.*, *13*: 2373–85.

Shelley, I.J., Takahashi-Nosaka, M., Kano-Nakata, M., Haque, M.S. and Inukai, Y. (2016). Rice Cultivation in Bangladesh: Present Scenarios, Problems, and Prospects. *J. Intern. Coopera. Agricul. Develop.*, *14*: 20–29.

Singh, P., Adhale, P., Guleria, A., Bhoi, P.B., Bhoi, A.K., Bacco, M. and Barsocchi, P. (2022). Crop Diversification in South Asia: A Panel Regression Approach. *Sustain.*, *14*: 9363.

Smith, K.A., Mosier, A.R., Crutzen, P.J. and Winiwarter, W. (2012). The role of N_2O derived from crop-based biofuels, and from agriculture in general, in Earth's climate. *Philos. Trans. R. Soc. Lond. B.*, *367*: 1169–74.

SRDI. (2010). *Saline Soils of Bangladesh*. Dhaka: Soil Resource Development Institute (SRDI). Ministry of Agriculture.

Srour, A.Y., Ammar, H.A., Subedi, A., Pimentel, M., Cook, R.L., Bond, J., Fakhoury, A.M. (2020). Microbial communities associated with long-term tillage and fertility treatments in a corn-soybean cropping system. *Front Microbiol.*, 11. http://dx.doi.org/10.3389/fmicb.2020.01363

Sun, R., Dsouza, M., Gilbert, J.A., Guo, X., Wang, D., Guo, Z., Ni, Y. and Chu, H. (2016). Fungal community composition in soils subjected to long-term chemical fertilization is most influenced by the type of organic matter. *Environ. Microbiol.*, *18*: 5137–50.

Thakuria, D., Talukdar, N.C., Goswami, C., Hazarika, S., Kalita, M.C. and Bending, G.D. (2009). Evaluation of rice-legume-rice cropping system on grain yield, nutrient uptake, nitrogen fixation, and chemical, physical, and biological properties of soil. *Biol. Fert. Soils*, *45*: 237–51.

Timmusk, S. (2017). Perspectives and challenges of microbial application for crop improvement. *Front. Plant Sci.*, *8*: 49.

UNFCCC. (2015). Strengthening Gender Considerations. *In*: *Adaptation Planning And Implementation in the Least Developed Countries*. Available at https://unfccc.int/files/adaptation/application/pdf/21673_unfccc_leg_gender_low_v5.pdf.

UN RC Bangladesh. (2022). *Bangladesh: Severe Flash Flood*. Office of the UN Resident Coordinator Situation Update #2.

World Bank. (2021). *Groundswell Part 2: Acting on Internal Climate Migration*. World Bank, Washington, DC. https://openknowledge.worldbank.org/handle/10986/36248.

World Food Program (WFP). (2021). *Bangladesh: Food Security Monitoring Bulletin*. https://documents.wfp.org/stellent/

WFO (World Food Program). (2023). *Bangladesh Food Security Monitoring: Remote Household Food Security Survey Brief*. This brief was developed by WFP in January 2023, based on remote household food security surveys conducted in early January 2023.

WRI. (2013). *Climate Analysis Indicator Tools (CAIT) Climate Data Explorer*. Bangladesh Profile. Washington D.C.: World Resources Institute. http://cait.wri.org/profile/Bangladesh.

Yimam, M.M. (2022). Conservation Agriculture as Climate Change Mitigation. *Int. J. Adv. Res. Biol. Sci.*, *9*: 144–55.

CHAPTER 17

Productivity Improvement Opportunities in the Polders of the Climate-vulnerable Coastal Zone to Address Future Food Security Challenges of Bangladesh

M.K. Mondal[1], *Z.H. Khan*[2], *A. Sutradhar*[2], *F.H. Kamal*[2], *S.P. Ritu*[3,4], *S. Parvin*[3] and *S. Yadav*[1,5]

Introduction

The coastal zone of Bangladesh, occupying 30% of the country's cultivable land (SRDI, 2010) is characterized by extreme climatic and environmental challenges limiting agricultural productivity and consequently, influencing the livelihood of the communities (PDO-ICZMP, 2003a; Paul, 2009; Shamsuddoha and Chowdhury, 2007). This zone can broadly be classified into two broad categories: interior cost and exterior cost. Out of 147 Upazilas in 19 coastal districts, 48 upazilas in 12 districts (23,900 km^2) are exposed to the sea and/or lower estuaries and defined as the exposed coast and the remaining 99 upazilas are termed as interior coast with a total area of 47,200 km^2 (PDO-ICZMP, 2003b). Morphologically it is divided into southeast, south-central, and southwest regions of Bangladesh (Islam, 2001; Maniruzzaman et al., 2015). This unique ecosystem comprises of low-lying lands with a dense network of rivers and canals (Islam, 2004; Dasgupta et al., 2014; Haque et al., 2019). The rivers are tidal, and this effect extends up to about 150 km inland (Chowdhury et al., 1997;

1 International Rice Research Institute, Bangladesh and Philippines.
2 Institute of Water Modelling, Bangladesh.
3 Bangladesh Rice Research Institute, Bangladesh.
4 Sylhet Agricultural University, Bangladesh.
5 The University of Queensland, Australia.
* Corresponding author: m.mondal@contractors.irri.org

IWM, 2014; Khan et al., 2015a). High and low tides occur twice daily, and the water level in the rivers at high tide in the monsoon remains higher than the land elevation and at low tide remains lower than the land elevation throughout the year (Khan et al., 2015a; Haque et al., 2019).

The water resources in the coastal zone are vital for agriculture, ecosystem sustenance, and livelihood. Recognizing the importance, about 1.2 million ha lands of the coastal zone were poldered by the government during the 1960s and 1970s to reduce the loss of life and enable the production of Aman rice that does not need appreciable drainage and fertilizer application for optimum yield. Poldering involved building large embankments around the perimeter of the islands between the rivers. Inside the polders are found also dense natural canal networks, and some of the larger internal canals are connected to the surrounding rivers by sluice gates installed in the polder embankments. Thus, with judicious management of the sluice gates, water can be taken into the polders at high tides and drained from the polders at low tides. Despite huge investment and efforts from the government, non-government and international organizations, the cropping intensity and productivity in this region are much lower than in other parts of the country (MoA-FAO, 2013). However, over the past 50 years, there have been tremendous improvements in agricultural technologies and practices. Bangladesh has achieved self-sufficiency in rice using these green revolution technologies. But the coastal zone, especially the polders is deprived of these technological advancements (MoA-FAO, 2013; Tuong et al., 2014).

Unlike the rest of Bangladesh, the farmers in the coastal zone have not widely adopted modern high yielding varieties (HYV) of rice because unfavourable water environment often prevails in the polder ecosystem during the monsoon. They usually cultivate traditional low yielding *Aman* (wet season) rice, sometimes followed by late sown low yielding *rabi* (dry season) crops and/or preceded by a pre-monsoon *aus* rice (Mondal et al., 2015a). The rabi crops are often damaged by the pre-monsoon rains and/or by cyclones at or near maturity. Moreover, late harvest of traditional rice and waterloggged soil thereafter limits the cultivation of high-yielding rabi crops. Due to these vulnerabilities, large areas of the coastal land lie fallow during the dry season (Hasan et al., 2013). However, there are tremendous opportunities to increase the productivity through the use of improved crops and water management practices (Mondal et al., 2015a, b; Ritu et al., 2015; Saha et al., 2015; Bhattacharya et al., 2019; Mondal et al., 2020; Yadav et al., 2020; Asefa et al., 2021). But the cropping system intensification and adoption of improved production systems in coastal polder zone has been limited by the misperception that river water is too saline. In reality, river water in all the coastal rivers is not saline all the time (Mondal et al., 2006; Khan et al., 2015a). Most rivers in the south-central coastal zone remain non-saline throughout the year and will remain so until 2030 with moderate climate change scenario A1B with 22 cm sea level rise and a moderate precipitation change. Some river water has salinity part of the year, and some carry it throughout the year. Like the water salinity, soil salinity showed a large temporal and spatial variation across the coastal zone depending on the rainfall pattern, the intensity of tidal flooding and the nature of the movement of saline groundwater (SRDI, 2010; Burman et al., 2019; Yadav et al., 2020). In addition, residual soil water in the soil profile retained after monsoon is seen as a disadvantage for *rabi* crops cultivation in the coastal zone, which could potentially be utilized for cultivation and safe harvest of high-value *rabi* crops

(Mondal et al., 2015a, 2020). The present study was, therefore conducted to take the next step in farmer-participatory cropping system intensification with improved water management to demonstrate the potential of the low and medium saline coastal zones of Bangladesh.

Materials and Methods

Study Site

The study was conducted in the low and medium saline coastal zones of Bangladesh. The farmer-participatory studies of different cropping patterns were conducted in polder 43/2F in the low saline south-central and in polder 30 in the medium saline southwest region of Bangladesh (Fig. 1).

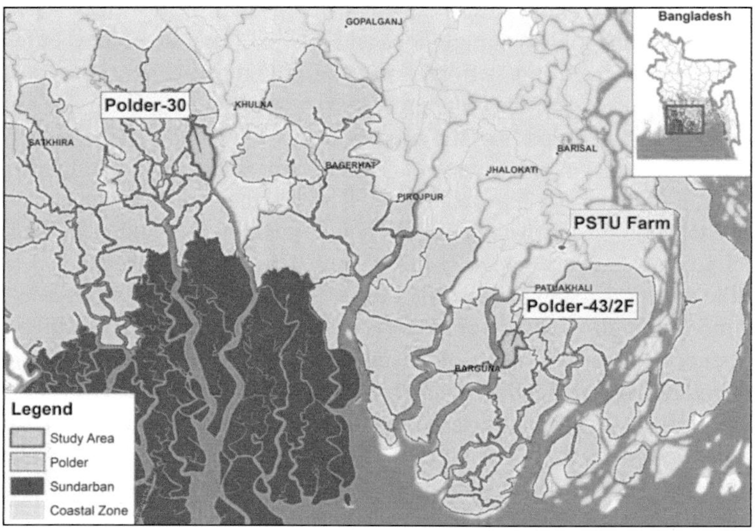

Fig. 1 Map of the southwest and south-central coastal zone of Bangladesh showing the study polders.

Crop and Water Management

Water in the rivers and canals in polder 43/2F remains non-saline throughout the year; whereas scarcity of freshwater observed in the second half of the dry season in the polder 30 of southwest coastal zone (Khan et al., 2015a; Mondal et al., 2015a, 2010). Utilizing these water resources, *aus-aman-boro* and *aus-aman-rabi* cropping systems were demonstrated and evaluated in the low saline zone from 2011–2014; and *aman-boro* were evaluated from 2005–2007 and 2011–2014, and *aman-rabi* from 2011–2016 in the medium saline coastal zones of Bangladesh.

High yielding varieties of rice and *rabi* (dry season) crops were used in all the tested cropping patterns in the low and medium saline zones. For rice, short duration, submergence tolerant and zinc biofortified varieties were chosen to fit two to three crops per year; and for *rabi* crops, hybrid varieties of maize and sunflower were evaluated in different cropping systems. The crops were cultivated within the recommended period and by adopting recommended management practices (BARI, 2006; BRRI, 2004, 2012). In rice, all fertilizers (P, K, S, Zn) except nitrogen (N) were broadcast just before the final land preparation and N was broadcast in two to three equal splits at mid-tillering, before panicle initiation and at flowering. Excess water from the *aus* and *aman* fields was drained out by gravity opening the sluice gates at low tides in the rivers before topdressing N fertilizer. River water was taken in after N top dressing N fertilizer, and in case of drought in *aus* and *aman* seasons. Terminal drainage was done two weeks before rice harvest opening the sluice gate during low tides. For *rabi* crops, 50% of the N and all the P, K, S, Zn, and B were broadcast just before final land preparation; the rest amount of N was top-dressed twice in maize and sunflower (Mondal et al., 2015a). Irrigation was provided to all the dry season crops in low saline coastal zone, but only *boro* rice was grown under irrigated condition and the *rabi* crops were cultivated under rain-fed environment in the medium saline coastal zone.

Crop Yield

Rice yield was determined by harvesting a 5 m × 2 m harvest area from the middle of the plot. Grain moisture content was determined at harvest and the yield was calculated at 14% moisture content. The grain yield of the *rabi* crops was determined from a 5 m × 4 m harvest area in the middle of each plot and the yield was converted to 12% moisture content. Rice equivalent yield (REY) of the *rabi* crops was calculated taking farm gate price of rice and *rabi* crops using the formula:

$$\text{REY (kg/ha)} = \text{rabi yield (kg/ha)} * \text{rabi price (Tk/kg)} / \text{rice price (Tk/kg)}$$

The price of rice and rabi crops was collected by survey when the farmers sold in the local market. The average price of HYV rice was Tk. 16.25 per kg at low saline sites and Tk. 18.75 at medium saline site (Mondal et al., 2015a). The prices of *rabi* crops were almost similar at both low and medium saline sites. The farm gate prices of maize and sunflower was Tk. 16.25 and 33.75 per kg, respectively were used to determine the REY of *rabi* crops.

Water Salinity

The salinity (electrical conductivity, EC) of the river water was measured using a portable EC meter in both low (2011–2012, and 2020–2023) and medium (2005–2008, and 2022–2023) saline coastal zones of Bangladesh. During 2005–2012, EC was measured daily, and it was measured weekly during recent years (2020 onwards). The EC values were averaged to document the monthly salinity dynamics of Payra and Passur Rivers flowing along the periphery of polder 43/2F and polder 30, respectively.

Results

Salinity and Water Management

Water salinity of the Payra River flowing along the periphery of polder 43/2F in low saline polder zone varied from 0.1 ppt to 0.3 ppt during 2011–2012, showed increasing

(EC_{max} ~4.0 dS/m in April) trends in the recent years (2020–2023) compared to last decade. Despite water of the Payra River remains suitable for irrigation throughout the year. The EC values of river water in the medium saline coastal zone also showed an increasing trend. Salinity of Kazibacha-Passur River system flowing along the east side of polder 30 started to increase from January, reached the maximum in May (~26 dS/m) and declined to <1.0 dS/m during the monsoon. Unlike in the low saline coastal zone, river water EC in the medium saline zone remains unsuitable for irrigation in the dry season spanning from January to June. IRRI (2017, 2023), Mondal et al. (2010, 2016), and Khan et al. (2015a) reported similar river water salinity in the low and medium saline coastal zones of Bangladesh (Fig. 2).

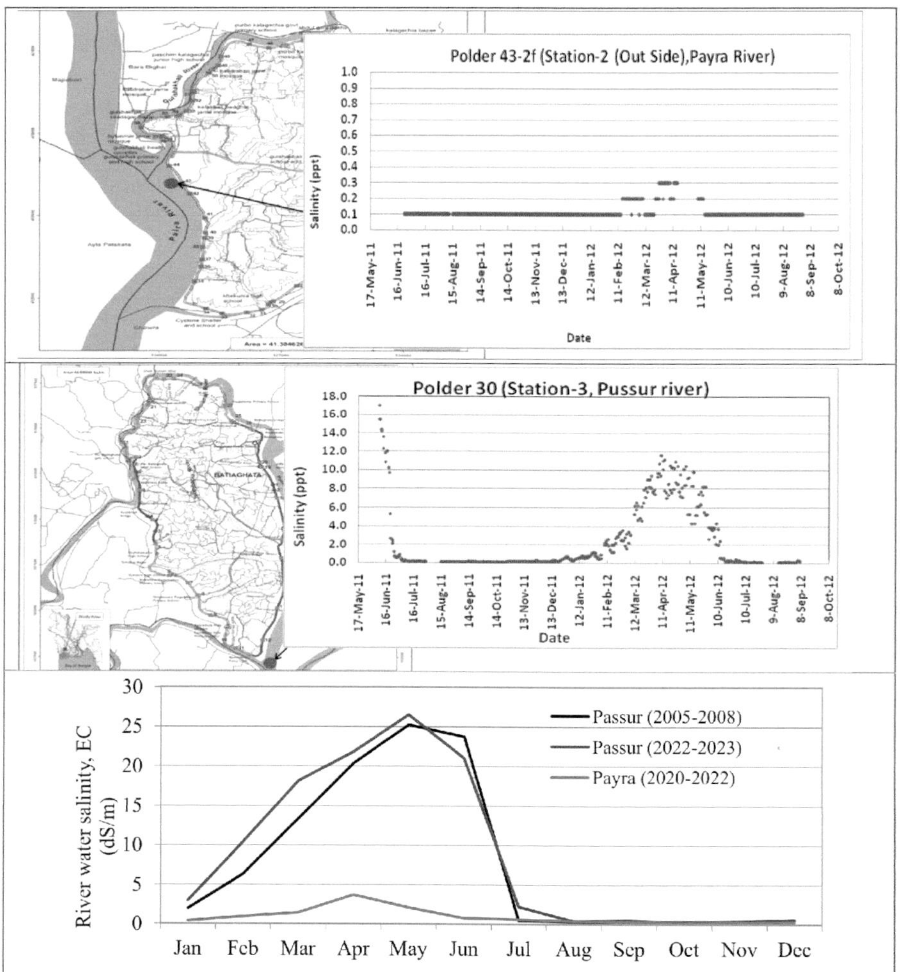

Fig. 2 River water salinity at low (Payra River - top) and medium (Passur River - middle) saline coastal zone (2011–2012) and in the recent years (2020–2023 for both rivers - bottom) (*Source:* Khan et al., 2015a; Mondal et al., 2006; IRRI, 2023; SRDI, 2023).

Rainfall, river water, and residual soil water was judiciously utilized for crop cultivation in both wet and dry seasons. Topography plays a vital role in water management in the coastal polder zone and the improved water management practices adopted for cropping intensification in the polder zone. Land elevation in the polders of the coastal zone is lower than water level elevation at high tides in the peripheral rivers of the polder ecosystem in the wet season (Mondal et al., 2022). Utilizing these opportunities of the polder ecosystem, tidal river and rainfall were effectively utilized to cultivate HYV rice in both *aus* and *aman* seasons. When rainfall was insufficient, tidal water during high tides was taken in through sluice gates by gravity and pumping was not required. And excess rainwater was drained out to the peripheral rivers of polders 30 and 43/2F by opening the sluice gates at low tides to facilitate fertilizer management for higher yield, timely rice harvest, and early establishment and safe harvest of *rabi* crops (Fig. 3). In the dry season, water level elevation at high tides remains lower than the land elevation. For that, irrigation is necessary for higher productivity of dry season crops along with drainage to save the *rabi* crops in case of excess rainfall. Accordingly, the sluice gates of the polder ecosystem were operated to take river water in for dry season cropping in the low saline coastal zone, and shut-down the gates to prevent entry of saline river water in the medium saline polder lands. In polder 43/2F, the river water was taken in through the sluice gate and stored in the polder canal networks to irrigate dry season crops (*boro* rice and *rabi* crops). In contrast, the sluice gates in polder 30 of the medium saline coastal zone were kept closed between February–June to curb saline water intrusion. In medium saline zone, residual soil water and rainfall were effectively utilized to grow *rabi* crops. The only exception was that the canals inside polder 30 were filled-up with non-saline river water by opening the sluice gates during high tides until January to irrigate *boro* rice. It was observed that with the stored water only 15–20% of the lands can be brought under *boro* rice cultivation (Mondal et al., 2010) (Fig. 3).

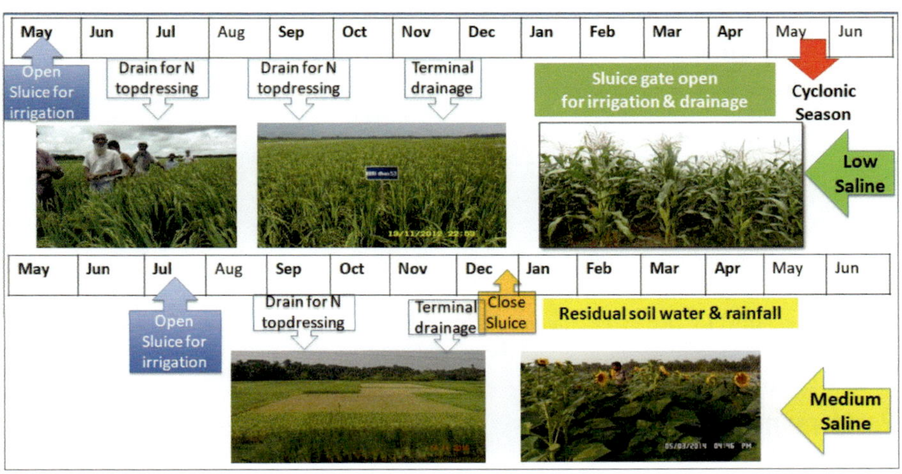

Fig. 3 Scheduling sluice gate operation for improved cropping in the low (top) and medium (bottom) saline coastal zone.

Performance of Rice

Traditional rice was popularly cultivated by the farmers in the polders of the coastal zone of Bangladesh, the productivity of which is low and varied from 1.5–3.5 t/ha depending on the varieties grown in *aman* season (Mondal et al., 2006, 2015a). The performance HYV rice was evaluated in *aus, aman,* and *boro* in low (polder 43/2F) and *aman* and *boro* seasons in medium (polder 30) saline coastal zone. The yield of *aus* rice varied from 3.6–5.2 t/ha, *Aman* rice ranged from 4.2–5.5 t/ha and *Boro* rice varied from 2.4–7.9 t/ha depending on the sowing time, varietal characteristics, and study sites. The mean yield of all the HYVs tested in *aman* season was similar (4.6–4.8 t/ha) (Fig. 4). The yield of *Boro* rice in low saline coastal zone ranged from 4.4–7.9 t/ha (av. 6.4 t/ha) and it varied from 2.4 (early crop damaged due to cold stress at panicle initiation, Mondal et al., 2015b) to 6.3 t/ha in medium saline coastal zone, average of which was 4.6 t/ha (Fig. 4). Yield of *Boro* rice drastically reduced when the crop was sown in the latter part of October due to cold injury at the panicle initiation stage (Mondal et al., 2015b). Moreover, salinity of the stored canal water used to irrigate *Boro* rice increased as the dry season progressed (Mondal et al., 2010). Therefore, the main cause of lower yields in the medium saline zone may be the combination of high-water salinity and adverse effects of low temperature at the panicle initiation stage of *Boro* rice compared to low saline coastal zone. This study provides strong evidence of the feasibility of HYV rice cultivation and increasing land productivity in the low and medium saline coastal zone with improved management of tidal river water through systematic operation of the sluice gates of the polder ecosystem. The study also exhibited the potential of *Boro* rice cultivation in the low saline coastal zone of Bangladesh, where temperature and river water salinity remain within the permissible limits for HYV rice (Mondal et al., 2015b). Introduction of recently developed salt and heat tolerant HYV rice varieties might increase the productivity further.

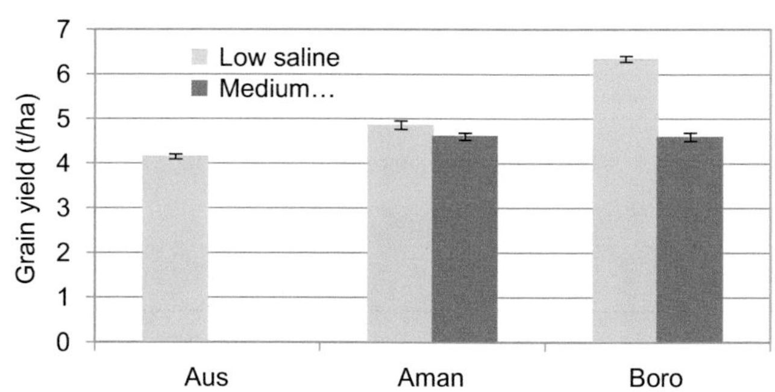

Fig. 4 Yield of HYV rice in *aus*, *aman*, and *boro* seasons in the low and medium saline polders of the coastal zone.

Performance of Rabi Crops

The performance of *rabi* crops varied across study sites and study periods (Mondal et al., 2015a). Overall, yields of maize and sunflower were higher in low saline than in the medium saline coastal zone (Fig. 5). The yield of maize in low saline polder 43/2F ranged from 7.5–9.5 t/ha, and sunflower from 2.9–3.8 t/ha (Mondal et al., 2015a). Initially the performance of mungbean and sesame were evaluated in polder 30, yield of which was poor (0.4–0.9 t/ha) due to poor germination, lack of irrigation, and crop damage either partially or fully at the later growth stage by pre-monsoon rains (Mondal et al., 2015a). Therefore, maize and sunflower were evaluated during the dry seasons of 2011–2012 to 2015–2016 in the medium saline polder 30 to evaluate the feasibility of the crops under rain-fed environment. Maize and sunflower cultivation were also affected by both waterlogging (due to high rainfall) and drought (due to lack or no rainfall) to different degrees over the study period (Mondal et al., 2015a). Despite such environmental limitations (salinity, drought, and excess rain), maize (5.7–9.5 t/ha) and sunflower (1.0–3.8 t/ha) produced a reasonable yield, the mean of which was 6.1 t/ha and 2.2 t/ha, respectively (Fig. 5), indicating the crops may be climate-resilient and suitable for the coastal zones of Bangladesh.

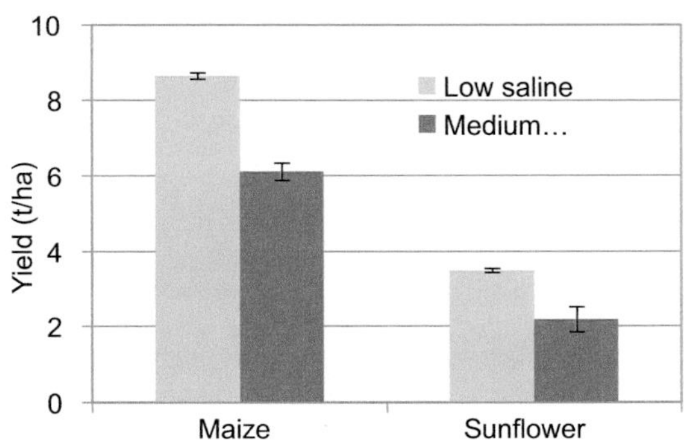

Fig. 5 Mean yield of maize and sunflower in low and medium saline coastal polder zone.

System Productivity

Four major cropping patterns were tested in the low and medium saline polders of the coastal zone. The patterns were: (1) *Aus-aman-boro* and (2) *Aus-aman-rabi* in the low saline coastal zone; and (3) *Aman-boro*, (4) *Aman-rabi* in the medium saline coastal zone.

In the low saline coastal zone, total grain yield, expressed in rice equivalent yield (REY) of *aus-aman-boro* cropping pattern ranged from 13.2–16.6 t/ha/yr (Mondal et al., 2015a). The mean REY of this pattern was 15.2 t/ha rice per year (Fig. 6), which was 2–3 times higher than the farmers' production (Mondal et al., 2015a). The system productivity of *aus-aman-rabi* cropping pattern in low saline coastal zone also varied as observed in triple rice pattern (Mondal et al., 2015a). The mean REY of maize-based triple cropping was 17.8 t/ha/yr and that of sunflower-based pattern was 16.4 t/ha/yr (Fig. 6). The REY of *aus-aman-rabi* cropping patterns was higher than triple rice system, due to the higher yield of maize and higher price of sunflower seed (Mondal et al., 2015a) (Fig. 6).

Only two crops were grown in the medium saline coastal polder zone. The REY of *aman-boro* cropping pattern varied from 4.7–10.5 t/ha/yr and was strongly affected by sowing date of *boro* rice (Mondal et al., 2015b). The mean REY of double

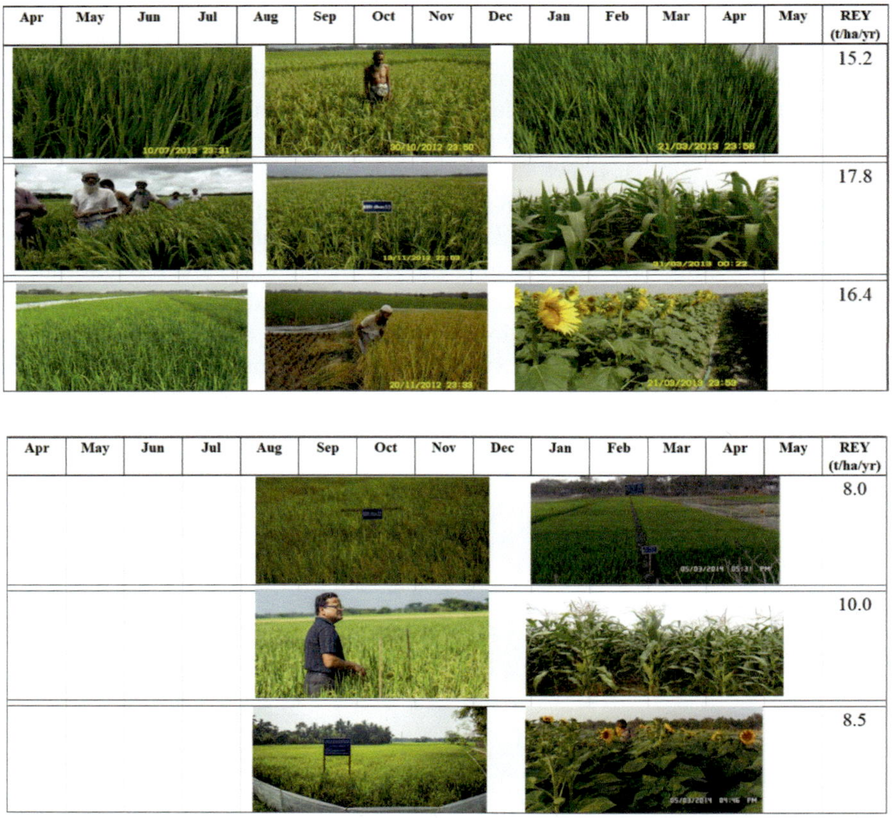

Fig. 6 System productivity (expressed in rice equivalent yield, REY) of improved production systems in low (top) and medium (bottom) saline coastal zones of Bangladesh.

rice system was 8 t/ha/yr (Fig. 6). In contrast, the system productivity of *aman-rabi* cropping pattern was higher than *aman-boro* pattern as observed in the low saline environment. REY of the *aman-rabi* cropping pattern varied from 6.6–10.0 t/ha/yr (Mondal et al., 2015a). The mean REY of *aman*-maize and *aman*-sunflower cropping patterns was 10.0 t/ha/yr and 8.5 t/ha/yr, respectively (Fig. 6). The systems with maize produced the highest yield at both low and medium salinity sites. The productivity of the improved cropping systems was at least double than of the farmers' practice (rice-mungbean and rice-sesame) in the medium saline coastal zone of Bangladesh (Mondal et al., 2015a, b, 2020; Asefa et al., 2021).

Discussion

The general perception is that improved agricultural technologies cannot be adopted in the coastal zone due to soil and water salinity problem. Thus, the cropping system intensification in the coastal zone of Bangladesh has been limited by this misperception that the river water is too saline for irrigation. In reality, most of the rivers in the south-central coastal zone remain non-saline throughout the year, while the river water salinity in most parts of the southwest region (except Satkhira district) lies within the permissible limit for crop irrigation from July to January (Mondal et al., 2016, 2010; Khan et al., 2015a). Khan et al. (2015a) showed that most rivers in the south-central coastal zone remain non-saline throughout the year; will remain so until 2030 with a moderate climate change scenario A1B with 22 cm sea level rise and a moderate precipitation change. Our findings on water salinity in polders 30 and 43/2F were consistent with the above findings. Although in the recent years, river water salinity is exhibiting an increasing trend, still it remains suitable for irrigation during most part of the year in the south-central region of Bangladesh.

There is also a misperception that improved agricultural technologies cannot be adopted in the coastal zone due to salinity and unfavorable water environment. We proved that the water environment can significantly be improved through the systematic operation of the sluice gates synchronizing with the tidal phenomenon of the coastal river systems, which will facilitate wide-scale adoption of improved production systems in the coastal polder zone. In the wet season, HYV rice can be grown with at least 2.0 t/ha higher yield than farmers' practice by draining out excess water at low tides that occurred twice daily in the coastal rivers. Possibilities of HYV rice cultivation in *aman* season through improved water management in the polders were also reported by Mondal et al. (2022). Although HYV rice yield was similar at both south-central and southwest coastal zone, the dry season productivity in the medium saline coastal zone was lower than that observed in low saline coastal zone, and it varied from year to year depending on the amount and frequency of rainfall. Excess rainfall of around 60 mm in a day destroys the crop (zero yield), lack or little rain hampers crop growth and development (30–50% of achievable yield) and the evenly distributed rainfall throughout the dry season provided beneficial effects (good yield) as irrigation does (Mondal et al., 2020, 2015a; Asefa et al., 2021).

The farmer-participatory research clearly demonstrated the feasibility of year-round cropping in the south-central coastal zone of Bangladesh through judicious management of polder infrastructure (particularly sluice gates) and tidal river water with a productivity of at least 15.0 t/ha rice per year in comparison to the farmers' production of 5–9 t/ha/yr (Mondal et al., 2015a, 2020). In contrast, only two crops can be grown in the medium saline coastal zone with a productivity of at least 8.0 t/ha/yr. Our findings are similar with the findings of other researchers (Bhattacharya et al., 2019, Mondal et al., 2020, 2019, 2015a; Saha et al., 2015; and Ritu et al., 2015). The system productivity can be further increased with the adoption of recently developed salt-tolerant crop varieties. But early establishment of dry season crops in this region is essential for resilient cropping and consistently higher productivity compared to the farmers' production of 2–3 t/ha of *Aman* rice and less than 1.0 t/ha of *rabi* crops (Mondal et al., 2015a).

The key prerequisite for sustainable cropping intensification in the coastal polder zone of Bangladesh is early (by November) harvest of the *Aman* rice, draining out excess water from the paddy field by mid-November and drain-out excess rainfall, if any in the dry season (Mondal et al., 2022, 2019, 2016, 2015a,b). Drainage is essential not only to create a favorable environment for *aman* harvest, but also to facilitate early establishment and safe harvest (before onset of the cyclonic season) of *rabi* crops. Although residual water in the soil profile is seen as a disadvantage for the successful cultivation of *rabi* crops in the coastal polder zone, our studies demonstrated that this water is a valuable resource in a saline-prone and water-scarce environment limiting the dry season productivity (as in medium saline coastal zone). In addition, introduction of maize and sunflower not only increased the productivity; it also exhibited resilience to the adverse effects of climate change in improving and sustaining dry season productivity.

Conclusions

Our studies revealed the potential for increased land productivity in Bangladesh's low and medium saline coastal zone through enhanced crop and water management practices. Key strategies include drainage during the monsoon for high-yielding Aman rice cultivation, mid-November terminal drainage for rice harvest and early rabi crop establishment, and periodic drainage in the dry season for safe rabi crop harvesting. In the south-central coastal zone, where non-saline river water is available year-round, triple cropping achieved a system productivity of 15–18 t/ha/yr. in farmers' fields. In the medium saline coastal zone, intelligent management of residual soil water and rainfall enabled the harvest of 8–10 t/ha/yr. of rice. With challenges such as increased natural calamities and a declining land-man ratio, Bangladesh faces significant hurdles in maintaining food sufficiency. Given the high cropping intensity and productivity in other regions, the underutilized 1.2 M ha lands in coastal polders present a crucial opportunity for increasing food production. Judicious water resource management in the coastal polder zone could contribute significantly to addressing future food security challenges in Bangladesh and achieving sustainable development goals within the designated time.

References

Asefa, Y., Yadav, S., Mondal, M.K., Bhattacrarya, J., Parvin, R., Sarker, S.R., Rahman, M., Sutradhar, A., Vara Prasad, P.V., Bhandari, H., Shew, A.M. and Krishna Jagadish, S.V. (2021). Crop diversification in the polders of Bangladesh: Yield stability, profitability, and associated risk. *Agricultural Systems*, *187*.

Bangladesh Agricultural Research Institute (BARI). (2006). *Krishi Projucti Hatboi* (in Bangla) 4th Edn. BARI, Gazipur, Bangladesh.

Bangladesh Rice Research Institute (BRRI). (2004, 2012). *Adhunik Dhaner Chas* (in Bangla). BRRI, Gazipur, Bangladesh.

Burman, D., Mondal, M.K., Khan, Z.H., Sutradhar, A.K. and Kamal, F.A. (2019). Soil and water resources of Sundarbans. *In*: H.S. Sen (Ed.), *The Sundarbans: A Disaster-Prone Eco-Region*, 165–96. Springer.

Bhattacharya, J., Saha, N.K., Mondal, M.K., Bhandari, H. and Humphreys, E. (2018). The feasibility of high yielding Aus-aman-rabi cropping systems in the polders of the low saline coastal zone of Bangladesh. *Field Crops Research*, *234*: 33–46.

Chowdhury J.U., Rahman, M.R. and Salehin, M. (1997). *Flood control in a floodplain country: Experiences of Bangladesh*. Islamic Educational, Scientific, and Cultural Organization (ISESCO), Rabat, Morocco.

Dasgupta, S., Huq, M., Khan, Z.H., Ahmed, M.M.Z., Mukherjee, N., Khan, M.F. and Pandey, K. (2014). Cyclones in a changing climate: The case of Bangladesh. *Clim. Dev.*, *6*: 96–110.

Hasan, M.N., Hossain, M.S., Islam, M.R. and Bari, M.A. (2013). *Trends in the Availability of Agricultural Lands in Bangladesh. National Food Policy Capacity Strengthening Programme*. Soil Resources Development Institute, Dhaka, Bangladesh.

Haque, A., Rahman, M.H. and Rahman, D. (2019). An evaluation of sea level rise vulnerability and resilience strategy to climate change in the coastline of Bangladesh. *Int. J. Environ. Sci. Nat. Res. 18*(1): 555983.

International Rice Research Institute (IRRI). (2017). *Community Water Management for Improved Food Security, Nutrition, and Livelihoods in the Polders of the Coastal Zone of Bangladesh.* Technical report submitted to the International Water Management Institute, IRRI, Philippines.

International Rice Research Institute (IRRI). (2023). *Pathways of Scaling Agricultural Innovations for Sustainable Intensification in the Polders of Coastal Bangladesh.* Bi-Annual Report submitted to the Kansas State University (USA), IRRI, Philippines.

Islam, M.S. (2001). *Sea-level Changes in Bangladesh: The Last Ten Thousand Years.* Asiatic Society of Bangladesh.

Islam, M.R. (2004). *Where Land Meets the Sea: A Profile of the Coastal Zone of Bangladesh.* The University Press Limited, Dhaka, Bangladesh.

Institute of Water Modelling (IWM). (2014). *Bangla-Dutch Research Initiative on Flood Risk Management and Morphological Assessment.* Final report, December 2014.

Khan, Z.H., Kamal, F.A., Khan, N.A.A., Khan, S.H. and Khan, M.S.A. (2015a). Present surface water resources of the Ganges coastal zone of Bangladesh. *In*: Humphreys, E., T.P. Tuong, M.C. Buisson, I. Pukinskis, M. Philipps (Eds.), *Revitalizing the Ganges Coastal Zone: Turning Science into Policy and Practices*, 14–26. *Conference Proceedings*. Colombo, Sri Lanka: CGIAR Challenge Program on Water and Food.

Maniruzzaman, M., Biswas, J.C., Khan, M.A.I., Sarker, G.W., Haque, S.S., Biswas, J.K., Sarker, M.H., Rashir, M.A. Sekhar, N.U., Nemes, A., Xenarios, S. and Deelstra, J. (2015). Climate change impact on crop production and water requirement in southern Barisal, Bangladesh. *In*: Humphreys, E., T.P. Tuong, M.C. Buisson, I. Pukinskis, M. Philipps (Eds.), *Revitalizing the Ganges Coastal Zone: Turning Science into Policy and Practices*, 39–52. *Conference Proceedings*. Colombo, Sri Lanka: CGIAR Challenge Program on Water and Food (CPWF),

Ministry of Agriculture-Food and Agriculture Organization (MoA-FAO). (2013). *Master Plan for Agricultural Development in the Southern Region of Bangladesh.* Prepared by the Ministry of Agriculture of the Government of Bangladesh and Food and Agriculture Organization of the United Nations, Dhaka, Bangladesh.

Mondal, M.K., Tuong, T.P., Ritu, S.P., Choudhury, M.H.K., Chasi, A.M., Majumder, P.K., Islam, M.M. and Adhikary, S.K. (2006). Coastal water resource use for higher productivity: Participatory research for increasing cropping intensity in Bangladesh. *In*: Hoanh, C.T, Tuong, T.P, Gowing, J.W., Hardy, B. (Eds.), *Environmental and Livelihoods in Tropical Coastal Zones: Managing Agriculture-fishery-aquaculture Conflicts*, 72–85. Comprehensive Assessment of Water Management in Agriculture Series, No. 2. CABI Publishing.

Mondal, M.K., Tuong, T.P., Sharifullah, A.K.M. and Sattar, M.A. (2010). Water supply and demand for dry-season rice in the coastal polders of Bangladesh. *In*: Hoanh, C.T., Szuster, B.W., Suan-Pheng, K., Ismail, A.M., Noble, A.D. (Eds.), *Tropical Deltas and Coastal Zones: Food Production, Communities, and Environment at the Land-Water Interface*, 264–78. Comprehensive Assessment of Water Management in Agriculture Series, No. 9. CABI Publication.

Mondal, M.K., Paul, P.L.C., Humphreys, E., Tuong, T.P., Ritu, S.P. and Rashid, M.A. (2015a). Opportunities for cropping system intensification in coastal zone of Bangladesh. *In*: Humphreys, E., T.P. Tuong, M.C. Buisson, I. Pukinskis, M. Philipps (Eds.), *Revitalizing the Ganges Coastal Zone: Turning Science into Policy and Practices*, 449–76. *Conference Proceedings*. Colombo, Sri Lanka: CGIAR Challenge Program on Water and Food (CPWF).

Mondal, M.K., Saha, N.K., Ritu, S.P., Paul, P.L.C., Sharifullah, A.K.M., Humphreys. E., Tuong, T.P. and Rashid, M.A. (2015b). Optimum sowing window for *Boro* cultivation in the coastal zone of Bangladesh. *In*: Humphreys, E., T.P. Tuong, M.C. Buisson, I. Pukinskis, M. Philipps (Eds.), *Revitalizing the Ganges Coastal Zone: Turning Science into Policy and Practices*, 342–60. *Conference Proceedings*. Colombo, Sri Lanka: CGIAR Challenge Program on Water and Food (CPWF).,

Mondal, M., Humphreys and E., Sudhir-Yadav. (2016). *An Innovative Water Management Approach for Food Security of Coastal Zone Communities in Bangladesh.* Paper presented at the 2nd World Irrigation Forum (WIF2), Chiang Mai, Thailand, 6–8 Nov.

Mondal, M., Sudhir-Yadav, Krishna Jagadish, S.V., Islam, S. and Masud, J. (2019). *Cropping System Options in the Polders of Coastal Bangladesh – Synergies and Tradeoffs*. Poster presented in the SIIL Annual Meeting held at Saly, Senegal, 7–9 April.

Mondal, M.K., Saha, N.K., Bhattacharya, J., Ritu, S.P., Paul, P.L.C., Parvin, S., Khan, Z. H., Sutradhar, A. and Kamal, F.A. (2020). Productivity improvement opportunities in the coastal polders to address food security challenges of Bangladesh. *J. Indian Society of Coastal Agricultural Research, 38*: 23–37.

Mondal, M.K., Yadav, S., Baidya, B., Khan, Z.H., Sutradhar, A., Humphreys, E., Kamal, F.A. and Krishna Jagadish, S.V. (2022). Evaluation of gravity-led and energy-fed drainage for sustaining food security in the polders of the coastal zone of Bangladesh. *J. Irrigation and Drainage*, 1–14. https://doi.org/10.1002/ird269.

Paul, B.K. (2009). Why relatively fewer people died? The case of Bangladesh's cyclone Sidr. *Nat. Hazard*, *50*(2): 289–304.

Project Development Office for Integrated Coastal Zone Management Plan (PDO-ICZMP). (2003a). *Coastal Zone Management: An Analysis of Different Policy Documents*. WP009, Dhaka, Bangladesh.

Project Development Office for Integrated Coastal Zone Management Plan (PDO-ICZMP). (2003b). *Delineation of the Coastal Zone*. PDO-ICZMP, Dhaka, Bangladesh.

Ritu, S.P., Mondal, M.K., Tuong, T.P., Talukdar, S.U. and Humphreys, E. (2015). An a*us-aman* system for increasing the productivity of a moderately saline region of the coastal zone of Bangladesh. *In*: Humphreys, E., T.P. Tuong, M.C. Buisson, I. Pukinskis, M. Philipps (Eds.), *Revitalizing the Ganges Coastal Zone: Turning Science into Policy and Practices*, 361–88. *Conference Proceedings*. Colombo, Sri Lanka: CGIAR Challenge Program on Water and Food (CPWF).

Saha, N.K., Mondal, M.K., Humphreys, E., Bhattacharya, J., Rashid, M.H., Paul, P.C. and Ritu, S.P. (2015). Triple rice in a year: Is it a feasible option for the low salinity areas of the coastal zone of Bangladesh? *In*: Humphreys, E., T.P. Tuong, M.C. Buisson, I. Pukinskis, M. Philipps (Eds.), *Revitalizing the Ganges Coastal Zone: Turning Science into Policy and Practices*, 421–35. *Conference Proceedings*. Colombo, Sri Lanka: CGIAR Challenge Program on Water and Food (CPWF).,

Shamsuddoha, M. and Chowdhury, R.K. (2007). *Climate Change Impacts and Disaster Vulnerabilities in the Coastal Areas of Bangladesh.* COAST Trust, Dhaka, Bangladesh.

Soil Resources Development Institute (SRDI). (2010). *Saline Soils of Bangladesh.* SRMAF project, Ministry of Agriculture, Govt. of Bangladesh. Dhaka, Bangladesh.

Soil Resources Development Institute (SRDI). (2023). *Personal communication: Monthly Soil and Water Salinity Bulletin.* SRDI, Barishal, Bangladesh.

Tuong, T.P., Humphreys, E., Khan, Z.H., Nelson, A., Mondal, M., Buisson, M.C. and George, P. (2014). *Messages from the Ganges Basin Development Challenge: Unlocking the Production Potential of the Polders of the Coastal Zone of Bangladesh through Water Management Investment and Reform.* CPWF Research for Development Series 9. https://cgspace.cgiar.org/bitstream/handle/10568/41708/CPWF%20Ganges%20basin%20messages%20Sept%2014.pdf?sequence=5.

Yadav, S., Mondal, M.K., Shew, A., Krishna Jagadish, S.V., Khan, Z.H., Sutradhar, A., Bhandari, H., Humphreys, E., Bhattacharya, J., Parvin, R., Rahman, M. and Chandna, P. (2020). Community water management to intensify agricultural productivity in the polders of the coastal zone of Bangladesh. *Paddy and Water Environment*, *18*(2): 331–43.

Chapter 18

Flood Management and Livelihood Improvement in Wetland Ecosystems (Haor Region) of Bangladesh

S.M. Rezaul Mostofa Kamal[1], *Muhammad Masood*[2] and *Shamal Chandra Das*[2]

Introduction

The northeastern region of Bangladesh (Fig. 1) is blessed with a unique natural feature, locally known as 'Haor'. These Haors are expansive bowl-shaped floodplain depressions, typically surrounded by river levees. Covering an impressive area of 8,58,461 hectares, which accounts for 44% of the total area of seven districts, namely Sunamganj, Sylhet, Moulavibazar, Habiganj, Kishorganj, Netrokona, and Brahmanbaria. In total, these seven districts span a vast land area of 19,66,907 hectares. The region's abundant Haors, numbering 373, play a vital role in Bangladesh's agricultural sector, contributing significantly to rice production. Annually, these Haors yield an impressive 53 lakh tons of rice, amounting to approximately 16% of the country's total rice production. These Haors are not only a natural wonder but also a crucial factor in supporting the livelihoods and food security of countless people in the region.

The Haor area is characterized by its complex network of rivers, with the Surma, Kushiyara, old Surma, Baulai, Ghorautra, Kalni, Dhaleswari, Dhala gang, Chela, Nawa gang, Chalti, Jadukata, Bhugai, Kangsha, Manu, Juri, Khowai, and upper Meghna rivers flowing through its expanse.

This region houses a total population of 20 M people. As outlined in the Haor Master Plan, this population is projected to grow to 23 million by the year 2030 (CEGIS, 2012). A substantial 86% of the local population is engaged in agriculture

[1] Ministry of Home Affairs, Government of Bangladesh.

[2] Bangladesh Water Development Board, Bangladesh.

* Corresponding author: mostafa.banna@gmail.com

Hoar area briefly

- Natural depression of unique hydro-ecological characteristics
- Total number of hoar: 373
- Area coverage: 1.99 million ha (13.5% of country's total land)
- Total population: 19.37 million (about 12% of national population)
- Major contribution to national economy: estimated annual production of 5.25 million MT of rice, country's 20% inland fish production, 22% cattle production, and 22% duck production.
- Biodiversity niche including home for migratory birds, aquatic wildlife,
- Hakaluki Hoar (Moulvibazar district) and Tanguar Hoar (Sunamganj district) are declared as Ecologically Critical Area (ECA) and RAMSAR site, respectively.
- The only wetlands permanent sanctuary (Baikka Beel) of Bangladesh is in Hail Hoar (Moulvibazar district)
- Reservoir for rain and flood water and controller of siltation
- Source of recharging ground water
- Source of drinking water

(*Source:* reproduced and updated from HCTT, 2017)

as their primary occupation, while fisheries serve as a secondary choice for many. Within the Haor area, approximately 30% of the residents find themselves living in conditions classified as lower poverty level, emphasizing the complex socio-economic landscape of the region (CEGIS, 2012).

The Haor areas of the Meghna River basin is centred on the Surma-Baulai River and the Kushiyara-Kalni River system. There are three distinct river systems in the northeast region, Kushiyara-Kalni-Dhaleswari-Upper Meghna River System, Surma Baulai River System, and Bhugai-Kangsha River System. Additionally, there exists a network of smaller rivers and khals that interconnect the Haors with the principal rivers. Notably, all these rivers exhibit remarkably rapid fluctuations in water levels.

Haor areas of Bangladesh experiences both flash flood and monsoon (riverine) flood. Flash flood is the common natural hazard in this area, because of its upstream hilly topography and steep slope of the river flowing through the area (Mahtab et al., 2018) and by excessive rainfall in upper mountainous area of India as well as within the country. It occurs in the pre-monsoon period (April-May) causes severe damages to Boro crops just before harvesting in the area. It also damages infrastructures by erosion and channel shifting and often causes loss of lives and properties including houses, livestock, food etc (Choudhury, 2005; Mahtab et al., 2018). As a result, it increases the threat to the economic, social, and environmental fields of the Haor areas (IWM, 2007).

Concept of Submersible Embankment

The hydrological regime in the Hoar basin is conducive to growing only Boro rice, which is vulnerable to pre-monsoon flash floods. The concept of protecting Boro rice from pre-monsoon flash flood through constructing submersible embankment was originally initiated by Zamindars (landlords) in 1920s especially in Shonir Hoar

The Bangladesh Delta Plan-2100 (2018) is the world's first comprehensive delta plan using Adaptive Delta Management (ADM), integrating medium to long-term aspirations for income growth, poverty elimination, and ecological sustainability. It includes 65 physical projects and 15 institutional projects categorized under geographical hotspots, addressing diverse challenges across the nation.

The Eighth Five-Year Plan (8th FYP) for 2021–2025 aligned with the long-term development vision of Vision 2041. This plan recognized the isolation of Haor areas from mainstream services and proposed utilizing idle Haor areas for increased agricultural productivity. Meanwhile, the Bangladesh Water Development Board (BWDB) formulated a Five-Year Strategic Plan based on FAP 6, NWP, and NWMP, focusing on river improvement, erosion control, and flood management in the Upper Meghna River basin.

Flash Flood in Haor Area

Flash floods, swift and destructive in nature, are primarily caused by a complex interplay of factors, prominently among them being excessive rainfall. In the pre-monsoon period, the upper mountainous regions of India witness intense precipitation, resulting in an influx of water downstream. This phenomenon contributes to around 40% of the water in the Haor region, saturating the landscape and elevating the risk of flash floods. Furthermore, a significant portion, approximately 60%, of the water that engulfs the Haor area originates as external inflows from the catchments of India, magnifying the potential for flood events. An astonishing illustration of the scale of this rain-induced vulnerability is Cherrapunji, situated upstream from the Haor region, known to receive the world's highest average annual rainfall, reaching a staggering 12,000 mm. This remarkable statistic underscores the challenging force of nature and the cascading effect that excessive rainfall, especially in upstream regions, can have on downstream areas like the Haor region, resulting in the devastating flash floods that have come to characterize this region. For example, a remarkable rainfall accumulation of 2,458 mm occurred over a mere span of three days, from June 15 to 17, 2022, in Cherrapunji. This extraordinary precipitation figure stands as the highest recorded in the past 122 years since the commencement of rainfall recording, as documented by PDNA in 2023 (PDNA, 2023).

Recent Flash-flood and Its Impact

Flood in 2017

In 2017, extensive rainfall and the surge of water from the upstream Meghalaya hills in India combined to result in the submersion of vast stretches of cropland in the Haor region and low-lying areas of the northeastern part of Bangladesh. The flood event, which commenced on March 28th, affected six districts – Sylhet, Moulavibazar, Sunamganj, Habiganj, Netrokona, and Kishoreganj. As the water levels continued to rise, embankments were breached, and overflowing water engulfed considerable cropland expanses. Notably, the flood inflicted severe damage on approximately 160,170 ha of nearly-mature Boro rice fields, disrupting agricultural activities in the process (Mondal et al., 2019; Nirapad, 2017) (details information in Table 1).

Table 1 Selected districtwise affected land in 2017 flood.

District	*Haor Area (ha)*	*Submerged Land (ha)*
Sunamganj	1666000	113000
Sylhet	76835	34369
Habiganj	116510	27010
Moulvibazar	53426	17432
Kishorganj	165515	24100
Netrokona	184320	48670

In the wake of the devastating flash floods of 2017, the government has undertaken significant initiatives to refine and enhance flood management strategies. A pivotal change has been observed in the KABITA (Taka for works) policy, originally established in 2010. Under the previous KABITA Policy-2010, the responsibility for implementing maintenance and rehabilitation works of embankments rested with Bangladesh Water Development Board (BWDB) through a Project Implementation Committee (PIC). This committee comprised a diverse array of members, including elected officials, women representatives chosen by the Honorable Member of Parliament, Ansar/VDP nominees by the Upazilla Chairman, schoolteachers, and NGO representatives.

Subsequently, the government introduced the KABITA Policy-2017, signifying a fundamental shift in the implementation approach. Under this new policy, the maintenance and rehabilitation works of embankments are now delegated to Upazilla Committees headed by the Upazilla Nirbahi Officer (UNO), who in turn establish the Project Implementation Committees (PIC). These committees consist of five to seven members, focusing on localized expertise and engagement. Remarkably, the PIC members are drawn from the ranks of local farmers residing in the adjacent embankment areas and directly involved in cultivation. This strategic alteration reflects a more community-centric approach, acknowledging the vital role of those with an intimate connection to the land and its management. These changes underscore the government's commitment to evolving flood management strategies by fostering localized knowledge, engagement, and empowerment in the aftermath of the 2017's flash flood.

Flood in 2022

The flash flood of 2022 had a significant impact on the Haor area, particularly in Sylhet, Netrokona, and Sunamganj. The situation deteriorated on April 16th when water levels at the stations in Sylhet exceeded the pre-monsoon danger level (PMDL). Subsequently, stations in Netrokona followed, while Sunamganj floated close to the PMDL. However, post-April 16th, water levels gradually receded, and by the end of April, the floodwaters had subsided.

The effects of this flash flood were wide-reaching. A total of 4,398 ha of crops were damaged out of the vast agricultural expanse covering 472,766 ha. The flood also took a toll on infrastructure, damaging 15 km of embankments out of the 862 km that had been rehabilitated in 2022. Moreover, the flood impacted the functionality of 28 PIC out of a total of 1,049 established in that year.

While the losses were substantial, they were comparatively lower than the losses experienced during the devastating flash flood of 2017 in the Haor area. The consequences extended beyond just physical damages; properties including houses, livestock, and food supplies were also adversely affected. The losses triggered by the flood extended to homestead lands, intensifying the struggles of affected communities. The interruption of economic activities compounded the difficulties, resulting in widespread malnutrition, diseases, and compromised sanitary and hygiene conditions. This situation was further worsened by the low regional literacy rate, impeding effective communication and response efforts. The flood's impact reverberated through community activities, eroding both their development and cohesion. In conclusion, the Haor area's vulnerabilities were laid bare during the flash flood of 2022, underscoring the pressing need for comprehensive disaster management strategies and sustainable development initiatives.

Intervention for Flash Flood Management

Bangladesh Water Development Board (BWDB) has undertaken substantial interventions aimed at mitigating the impacts of flash floods. Among these measures, the construction of 2900 km of submersible embankments stands as a crucial defense against floodwaters. The strategic placement of 128 regulators, 65 sluices, 28 closures, and 3 cross dams further enhances flood management capabilities (BWDB, 2022). Complementing these physical structures, BWDB has embraced technological advancements, utilizing qualitative forecasts based on satellite images from organizations like BMD, IMD, and NOAA, which allow for informed decisions and timely responses. The inclusion of long-range weather forecasts, disseminated nationwide, enables communities to prepare for potential flash flood scenarios. BWDB's current focus on providing a 3-day flash flood forecast specifically for the Northeast region (Sylhet region) employs a Deterministic Flash Flood Model. These multi-faceted initiatives underscore the board's dedication to comprehensive flash flood management, employing both physical infrastructure and cutting-edge forecasting technologies to safeguard communities and assets from the devastating impacts of flash floods.

In response to the devastating flash flood in 2022, the Ministry of Water Resources took action by assembling a 9-member technical committee. This committee was tasked with investigating the potential causes behind the delayed release of floodwaters and proposing practical solutions. The committee's efforts culminated in several recommendations aimed at enhancing early flash flood management in the Haor region. Two notable suggestions include: (a) reserving a few Haors for non-crop purposes and transforming them into permanent reservoirs, and (b) pinpointing suitable Haors where early-maturing rice varieties or other crops could be cultivated, with the flexibility to intentionally allow flooding in emergency situations, thereby safeguarding other areas from the threat of early flash floods. In essence, by converting select Haors into reservoirs, the committee proposed a strategic approach to mitigate the impact of early flash floods in the region (Masood et al., 2022).

Livelihood Management in Haor Area

Livelihood management in the Haor area necessitates a strategic approach that aligns with the unique challenges and opportunities of the region. To diversify income sources and enhance resilience, promoting poultry farms presents an effective avenue. Additionally, bolstering duck cultivation can offer a sustainable source of income, as ducks thrive in wetland environments. Livestock rearing, another viable option, provides both meat and dairy products. Recognizing the role of women in household economies, providing tailored training in sewing machine operation, duck cultivation, and vegetable farming can empower them as vital contributors.

Considering the environmental conditions of the Haor area, cultivating short-duration rice varieties becomes imperative. These varieties ensure quicker harvests, mitigating risks posed by unexpected floods. Furthermore, prioritizing cold-tolerant rice varieties is crucial, given the region's vulnerability to temperature fluctuations. This dual-pronged rice cultivation approach not only improves food security but also bolsters the resilience of communities against climate-related challenges. By carefully aligning livelihood strategies with the unique characteristics of the Haor area, a holistic and sustainable approach can be realized, fostering economic stability and enhancing the well-being of its inhabitants.

Recommendations

This compilation of recommendations strives to offer practical solutions for both flood control and sustaining livelihoods in the Haor area. By taking into account the distinct environment and involving local communities, these recommendations aim to foster a safer, more sustainable future for the Haor area while also safeguarding the means of people's livelihoods.

- A comprehensive approach to the rehabilitation and sustainable management of the Haor area involves a range of strategic measures. Prioritizing the completion of submerged embankment rehabilitation by February is essential to fortify flood defenses ahead of the monsoon season. Additionally, consistent dredging of major rivers like Surma, Kalni, Kushiyara, and others is crucial for maintaining proper drainage and preventing waterlogging.
- To counter flash flood challenges, the cultivation of short-duration rice varieties is recommended, ensuring timely harvests despite unpredictable weather patterns. Embracing advanced agricultural technologies, including Integrated Pest Management (IPM), Integrated Crop Management (ICM), Good Agricultural Practice (GAP), and Crop Diversification, can enhance productivity while minimizing environmental impact.
- Recognizing the region's vulnerability to temperature fluctuations, cultivating cold-tolerant rice varieties remains a priority. Enhancing green cover is imperative for ecological balance; afforestation and grass plantation initiatives should be implemented to prevent soil erosion and bolster the ecosystem. For example, suitable species of trees (i.e., Hijal, Karach, and Barun) should be planted for afforestation at the base (along the toe) of the embankment.

- Permanent nature of infrastructure is to be constructed in critical section of embankment to withstand the forces of nature. By systematically integrating these strategies, the Haor area can progress towards a resilient and sustainable future, better equipped to manage its unique challenges while safeguarding the livelihoods and well-being of its residents.
- Ensuring the sustainable development of the Haor area necessitates the coordination of eco-friendly measures across planning, design, and implementation efforts by key agencies such as BWDB, LGED, and RHD.
- To enhance embankment effectiveness, a comprehensive review of embankment design based on model studies should be a priority, and maintaining a proper slope (at least 1:3) is crucial for stability. It's imperative to maintain a 30-meter buffer zone around embankments, refraining from excavating khals (canals), ponds, or borrow pits to prevent compromising embankment integrity.
- Incorporating local expertise is vital. The participation of local farmers who are directly involved in cultivation within Project Implementation Committees (PIC) should be strictly enforced to ensure community engagement and effective decision-making. Road construction in the Haor region should align with sustainable practices, avoiding all-season roads that might disrupt local ecosystems.
- Empowering the people of the Haor region is essential. Comprehensive training programs for poultry farming, livestock rearing, vegetable cultivation, sewing, and other viable livelihood options should be provided to enhance economic resilience. Proper compaction of embankments is essential to ensure their durability and effectiveness in flood management.
- Lastly, efficient allocation and timely disbursement of funds are critical to the success of development initiatives. By adhering to these principles and adopting a holistic, sustainable approach, the agencies responsible for the development of the Haor area can promote long-term resilience, community well-being, and environmental preservation.

Conclusions

To encapsulate, the Haor region in northeastern Bangladesh is a distinct hydro-ecological area vulnerable to flash floods. These floods stem from heavy rains and impact agriculture, particularly Boro rice cultivation, threatening livelihoods. Contributing factors include excessive upstream rainfall, sedimentation, infrastructure issues, and climate change effects. Flood management challenges, such as pre-monsoon floods and transboundary cooperation, have been discussed alongside significant directives like FAP, NWP, NWMP, FYP, and BDP2100. The region's ecological significance, livelihood strategies, recent flash floods, interventions, and BWDB initiatives have also been examined. Recommendations focus on integrated flood risk management and improving livelihoods. A recurring cycle of flash floods every five years underscores the need for preparation and public awareness to address this challenge effectively.

References

Biswas, R.K., Das, S.C. and Islam, A.K.M.S. (2013). Integration of Environmental Concept for Water Resources Management in Bangladesh Chapter-5, (98-119). *Impact of Climate Change on Socio-economic Conditions of Bangladesh.* Alumni Association of German Universities in Bangladesh (AAGUB).

BWDB. (2022). *Annual Report 2021–2022*, 120 pp. Bangladesh Water Development Board, 72 Green Road, Dhaka, Bangladesh.

CEGIS. (2012). *Master Plan of Haor Area.* Summary Report, Vol. 1, Bangladesh Haor and Wetland Development Board.

Choudhury, A.G. (2005). *Holistic Development of Haor: A Challenge for the Millennium.* Unpublished Document, CEGIS, Gulshan-1, Dhaka.

FAP-6. (1993). *Northeast Regional Water Management Plan.* Government of Bangladesh, FPCO, September 1993.

HCTT. (2017). Coordinated Needs Assessment (CNA). *Floods in Northeast (Hoar) Areas of Bangladesh, April–May 2017.* Government of Bangladesh.

IWM. (2007). *Mathematical Modelling along with Hydrological Studies and Terrestrial Survey under the Haor Rehabilitation Scheme.* Final Report. March, Institute of Water Modelling, Dhaka, Bangladesh.

Khan A.A. (2010). *Impact of a Submersible Embankment Project on Hoar-based Livelihood.* M.Sc. Engineering thesis, Institute of Water and Flood Management, Bangladesh University of Engineering and Technology.

Mahtab, M.H., M. Ohara and M. Rasmy. (2018). The impact of rainfall variations on flash flooding in Haor areas in Bangladesh. *Water Conserv. Manag.*, *2*(2): 6–10.

Masood, M., Ullah, M.E. and Das, S.C. (2022). Causes of Early Flash-Flood in Haors of Bangladesh and Recommendations for Flood Management to Save Crops and to Maintain Biodiversity. *In*: *Book of Abstract, 11th International Perspective on Water Resources and the Environment (IPWE-2023).* Dhaka, Bangladesh, Reston, VA, USA, 4–6 January 2023, AbstractID_100692, ISBN: 978-984-35-3438-5.

Masood, M. and Takeuchi, K. (2016). Climate change impacts and its implications on future water resource management in the Meghna Basin. *Futures*, *78*: 1–18.

Mondal, S., Akter, L., et al. (2019). Effects of 2017 Early Flash Flooding on Agriculture in Haor Areas of Sunamganj. *J. Environ. Sci. and Natural Resources*, *12*(1&2): 117–25.

Nirapad. (2017). Flash Flood Situation., reliefweb, https://reliefweb.int/report/bangladesh/ bangladesh-flash-flood-situation-april19-2017.

PDNA. (2023). *Post Disaster Needs Assessment Bangladesh: Floods 2022*, 50 pp.. Ministry of Disaster Management and Relief, Government of People's Republic of Bangladesh.

Chapter 19

Re-adjustment of Transplanting Time of Rice Seedlings

A Simple and Effective Way to Ensure Food Security for Haor Communities

A.K.M. Golam Sarwar[1*]

Introduction

The Haor, also known as a back swamp, is a bowl- or saucer-shaped tectonic downturn or freshwater wetland environment in the northeastern region of Bangladesh and flooded to a depth of at least 5 m for 7–8 months during the rainy season (Fig. 1; Chakraborty et al., 2021). It is made up of a variety of wetland environments, including rivers, streams, canals, vast stretches of agricultural plains that are periodically inundated, and beels (wetlands). A total of 373 Haor/wetlands cover an area of around 0.86 m ha, or almost 43% of the total area of the Haor districts, viz., Sunamganj, Sylhet, Habiganj, Maulvibazar, Netrakona, Kishoreganj, and Brahmanbaria (CEGIS, 2012). The haors contribute about 20% of the total Boro rice production in the dry season and also provide a rich source variety of fish in the wet season of Bangladesh. Additionally, domestic ducks have recently been raised in the marshes. Haor's economic activities contribute 2% of the country's GDP (Barkat et al., 2019). The principal causes of food insecurity in Haor households were lack of year-round access to land, monoculture (Boro rice), seasonal employment, and natural disasters (Kazal et al., 2010).

In many rice-growing nations, like Bangladesh, rice security is synonymous with food security, which is also a crucial factor in influencing social and political stability (Brolley, 2015; Rashid and Yasmeen, 2017). Rice production is, therefore, the principal agricultural activity in Bangladesh as self-sufficiency in cereals receives priority in national development plans (Anonymous, 2013). On the other hand, the

[1] Department of Crop Botany, Bangladesh Agricultural University, Mymensingh, Bangladesh.
* Corresponding author: drsarwar@bau.edu.bd

changing climate has been adding serious problems to Bangladesh's agriculture. So, the development of climate resilient technologies is a demand of the hour. The adverse environmental conditions that affect Boro-rice cultivation are the low temperature at the beginning of the cropping season and the early flash flood, which coincides with the harvesting time of Boro-crops, at the end (Rashid and Yasmeen, 2017).

Fig. 1 Haor area in Bangladesh. A-B. During rainy season; C-D. During Boro season.

Temperature and precipitation patterns on a global scale are impacted by climate change. The severity and the frequency of extreme environmental events including hurricanes, heat waves, flash floods, droughts, storms, and forest fires are all influenced by these effects. The occurrence of flash floods is a common phenomenon and threatens the food security and livelihood of Haor communities of Bangladesh. For example, the Sunamganj district had significant flash floods in 1998, 2004, 2007, 2010, 2012, 2015, and 2017, showing that catastrophic flash floods typically recur every three years. Heavy rainfalls in the upstream (the mountainous regions of India) are the main cause of (early) flash floods and the occurrence of early monsoon aggravates the situation (CEGIS, 2012). In 2017, flash floods in Haor districts caused colossal damage to Boro-rice production due to poor management of embankments, and Haor farmers were the worst sufferers of the flooding situation (Seraj, 2018; Chakraborty et al., 2021). According to the hazard calendar (Rashid, 1977 *cf.* Chakraborty et al., 2021), the first week of May to June is the peak time for the occurrence of flash floods in Bangladesh. However, due to climate change variabilities, the flash flood recently occurred in April, even as early as mid-March (Table 1). In addition, flooding in these wetlands in the northeast of India is predicted to increase soon due to climate change, making the future viability of farming and production systems even more critical (Dey et al., 2021). This necessitates the identification of appropriate crop cultivars and production systems to improve food security in the Haor areas.

The remedial measures to avoid crop damage from flash floods might be to cultivate short-duration rice cultivars, and/or the early transplanting of HYV rice cultivars. But short-duration or low-temperature tolerant rice cultivars with high

Table 1 Date of flood water flow enters into the Haor and consequent flash flood from 1990–2023 in the study area.

Year	*Entrance of Flood Water*	*Declaration of Flash Flood*
1990	18 March	21 March
1991	22 May	25 May
1992	8 May	10 May
1993	27 April	30 April
1994	30 May	2 June
1995	25 May	27 May
1996	16 March	18 March
1997	22 May	24 May
1998	20 May	23 May
1999	3 May	6 May
2000	28 April	30 April
2001	27 April	30 April
2002	14 April	18 April
2003	27 May	30 May
2004	13 April	15 May
2005	22 May	25 May
2006	17 April	19 April
2007	25 April	28 April
2008	2 May	5 May
2009	16 April	18 April
2010	17 April	19 April
2016	–	3 May
2017	12 March	28 March
2022	1 April	6 April

(Adapted and modified from Awal, 2022).

yield potential is yet to be developed (Rashid and Yasmeen, 2017). On the contrary, early transplanting especially in Haor areas to avoid flash floods often causes chilling injury resulting in grain sterility during Boro season (BRRI, 2007; Biswas, 2011). Low temperature (as low as 12°C), prevails for 6 days or more during the season, may induce almost 100% grain sterility in rice (Kabir et al., 2008). Results of our previous studies have shown that the hybrid rice cultivars were more tolerant to low-temperature stress-induced spikelet sterility compared to inbred rice cultivars (Sarwar et al., 2017). Therefore, this study was designed to find out low-temperature tolerant cultivar(s) from among the available hybrid rice cultivars in Bangladesh through their early transplanting in the Boro season in natural conditions.

Materials and Methods

The experiments were conducted at the farmer's field in Doarabazar, Sunamganj during the Boro season for two consecutive years 2017–2018 and 2018–2019. Ten hybrid rice cultivars, viz., BRRI hybrid dhan2, BRRI hybrid dhan3, BRRI hybrid dhan5, SL-8H, Agomony, Hybrid LP-70, Partex-1, Balia-2, Mota Modina, Nobin, were used as experimental materials in 2017–2018. While 12 hybrid rice cultivars, viz., BRRI hybrid dhan2, BRRI hybrid dhan3, BRRI hybrid dhan5, SL-8H, Hybrid LP-70, Agomoni JBS-17(4), Win-302, Winall Hybrid 2, Nobin IS-1, Ispahani 1, Durbar IS-2, and Ispahani 2, were used in the following year (2018–2019). The experiment was laid out in a randomized complete block design with three replications.

Thirty-days old healthy rice seedlings were transplanted in the experimental plot and the standard rice cultivation procedure (BRRI, 2017) was followed. The transplanting operation of rice seedlings was done following the farmer's practice in Boro season of 2017–2018; and on 21 November 2018 (almost one month before the traditional practice) in the 2018–2019 season. The crop was harvested at the maturity stage, 80% of grains attained golden color, and data regarding growth and yield contributing descriptors, viz., plant height, tiller number, number of effective and non-effective tillers, length and maximum width of flag leaf, panicle length, number of primary and secondary branches of panicle, number of filled and un-filled grains panicle^{-1}, 1000-grain weight, grain yield (t ha^{-1}).

The collected data were analyzed following the ANOVA using the statistical computer package program MSTAT-C. The significant mean differences of measured parameters among the cultivars were evaluated using Duncan's New Multiple Range Test (DMRT) at $p \leq 0.05$ unless otherwise mentioned (Gomez and Gomez, 1984).

Results and Discussion

In the 2017–2018 growing season, out of 10 cultivars, seeds of six cultivars including BRRI hybrid dhan5, Hybrid LP-70, Partex-1, Balia-2, Mota Modina, Nobin, failed to germinate and/or to produce healthy seedlings. It might be due to old seeds (previous years) supplied by the seed vendors or improper handling of seeds during storage. Most of the yield and yield contributing descriptors, e.g., plant height, number of effective tillers, panicle length, number of primary and secondary branches, etc. differed significantly (Table 2). These descriptors might be controlled by the genetic makeup of respective cultivars (BRRI, 2017). However, there was no significant difference observed in pollen morphological descriptors and grain yield, which might be due to the favorable weather conditions prevailing during the cropping season. All the cultivars studied may possess similar yield potential, though the BRRI hybrid dhan-3 produced the maximum grain yield (7.85 t ha^{-1}).

In the 2018–2019 Boro season, similar significant variations in all growth and yield contributing descriptors, viz., plant height, tiller number, number of effective and non-effective tillers, length and width of flag leaf, panicle length, number of primary and secondary branches, number of filled and un-filled grains panicle^{-1}, 1000-grain weight, grain yield (t ha^{-1}), among the cultivars studied were recorded (Table 3). The genetic potential and growing (environmental) conditions of the rice crop determine

Table 2 Growth and yield attributes of hybrid rice cultivars, 2017–2018.

Cultivars	*Plant Height (cm)*	*Tiller / Hill (no.)*	*Effective Tiller/ Hill (no.)*	*Non-effect Tiller/ Hill (no.)*	*Flag Leaf*		*Panicle Length (cm)*	*Primary Branch/ Panicle (no.)*	*Secondary Branch/ Panicle (no.)*	*Total Grain /Panicle (no.)*	*Filled grain/ Panicle (no.)*	*Unfilled Grain/ Panicle (no.)*	*1000 -seed Weight (g)*	*Grain Yield (t/ha)*
					Length (mm)	*Width (mm)*								
BRRI hybrid dhan2	100a	14.4a	13.8a	0.6b	208a	16.2a	25.0a	11.4bc	15.8ab	124.8a	109.4ab	15.0a	28.76	7.54
BRRI hybrid dhan3	97ab	14.8a	12.4ab	2.4ab	225a	17.8a	25.2a	17.4a	14.2ab	139.4a	132.0a	7.4b	28.08	7.85
SL8H	91.8bc	12.6a	10.4b	2.2ab	249a	18.2a	21.0b	13.6b	13.0b	132.0a	119.4ab	12.6ab	27.26	7.65
Agomany	88.4c	13.8a	11.4ab	2.6a	268a	18.0a	23.6a	10.6c	18.2a	112.4a	97.2b	15.2a	27.80	7.51

Note: Means under each column separated by same lower-case letter were not significantly ($p \leq 0.05$) different according to DMRT.

Table 3 Growth and yield attributes of hybrid rice cultivars 2018–2019.

Cultivars	*Plant Height (cm)*	*Tiller / Hill (no.)*	*Effective Tiller/ Hill (no.)*	*Non-effect Tiller/ Hill (no.)*	*Flag Leaf*		*Panicle Length (cm)*	*Primary Branch/ Panicle (no.)*	*Secondary Branch/ Panicle (no.)*	*Total grain /panicle (no.)*	*Filled Grain/ Panicle (no.)*	*Unfilled Grain/ Wanicle (no.)*	*1000 -seed weight (g)*	*Grain Yield (t/ha)*
					Length (mm)	*Width (mm)*								
BRRI hybrid dhan2	62.4 de	14.6 a	13.3 ab	1.3 bc	18.20 b-d	10.3 e	20.10 cd	8.5 bc	13.5 ef	89.5 de	25.9 d	28	26.76	3.85 e
BRRI hybrid dhan3	65.0 c-e	15.8 a	14.1 a	1.7 a-c	19.15 b-d	11.2 c-e	20.67 b-d	7.8 c	12.2 f	83.5 e	28.7 cd	34	29.24	4.14cd
BRRI hybrid dhan5	74.1 ab	11.0 c	10.0 d	1.0 bc	21.70 ab	11.9 bc	22.94 a	10.6 a	29.6 a	160.1 a	47.9 a	29	31.89	5.90 a
SL-8H	64.2 c-e	13.1 a-c	11.0 b-d	2.1 ab	19.31 b-d	12.0 a-c	17.85 f	9.0 bc	18.6 cd	92.6 c-e	34.1 a-c	37	25.97	4.05 d
Durbar IS-2	66.4 cd	14.1 ab	13.1 a-c	1.0 bc	17.20 cd	10.4 de	21.24 b	8.4 bc	14.8 d-f	105.5 cd	21.3 d	20	26.97	4.25 c
Nobin IS-1	75.4 a	12.7 a-c	12.1 a-d	0.4 c	23.65 a	13.2 a	23.63 a	10.5 a	26.0 ab	140.7 b	35.4 a-c	25	28.10	5.88 a
Hybrid LP-70	61.7 e	14.1 ab	12.7 a-c	1.4 bc	16.15 d	11.1 c-e	19.95 d	8.1 c	13.9 ef	93.1 c-e	29.4 cd	31	26.66	4.32 b
Ispahani 1	65.1 c-e	14.8 a	12.2 a-d	2.6 a	19.35 b-d	12.5 ab	20.10 cd	8.5 bc	16.4 d-f	93.5 c-e	30.8 a-d	33	28.24	4.52 b
Ispahani 2	66.8 cd	12.7 a-c	11.3 a-d	1.4 bc	20.40 a-c	11.5 b-d	18.58 ef	9.5 ab	22.9 bc	109.0 c	44.6 ab	41	26.96	3.44 g
Winall hybrid-2	63.0 de	14.1 ab	12.2 a-d	1.9 ab	18.05 b-d	10.5 de	21.01 bc	8.4 bc	15.6 d-f	90.6 de	32.9 a-d	36	29.36	3.68 f
Agomani JBS-17(4)	63.2 de	11.8 bc	10.4 cd	1.4 bc	19.45 b-d	12.2 a-c	20.25 cd	8.2 bc	17.2 de	105.4 cd	31.2 b-d	29	27.93	4.15cd
Win-302	69.9 bc	11.8 bc	11.3 a-d	0.5 c	20.20 a-c	10.7 de	19.02 e	8.6 bc	24.5 b	107.7 cd	30.4 b	28	28.51	4.06cd

Note: Means under each column separated by same lower-case letter were not significantly ($p \leq 0.05$) different according to DMRT.
(Adapted and modified from Sarwar et al., 2020)

its growth and yield (Yoshida, 1981; Sarwar and Chanda, 2020). According to Yoshida (1981), grain yield depends on panicle number m^{-2}, spikelet number panicle^{-1}, full spikelets (%), and 1,000-grain weight (g). With the exception of the number of unfilled grains (sterile spikelets) panicle^{-1}, all of the analyzed descriptors' numerical values decreased as a result of the early transplanting, which may be one of the causes of the low yield (Tables 2, 3). The highest grain yield among the cultivars was obtained by BRRI hybrid dhan5 (5.90 t ha^{-1}), followed by Nobin IS-1 (5.88 t ha^{-1}) with no statistically significant difference (Table 3). The largest flag leaf, the longest panicle length, the most primary and secondary branches, and the filled grain panicle^{-1} in the different cultivars may be related to the maximum grain output (Tables 2, 3).

The grain yield of some cultivars was occasionally as much as 48% lower than the yield of the previous year when all growth conditions were ideal, i.e., without any low-temperature stress and absence of the flash flood. The integration of low-temperature stress-related spikelet sterility, increased prevalence of insect and disease pests, hailstorms just a few days before harvest, and other factors that predominated over the growing season may be to blame for these decreased yields. However, this technology, early transplanting of (hybrid) rice seedlings of suitable cultivars, could be adapted in the field due to the grain yield of some cultivars, higher or equivalent to the national average yield of 4.8 t ha^{-1} (https://knoema.com/atlas), and/or complete destruction by early flash flood in several years (http://www.theindependentbd.com/home/printnews/90337). A secure harvest of Boro rice will be necessary to maintain a safe and secure rice harvest for the local population (Baishakhya et al., 2023). The marginal and smallholder farmers in the Haor wetlands are regrettably more susceptible to crop loss risk and typically have a low ability for adaptation and limited access to alternate production methods. Among the cultivars, BRRI hybrid dhan5 and Nobin IS-1 could be suggested for the highest grain yield at the earlier transplanting. Further experimentations are necessary to identify (more) cultivars suitable for early transplanting in the Boro season in the Haor regions.

Field Constraints

One of the main obstacles to the early transplanting of Boro rice in the Haor area was the mindset of the farmers. Motivating the farmer(s) to begin transplanting early was extremely tough. Even while the transplanting of rice seedlings into the experimental plots was taking place, other farmers were still criticizing the process. Bird attacks and hailstorms, particularly during the ripening stage when plants in all other rice fields were in the milk-dough or grain-filling stage, posed other difficulties. To get out of this condition, a community effort for early transplantation in a large land area would be helpful. Insect and disease pest incidence might also be decreased by doing this.

Acknowledgments

We thank the Department of Agricultural Extension personnel in Dowarabazar, Sunamganj for their active assistance. The Ministry of Science and Technology of the People's Republic of Bangladesh (grant #39.00.0000.09.06.79.2017/2/BS-379) and the Bangladesh Agricultural University Research System (grant #2017-252-BAU) provided partial funding for this work.

References

Anonymous. (2013). *National Sustainable Development Strategy. 2010–2021*, 123 pp.General Economics Division, Planning Commission, Govt. of the People's Republic of Bangladesh, Dhaka.

Awal, M.A. (2022). Rain and flood characteristics, agricultural damages and coping strategies in northeastern Haor region of Bangladesh. *Discovery*, *58*: 1274–96.

Baishakhya, S.D., Islam, M.A. and Kamruzzaman, M. (2023). Overcoming barriers to adapt rice farming to recurring flash floods in Haor wetlands of Bangladesh. *Heliyon*, *9*: e14011.

Barkat, A., Suhrawardy, G.M. and Rahman, M.I. (2019). *A Study on Haor Governance and Haor Dwellers' Rights in Bangladesh*. Final Report. Human Development Research Centre, Dhaka.

Biswas, J.K. (2011). *Cool-injury of Rice: Modern vs. Traditional Science*. (in Bangla). http://www.sangbad.com.bd/?view=details&type=gold&data=Entertainment&pub_no=597&menu_id=23&news_type_id=1&val=55338.

Brolley, M. (2015). *Rice Security is Food Security for Much of the World. Rice Today*, 30–32. International Rice Research Institute, Manila, Philippines.

BRRI (Bangladesh Rice Research Institute). (2007). *Boro-Dhane Chitar Karon O Protikar* (A booklet on Boro-rice sterility and its mitigation). Workshop of BRRI-DAE, 19 April. BRRI, Gazipur.

BRRI (Bangladesh Rice Research Institute). (2017). Modern Rice Cultivation (in Bangla), 1–80. 20th Edn., BRRI, Gazipur.

CEGIS (Center for Environmental and Geographic Information System). (2012). *Master Plan of Haor Area*. Vol. I, Summary Report. Bangladesh Haor and Wetland Development Board. Ministry of Water Resources, Govt. of Bangladesh.

Chakraborty, D., Mondal, K.P., Islam, S.T. and Roy, J. (2021). Flash Flood in Bangladesh: Lessons learnt. *In*: Shrestha, S., Djalante, R., Shaw, R., Pal, I. (Eds.), *Disaster Resilience and Sustainability Adaptation for Sustainable Development*, 591–10. Elsevier.

Dey, N.C., Parvez, M. and Islam, M.R. (2021). A study on the impact of the 2017 early monsoon flash flood: Potential measures to safeguard livelihoods from extreme climate events in the Haor area of Bangladesh. *Int. J. Disaster Risk Reduction*, *59*: 102247.

Gomez, K.A. and Gomez, A.A. (1984). *Statistical Procedures for Agricultural Research*. (2nd Edn.). John Wiley and Sons, New York.

Kabir, M.S., Howlader, M., Biswas, J.K., Mahbub, M.A.A. and Nur-e-Elahi, M. (2008). Probability of low temperature stress at different growth stages of Boro rice. Paper presented at *Intern. Symp. on Climate Change and Food Security in South Asia*, August 28–30. Dhaka. http://www.wamis.org/agm/meetings/rsama08/Bari203-Kabir_BRRI.pdf.

Kazal, M.M.H., Villinueva, C.C., Hossain, M.Z. and Das, T.K. (2010). *Food Security Strategies of the People Living in Haor Areas: Status and Prospects*. Final Report PR #3/08. Food Planning and Monitoring Unit, Ministry of Food, Govt. of Bangladesh.

Rashid, M.M. and Yasmeen, R. (2017). Cold injury and flash flood damage in Boro rice cultivation in Bangladesh: A review. *Bangladesh Rice J.*, *21*: 13–25.

Sarwar, A.K.M. Golam and Chanda, S.C. (2020). Growth and yield descriptors of rice influenced by transplanting date. *J. Agric. Food Environ.*, *1*: 48–52.

Sarwar, A.K.M. Golam, Chanda, S.C., Islam, M.M. and Jannat-E-Tajkia. (2020). Screening of rice cultivars suitable for early transplanting in Haor Agriculture. *J. Bangladesh Agric. Univ.*, *18*: 963–67.

Sarwar, A.K.M. Golam, Sohel, S.A., Islam, M.A. and Ashrafuzzaman, M. (2017). Palynological features and grain yield of rice as influenced by air temperature and relative humidity. *Bangladesh J. Bot.*, *46*: 947–54.

Seraj S. (2018). What protections for Haor crops? *The Daily Star*. Published on 29 March 2018. https://www.thedailystar.net/country/what-protections-haor-crops-1554760.

Yoshida S. (1981). *Fundamentals of Rice Crop Science*. International Rice Research Institute, Los Baños, Philippines.

CHAPTER 20

Bacillus thuringiensis (*Bt*) Biopesticide for the Control of Insect Pests of Agriculture Crops and Vectors of Human Diseases

Bangladesh Perspective

Mahbub Rahman[1] and *Mustafizur Rahman*[2]

Introduction

The world's urban population has grown exponentially over the last two decades. To feed this larger, more urban and richer population, food production must double by the year 2025. In addition, this massive urbanization creates the emergence of a severe risk of vector-borne diseases due to the destruction of insect habitats and rapid changes in the biology and ecology of insect vectors of human diseases associated climate change. Climate change and extreme weather events have a major impact on agricultural pests and insect vectors of human diseases. Climate changes can result in an expansion of insects' geographic distribution, increased survival during overwintering, increased number of generations, altered interspecific interaction, increased risk of invasion by migratory pests, increased incidence of insect-borne diseases, and reduced effectiveness of chemical and biological control agents. As a result, there is a serious risk of crop failure and economic losses, as well as a challenge to human food security, vector-borne diseases, and biosecurity.

Soil bacterium *Bacillus thuringiensis* producing insecticidal crystalline proteins, so-called Cry toxins (Bt cry toxins) are used widely as an alternative or supplement to synthetic chemical pesticides used in production agriculture, forest management, and control of vector insects (Ferre and Van Rie, 2002), as they are specific to their target pests, are safe to other animals and plants, and are biodegradable (Schnepf

[1] School of Agriculture, Food and Wine, The University of Adelaide, Australia.
[2] Department of Soil, Water and Environment, University of Dhaka, Bangladesh.
* Corresponding author: drmahbub.rge@gmail.com

et al., 1998). The Bt cry toxins are also a key source of genes for transgenic expression to provide pest resistance in plants, and thereby reduce reliance on conventional pesticides, providing economic viability, public health benefits, and environmental compatibility (Shelton et al., 2002). Therefore, the adoption of Bt transgenic crops by the farmers has exceeded all expectations, covering more than 80 M ha since 1996 (Bates et al., 2005). In the USA alone, Bt corn acreage grew from approximately 8% in 1997 to 19% in 2000, before climbing to 85% in 2023 (Economic Research Service, USDA).

In 2013, Bangladesh commercialized Bt eggplant (locally known as Bt brinjal), a genetically modified eggplant that is resistant to the fruit and shoot borer pests. This technology not only allows farmers to reduce their use of pesticides by 80% or more, but it also decreases yield loss due to pest damage (Shelton et al., 2018). Success of the Bt brinjal partnership is likely to affect the future of other genetically engineered (GE) crops in Bangladesh, as well as other parts of the world where biotechnology is needed for food security and environmental compatibility. In Bangladesh, mosquito-borne disease "dengue infection" is seeing a huge rise in the capital and has spread to other parts of the country for the last several months. This raised a considerable discussion amongst vector control experts and public in Bangladesh as Dhaka South City Corporation (DSCC) introduced *Bacillus thuringiensis subspecies israelensis* (Bti) formulation to control Aedes mosquitoes.

Bt crops and Bt biopesticides (Bti) are great tools to control insect and pest infestation of agriculture crops and mosquito vectors of human diseases, but the benefits may be very short-lived by the evolution of insect resistance. Understanding how Bt toxins work and how insects become resistant will provide the basis for taking measures to counter resistance. Without in-depth knowledge of "Bt mode of actions and resistant mechanisms and hands-on practical skills and proactive resistant management plan", indiscriminate use of Bti for the control of mosquitoes will allow the insect to develop resistant rapidly within very short period of time and within a couple of generations. If this happens, there will be no good use of Bt biopesticides to control the ever-growing mosquito problem in Dhaka City and Bangladesh, and neighboring South and Southeast Asian Countries. Considering public perception and general knowledge of Bt crops and Bt biopesticides amongst different stakeholders in Bangladesh, we provide a brief commentary about *Bacillus thuringiensis*, its molecular structure, mode of actions, and resistant mechanism.

Bt-endotoxins (Cry-proteins)

B. thuringiensis, commonly known as Bt, is a naturally occurring, spore-forming soil bacterium which produces large quantities of insecticidal crystal proteins, d-endotoxins or Cry-toxins, during the stationary phase as part of its natural life cycle (Agaisse and Lereclus, 1995). *B. thuringiensis* produces large quantities of Cry-toxins, each having defined host specificities and have gained widespread popularity for their ability to control a wide range of insect pests and nematodes in a natural, environmentally compatible manner (Ferre and Van Rie, 2002; Gould, 1998). Changes in a single amino acid in Cry-toxins have been shown to have dramatic effects on the range and specificity of activity against insect species (Chandra et al., 1999; Rajamohan et al.,

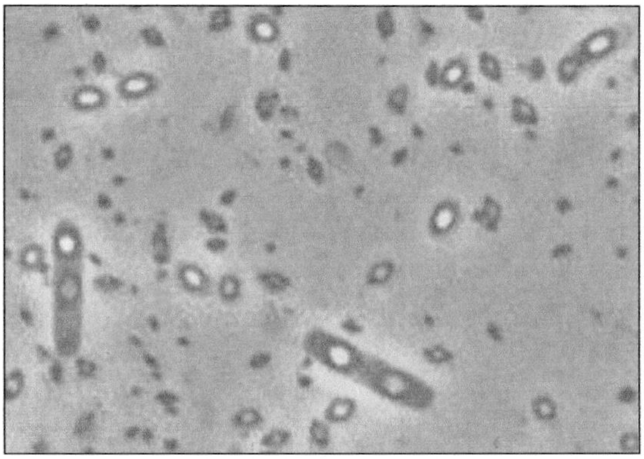

Fig. 1 *B. thuringiensis* viewed by phase contrast microscopy. The vegetative cells contain endospores (phase bright) and crystals of an insecticidal toxin (δ-endotoxin). Most cells have lysed and released the spores and toxin crystals (Deacon, 2003).

1996). With modern biotechnology, the genes coding for Bt Cry-toxins are readily isolated and introduced into plants and become stably inherited traits of new cultivars (Babu et al., 2003; Deacon, 2003). (Fig. 1)

A Brief History as Biopesticides

B. thuringiensis was first identified in grain storage facilities in Japan in 1901 (Ishawata, 1901). Then, in 1911, it was rediscovered as a pathogen in flour moths in the province of Thuringia, Germany (Berliner, 1911). In 1938, Bt Cry-toxins was first used as a commercial insecticide in France (Worthington, 1991), and then in the USA during the 1950s (Ghassemi, 1981). However, more effective Bt formulations were introduced in the 1960s with the discovery of various highly pathogenic strains, each of which produces its own unique insecticidal crystal protein, with particular activity against different types of insects (McGaughey and Johnson, 1994). Bt Cry-toxin gene was transferred and expressed in *E. coli* in 1981 (Schnepf and Whiteley, 1981) and in vegetable plants in 1987 (Vaeck et al., 1987). The first Bt corn was registered by US EPA in 1995 and commercially introduced in USA in 1996 (Shelton et al., 2002).

Three Bt crops are now commercially available: corn, cotton, and potato, and several others are in various stages of development including potato, rice, maize, wheat, canola, soybean, tobacco, tomato, apple, peanuts and broccoli (Babu et al., 2003; Shelton et al., 2002). Bt crops are currently grown in 17 countries and the adoption rates are rising rapidly with the global area of Bt crops grown increasing 24% between 2002–2003 (Bates et al., 2005). As of now, cotton is the most popular of the Bt crops and has led to a 3–27% increase in cotton yield in countries where it is grown since it was commercialized in Australia and USA in 1996 (Jarman, 2006). The adoption of Bt crops is predicted to continue its steady growth, influenced by an increasing confidence in the benefits of Bt crops and the availability of new Bt varieties that confer protection against additional crop pests (Bates et al., 2005).

Host Range and Commercial Products

Over the years, Bt Cry-toxins was available only for the control of Lepidoptera, using a highly potent strain (*B. thuringiensis* var *kurstaki*). Toxin from this strain still forms the basis of many Bt formulations (Meadows et al., 1992). Further screening of a large number of toxins from other Bt strains have shown some that are active against larvae of Coleoptera or Diptera. Most of these toxins have the same basic toxin structure, but differ in insect host range, perhaps because of different binding affinities to the toxin receptors in the insect gut (Deacon, 2003). Many commercial products from the toxin of *B. thuringiensis* var *kurstaki* for the control of Lepidoptera are currently marketed under various trade names such as Biobit®, Dipel®, and Javelin®.

The toxins produced by the strains of *B. thuringiensis* var *israelensis* are highly active against stimuli blackfly vectors of some tropical diseases, and also against fungus gnat larvae. This strain is also a key source of toxin for larval control of vector mosquito (Abdullah et al., 2003). The commercial products of this strain are Skeetal®, Vectobac®, and Mosquito Attack®. Toxins from *B. thuringiensis* var *san diego* or *B. thuringiensis* var *tenebrionis* are marketed for the control of some Coleoptera, especially for the control of the important Colorado potato beetle.

Classification of Cry-toxins

Toxins from *B. thuringiensis* are classified according to their size, crystal shape, and specificity: the major group is insect-specific Cry-toxins and the other one is cytolytic Cyt toxins (Schnepf et al., 1998) (Table 1). The Cry-toxins are encoded by different genes and have distinct three-domain structures with amphipathic a-helix (Li et al., 1991), where Cyt-toxins have only one b-domain that can enhance the effectiveness of Cry-toxins for insect control (Li et al., 1996). As of now, about 311 different genes that encode the Cry-toxins have been identified and sequenced (Crickmore et al., 2006), and this enables the toxins to be assigned to more than 15 groups on the basis of sequence similarities (Deacon, 2003).

Table 1 Bt-endotoxins and their host specificity (adopted from Deacon, 2003).

Bt Cry-toxins	*Protein Size (kDa)*	*Insect Activity*
Cry1 [several subgroups: A(a), A(b), A(c), B, C, D, E, F, G]	130– 138	Lepidoptera
Cry2 [subgroups A, B, C]	69– 71	Lepidoptera and Diptera
Cry3 [subgroups A, B, C]	73– 74	Coleoptera
Cry4 [subgroups A, B, C, D]	73– 134	Diptera
Cry 5–9	35– 129	Nematodes and others

Structure of Cry-toxins

The three dimensional structures of trypsin activated Lepidoptera specific Cry1Aa (Grochulski et al., 1995), Coleoptera specific Cry3A, and Cyt2A (Li et al., 1991) have been identified (Fig. 2). While Cry3A and Cry1Aa possess three domains and 36% amino acid sequence identity (Li et al., 1991), the Cyt2A consists of a single domain

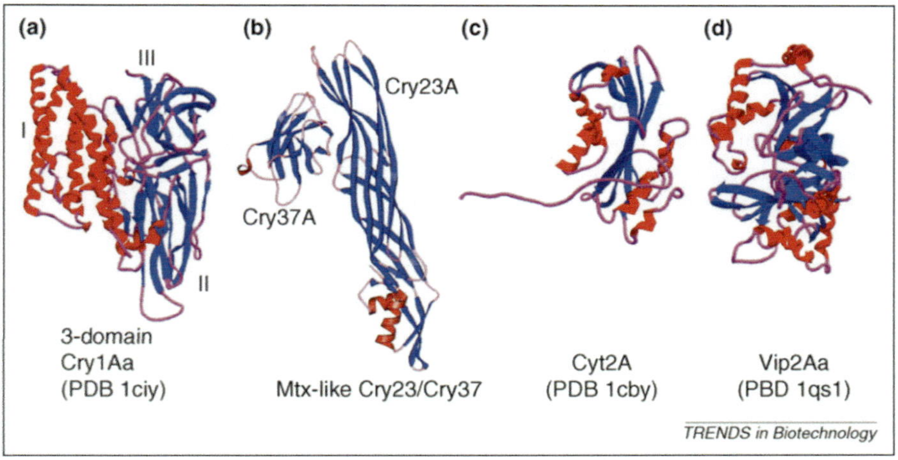

Fig. 2 Three-dimensional structure of representative toxins produced by *Bacillus thuringiensis* (Bravo et. al. 2008). (a) The Cry1Aa protein is shown as an example of the three domain family Cry proteins. Domain I (in red) is the pore-forming domain and domains II and III (in blue) are important for toxin-receptor interactions. (b) The Cry23/Cry37 as an example of binary Mtx-like Cry toxins. The Cry23 toxin shares homology with the Mtx toxins of *Bacillus sphaericus*, which can kill mosquitoes, but the Bt-Mtx-Cry toxins are active against coleopteran insects. Image kindly supplied by T.J. Rydel, Monsanto Co. (c) The Cyt2Aa toxin as an example of Cyt toxins. (d) The Vip2Aa as an example of the Vip toxins. These proteins are secreted during vegetative growth. Vip1 and Vip2 constitute a binary toxin active against some coleopterans.

and 20% amino acid sequence identity with Cry3A and Cry1Aa (Hofte and Whiteley, 1989; Li et al., 1996).

Domain I in Cry3A and Cry1Aa consists of seven antiparallel a-helices in which the 5th helix is encircled by the remaining helices and some, or all of the helices are thought to have pore-forming activity (Grochulski et al., 1995); Domain II consists of three antiparallel b-sheets, similar to the antigen-binding regions of immunoglobulins, and believed to bind to receptors in the gut lumen (Lee et al., 2000b); and Domain III consists of two twisted, antiparallel b-sheets forming a b-sandwich, which is thought to responsible for the toxicity (de Maagd et al., 1996) and to protect the exposed C-terminal region of the active toxin, preventing further cleavage by gut proteases (Aronson and Shai, 2001). The ability to enhance toxicity by domain III is due to its lectin-like fold in b-sheets, which can facilitate the toxin binding to receptors with specific glycodeterminants (Akao et al., 1999; Burton et al., 1999) and trigger the conformational change in Doman I for insertion into the membrane (Li et al., 2001).

Mode of Action of Cry-toxins in Insect Midgut

The Bt-crystals are b-sheet aggregates of a ca. 130 KDa protein that is actually a protoxin and must be activated by certain midgut proteases to become a 60–65 KDa active toxin (Hofte and Whiteley, 1989). The crystal protein is highly insoluble in normal and acid pH conditions, so it is entirely safe to humans, higher animals, and most insects (Schnepf et al., 1998). However, it is solubilized under reducing condition and high pH (above pH 9.5), the conditions commonly found in the mid-gut of lepidopteran larvae (Knowles and Dow, 1993).

In vitro studies of the mode of action of Bt Cry-toxins by using a variety of systems including insect cell lines, whole insect gut tissues, and brush border membrane vesicle preparations of insect midgut indicated that elucidation of the toxicity mechanism is made complicated by the number of subspecies of *B. thuringiensis* studied by different investigators, the production of more than one toxin by the same bacterium, a wide number of susceptible hosts, the interaction between crystals and spores in toxicity in some larvae, and multiple receptors for the same toxin in individual host. The molecular mechanisms of Cry-toxins have been discussed intensively, with several hypothesis based cascades of multi-step activities (de Maagd et al., 2001; Knowles and Dow, 1993).

A basic model for the mechanism of action of Bt Cry-toxins has been widely accepted and involves solubilization of the crystal in the insect midgut, proteolytic processing of the protoxin by midgut proteases, binding of the Cry-toxin to midgut receptors, and insertion of the monomeric toxin into the apical membrane to create tetrameric ion channels or pores (de Maagd et al., 2001). A susceptible insect digests the Cry-toxins with its diet. Once digested, the protoxin is solubilized under midgut alkaline condition and the protoxin is cleaved by a gut protease such as trypsin or chymotrypsin to produce an active toxin (Carroll and Ellar, 1993), where chymotrypsin has been shown to be necessary for solubilization at least for Cry3A (Carroll et al., 1997). During solubilization, a large segment of the protoxin is removed and leaves a protein-resistant active core toxin (Knowles, 1994). Genetic deletion studies have shown that C-terminal half of the 130 kDa Cry1 toxins is removed in this process, indicating that the first 600 amino acid residues are the insecticidal determinants in this group of toxins, while the Cry2 and Cry3 toxins could be truncated versions of Cry1 where proteolytic activation occurs primarily at the N-terminus (Bravo et al., 2002). Differences in the solubilization may affect the degree of toxicity among the Cry-toxins (Bora et al., 1994). For example, a reduction in solubility is thought to be involved for some insect resistance to the Cry-toxins (Crickmore et al., 1998).

After proteolytic activation, the active toxin passes through the peritrophic membrane (PM) and binds to specific receptors in the brush border membrane vesicles (BBMVs) of midgut epithelial cells (Gill et al., 1992; Hofmann et al., 1988; Van Rie et al., 1989). Therefore, the structure and functional properties of the PM for Bt-toxicity are of vital importance. However, the interaction of the Cry-toxins with the PM, a major component of the midgut lumen, is not well understood (Granados et al., 2001).

The PM is an extracellular thin sheet composed of proteins, glycoproteins, and chitin microfibrils in a proteoglycan matrix which, in one or more stages of the insect life, envelops the midgut lumen contents and lines on the microvilli of the underlying epithelial cells (Lehane, 1997). This separates the food from the midgut epithelium in most but not all the insects, and plays key roles in the intestinal biology of the insect, such as regulation of insect digestion (Terra, 2001). The PM is involved in several other functions of critical importance for the insect survival; it prevents mechanical lesions of the apical cell membrane and non-specific binding to cell surface and acts as a semipermeable membrane regulating passage of nutrients from the lumen to the intestinal epithelium (Binnington et al., 1998). It is also an important innate defense barrier against ingested parasites, pathogens, and toxins (Binnington et al.,1998; Wang and Granados, 1997; Wang and Granados, 2000).

Toxin binding to midgut receptors, which is mediated by domains II and III (Ballester et al., 1999a; Lee et al., 2000b; Li et al., 1991), and is dependent on the affinity of receptors in BBMVs of the midgut in susceptible larvae (Ballester et al., 1999b; Gill et al., 1992; Van Rie et al., 1990), is a crucial step for the toxicity of the Cry-toxins. A major conformational change is required to transform the initially soluble Cry-toxin into a structure able to insert into the membrane, where the conformational change was envisaged to expose a relatively non-polar helix hairpin from domain I to initiate membrane penetration (Li et al., 1991). The helices in domain I of Cry-toxins are largely amphipathic, but the α4-loop-α5 region is the most hydrophobic (Haider et al., 1986; Morse et al., 2001). The 'umbrella' model for pore formation by the Cry-toxins (Gazit et al., 1998; Knowles, 1994; Li et al., 1991) suggests that a pair of helices, probably α4 and α5 (Masson et al., 1999; Schwartz et al., 1997) insert into the membrane. The remaining helices were rearranged to open on the membrane surface like the ribs of an umbrella. Several studies have suggested that the toxin binding to the receptors depends on the toxin oligomeric structure (Gomez et al., 2002). While the helix a4 of domain I is involved in the toxin oligomerization process, the helix a5 plays an important role in toxin structure stabilization and pore formation (Rausell et al., 2004a; Tigue et al., 2001). However, formation of a pre-pore oligomeric structure, probably a tetramer, by intermonomeric contacts and subsequent insertion into the apical membrane to make active lytic pores (Gomez et al., 2002; Rausell et al., 2004b) has been postulated but how a water-soluble complex penetrates the lipid bilayer is not known yet. The pore formation in turn rapidly immobilizes the epithelial cells of the larvae with a gradual decrease in gut pH (Aronson and Shai, 2001; Knowles and Ellar, 1987), causing the larva to stop feeding. The acidic pH enables the bacterial spores to germinate, and the bacteria can then invade the host, causing its death.

As many pore-forming toxins, such as Bt Cry-proteins, are lectins, a mechanism of pore formation involving the membrane insertion by putative leverage-mediated (LM) uptake reactions have been proposed, where oligomeric adhesion molecules bent membrane-receptors around 'hinge'-like lipophorin particles (Schmidt and Theopold, 2004). In addition, recent studies describing alternative experimental approaches using the model invertebrate organism *Caenorhabditis elegans* have suggested a novel cellular receptor for the toxin, and have detailed a cellular response of the host towards intoxication (Griffitts et al., 2005). However, the knowledge of the Cry-toxins' structures and basic cellular uptake mechanisms is required to obtain a more critical understanding of their molecular basis of mode of action and toxicity.

Cry-toxin Receptors in Insect Midgut

Bt toxins kill insects by binding to specific target sites, the toxin receptors and disrupting midgut membranes (Schnepf et al., 1998). It is not surprising to have multiple receptors, as there are several different Bt toxins, but the science has become more complicated as Garner et al. (1999) suggested that the protein that binds Cry1A toxin is an aminopeptidase, while others reported that the protein that binds Cry1A toxin is a member of cadherin-like molecules (Keeton and Bulla, 1997; Gahan et al., 2001)

In *Manduca sexta*, the receptor for Cry1Aa, Cry1Ab, and Cry1Ac toxins is 210-kDa protein, known to be a member of the cadherin-like (BT-R_1) gene family (Francis et. al., 1997). The Cry1Aa, Cry1Ab, and Cry1Ac exhibit identical toxicities toward *M. sexta* larvae and show a high degree of sequence similarities and presumed structural identities. These similarities make it likely that there is a common mechanism of toxicity in these lepidopteran-specific toxins in terms of both mode of action and the receptor proteins through which these toxins exert their lepidopteran-specific toxicity. Midgut expressed cadherin-like cDNAs were isolated from *M. sexta* (Vadlamudi et al., 1995), *Bombyx mori* (Nagamatsu et al., 1998); *Heliothis virescens* (Gahan et al., 2001), and *Pectinophora gossypiella* (Morin et al., 2003). Competition assays with the cadherin-like BT-R_1 protein prepared from larval *M. sexta* midgut and transiently expressed in cultured cells reveal virtually identical affinities for the Cry1Aa, Cry1Ab, and Cry1Ac toxins (Keeton and Bulla, 1997; Keeton et al., 1998). This is an indication of a common Bt-receptor for toxins of the lepidopteran-specific Cry1A family, where a 210-kDa glycoprotein BT-R_1 binds all three Cry1A toxins. In some cases, the receptor for Cry1Ab in *M. sexta* is believed to be a cadherin-like 210-kDa membrane protein (Dorsch et al., 2002). Domain II of Cry1A binds lepidopteran midgut epithelial receptors (de Maagd et al., 2001) by interacting with two cadherin motifs (Gómez et al., 2003). Cadherin-like Bt-R1 protein has been suggested to induce a conformational change in Cry1Ab that allows the formation of a pre-pore toxin oligomer (Gomez et al., 2002). In *B. mori*, the cadherin-like protein BtR175 serves as a Cry1Aa receptor (Nagamatsu et al., 1998). Involvement of cadherin-superfamily gene disruption in resistance to Cry1Ac has been described for a laboratory strain of *H. virescens* (Gahan et al., 2001), although no cadherin-like proteins have been purified from this insect. (Fig. 3)

A 220-kDa cadherin-like peptide was identified as a midgut Cry1Ab receptor in *O. nubilalis* (Hua et al., 2001). Ligand blots demonstrate that the Cry1Ab binds to a 220-kDa cadherin-like glycoprotein, and 145 and 154-kDa aminopeptidase (APN) isoforms in *O. nubilalis* (Hua et al., 2001). Candidate Cry1Ab binding receptor genes have not yet been reported in *O. nubilalis*. Effects of cadherin toxin binding region variation on resistance traits are unknown.

Aminopeptidase was implicated as a midgut receptor (Gill et al., 1995; Knight et al., 1994), followed by cadherin. The N-terminal and internal amino acid sequences of protein from *H. virescens* show a high degree of similarity with the enzyme aminopeptidase N, an indicator for aminopeptidase N as the receptor for the CrylAc toxin in the brush border membrane (Heckel et al., 1997). Cloning and expression of a cDNA that encodes BT-R1 of *M. sexta* is a 210-kDa membrane glycoprotein that specifically binds the Cry1Ab (Vadlamudi et al., 1995). The BT-R1 shares sequence similarity with the cadherin superfamily of proteins. The Cry1Ac binds to a 120-kDa glycoprotein receptor in the larval midgut epithelia of the susceptible insect *M. sexta* identified as aminopeptidase N (Knight et al., 1995). In *H. virescens*, a 170-kDa aminopeptidase N, bound Cry1Aa, Cry1Ab, and Cry1Ac were identified, but not Cry1C or Cry1E, whereas *N*-acetylgalactosamine (GalNAc) inhibited the binding of Cry1Ac but not that of Cry1Aa or Cry1Ab (Luo et al., 1997). In *P. xylostella* and

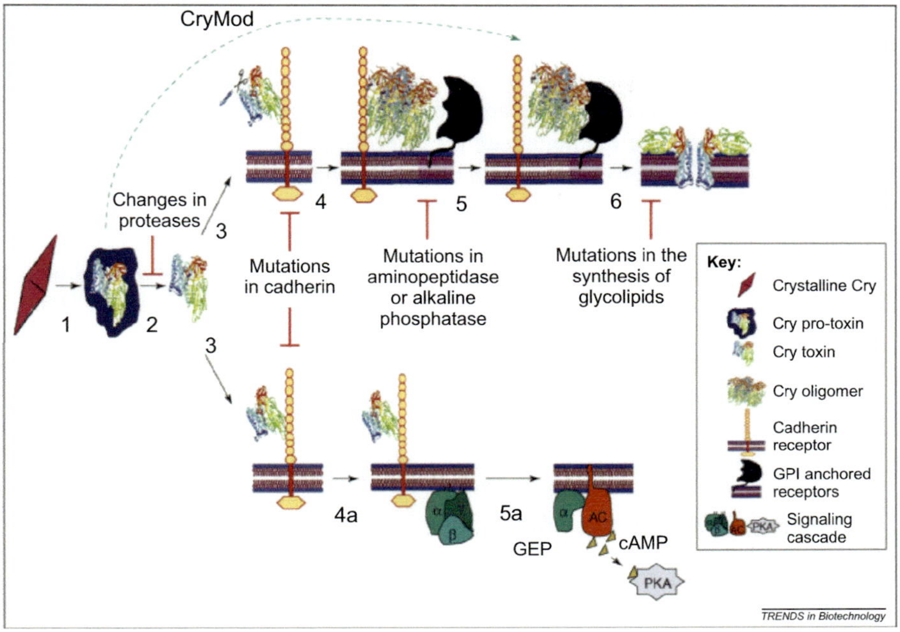

Fig. 3 Models of the mode of action of Cry-toxins and resulting mechanism for resistance (Bravo et. al., 2008). Two different mechanisms can be distinguished: the pore-formation model (top) and the signal transduction model (bottom), which both include similar initial steps for toxin solubilization in midgut lumen (1), activation by midgut proteases (2), and binding to primary receptor cadherin (3). In the pore-formation model (top), step 3 induces the cleavage of helix a-1 and triggers toxin oligomerization (4). The toxin oligomer then binds to a secondary receptor, such as aminopeptidase or alkaline phosphatase, which are anchored by a glycosylphosphatidylinositol anchor in the membrane (5). Finally, the toxin inserts itself into the membrane, thereby forming a pore that kills the insect cells (6). The signal transduction model (bottom) proposes that the interaction of the Cry-toxin with a cadherin receptor triggers an intracellular cascade pathway that is mediated by activation of protein G (4a), which, in a subsequent step (5a), activates adenylyl cyclase. This signal then increases the levels of cyclic adenosine monophosphate, which activates protein kinase A and leads to cell death.

B. mori, the Cry1Ac receptor also seems to be aminopeptidase N, while Cry1Aa and Cry1Ab bind to a 210-kDa brush border membrane vesicle (BBMV) protein (Crickmore et al., 1998). Membrane-bound alkaline phosphatase is also known to bind Cry1Ac receptor (Jurat-Fuentes and Adang, 2004; Ma et al., 2005). Recently, glycolipids being reported as Cry5B receptor in nematode *Caenorhabditis elegans* (Griffitts et al., 2005).

Pests Resistant to Cry-toxins

The success of Bt crops does not preclude resistance problems in the future, as at least seven resistant laboratory strains of three insect pests diamondback moth *Plutella xylostella*, pink bollworm *Pectinophora gossypiella* and cotton bollworm *Helicoverpa armigera* have completed development on Bt crops in both laboratory and greenhouse studies (Tabashnik et al., 2003). However, using either laboratory-adapted insects, or insects collected from wild populations, a number of insect populations of several insect species from different geographical areas, with different levels of resistance were obtained by laboratory selection with the Cry toxins. To date, laboratory-selected Cry toxin-resistant insects are the Indian mealmoth *Plodia interpunctella* (McGaughey, 1985), the diamondback moth *Plutella xylostella* (Tabashnik et al., 1991), the Colorado potato beetle *Leptinotarsa decemlineata* (Whalon et al., 1993), the cabbage looper *Trichoplusia ni* (Estada and Ferre, 1994), the cotton leafworm *Spodoptera littoralis* (Muller-Cohn et al, 1996), the beet armyworm *Spodoptera exigua* (Zheng et al., 2005), the tobacco budworm *Heliothis virescens* (Gould et al., 1992; Lee et al., 1995; MacIntosh et al., 1991; Forcada et al., 1996; Heckel et al., 1997), the European corn borer *Ostrinia nubilalis* (Huang et al., 1999), the mosquito *Culex quinquefasciatus* (Georghiou and Wirth, 1997), the cotton bollworm *Helicoverpa armigera* (Akhurst et al., 2003; Gunning et al., 2005; Ma et al., 2005), flour moth *Ephestia kuehnellia* (Rahman et al., 2004a) and a candidate member from nematodes *Caenorhabditis elegans* (Griffitts et al., 2005).

Mechanisms of Bt Cry-toxins Resistance in Insects

One or more genes may mediate lepidopteran resistance to the Cry toxins. Hypothetically, the mechanisms underlying the Cry toxins resistance may be found in any of the steps in the proposed mode of action (Schnepf et al., 1998). Therefore, any change in insect gut physiology that affects one or more steps in this process could prevent toxicity and potentially leads to the development of resistant pest populations (Li et al., 2004). To date, most studies for Bt resistance have focused on two steps in its mode of action: proteolytic activation of protoxin and binding of active toxin to receptors.

Several mechanisms of insect resistance to Bt toxins have been proposed (Gill et al., 1992). One of the best-known mechanisms of resistance is the reduced binding of toxin to midgut receptors. Modification of midgut receptor proteins that bind the Cry-toxins may reduce susceptibility of insects. This mechanism has been associated with the resistance to the Cry-toxins in many strains of *P. interpunctella* (Herrero et al., 2001; Van Rie et al., 1990), *P. xylostella* (Ferre et al., 1991; Escriche et al., 1995; Tabashnik, 1994; Tang et al., 1996), *H. virescens* (Gahan et al., 2001; Lee et al., 1995), and *S. exigua* (Moar et al., 1995). However, in some laboratory-selected resistant strains of *H. virescens*, reduced binding was not associated with the resistance to the Cry-toxins (Gould et al., 1992; MacIntosh et al., 1991).

The other mechanism of resistance may involve midgut proteinases that interact with Bt toxins. The resistant strain of *P. interpunctella* lacks a major gut proteinase, and has low Bt protoxin-hydrolyzing activity compared to the susceptible parent strains (Oppert et al., 1997). Changes in midgut protease activity and toxin binding together result in resistance in *L. decemlineata* (Loseva et al., 2002). Faster degradation of toxin by midgut proteases was associated with Bt resistance of *H. virescens* (Forcada et al., 1996; Forcada et al., 1999) and *Spodoptera littoralis* (Muller-Cohn et al., 1996). An increased rate of replacement or repair of damaged cells by Bt-resistant larvae compared to their susceptible counterparts is the basis of another mode of resistance in *H. virescens* (Martinez-Ramirez et al., 1999). However, two novel resistance mechanisms to the Cry-toxins have recently been identified in different species, one based on an elevated immune response in *E. kuehniella* (Rahman et al., 2004a) and *H. armigera* (Ma et al., 2005), a second mechanism was identified with glycolipids receptors in *C. elegans* mutants (Griffitts et al., 2005). Lower glycosphingolipids content in BBMV was also reported in high-level of resistance in a line of *P. xylostella* (Kumaraswami et al., 2001; Maruyama, 1999). Alternative mechanisms responsible for Bt-resistance or tolerance in individual insect species will be reviewed in this section.

The stored-grain pest *P. interpunctella* was one of the first insect species to evolve resistance to a *B. thuringiensis* subsp. *Kurstaki* formulation in the laboratory studies (McGaughey, 1985), where the resistance trait was proved to be recessive (Van Rie et al., 1990). The resistance was correlated with a 50-fold decrease in binding affinity of a receptor for the Cry1Ab protein but an increased susceptibility to Cry1Ca, a protein does not present in the selective formulation, and a corresponding increase in binding sites on the midgut for the Cry1Ca protein. In some other colonies, the level of resistance ranged from 24 to >2,000-folds after selection with Cry1Ac and Cry1Ab, but >10 for Cry1Aa, Cry1Ba, Cry1Ca, and Cry2Aa (McGaughey and Johnson, 1994). The diversity of resistance mechanisms to different toxins within species is common. For example, while resistance is associated with reduced binding of the Cry1Ac toxin to midgut cells, the same toxin shows unaltered binding to an 80-kDa BBMV protein (Mohammed et al., 1996). This strain has one of the two major trypsin-like proteases missing from midgut enzymes selected with Dipel or HD-133, and a reduced *in vitro* activation of Cry1Ac protoxin by midgut extracts when selected with HD-198, a toxin from *B. thuringiensis* subsp. *entomocidus* (Oppert et al., 1994).

Colonies of *H. virescens* with different levels of resistance and different resistance mechanisms have also been obtained in selection experiments with different Cry-toxins (Jurat-Fuentes et al., 2003). Selected with Cry1Ab protoxin for seven generations with Dipel, showed a 71-fold resistance and a lower binding affinity to a higher number of binding sites (MacIntosh et al., 1991). Larvae of *H. virescens* selected with the Cry1Ac protoxin could not survive on transgenic tobacco plants with moderate (0.01%) levels of Cry1Ab (Schnepf et al., 1998), but survived significantly better than the susceptible larvae with very high levels of resistance to Cry1Ac and to Cry1Ab, when selection was performed with Cry1Ac (Gould et al., 1995), an indication of the existence of multiple resistance mechanisms in this insect strain. Though Cry1Ac also binds to the Cry1Aa binding site (Van Rie et al., 1989), it is surprising that the binding of Cry1Ac and Cry1Ab was unchanged while the binding of Cry1Aa was abruptly reduced in this strain (Lee et al., 1995). However, it was proposed that the altered Cry1Aa binding site caused resistance to all three Cry1A toxins and that the additional binding sites recognized by Cry1Ab and Cry1Ac might not be involved in toxicity (Lee et al., 1995). Cadherin was implicated as a resistance factor to Cry1Ac by transposon knockout in *H. virescens* YHD2 colony resulting in a 40,000-fold resistance (Gahan et al., 2001), although no cadherin-like candidate proteins have been purified from this insect.

S. littoralis with more than a 500-fold resistant to Cry1Ca have shown partially recessive inheritance (Schnepf et al., 1998) and cross-resistance to Cry1Da and

Cry1Ea (MullerCohn et al., 1996). But in *S. exigua*, only an 11-fold resistance was observed after selection with Cry1C for seven generations, where the resistance is not entirely based on changes in toxin binding characteristics (Moar et al., 1995). This strain shows cross-resistance to Cry1Ab, Cry2Aa, Cry9C, and a Cry1Ea-Cry1Ca hybrid protein, where resistance to a single protein is more likely to occur first than resistance to spore-crystal mixtures (Bosch et al., 1994).

A common mechanism and a shared genetic basis of resistance to Cry1Ac have also been shown in some insects. For example, two independently-derived resistant strains of pink bollworm *P. gossypiella*, share a genetic locus at which three mutant alleles (r1, r2 and r3) confer resistance to Cry1Ac, where both strains are reported to show traits of "mode 1" resistance (Tabashnik et al., 2004). Cadherin was implicated as a resistance factor by phenotypic association in *P. gossypiella* AZP-R strain, where two copies of cadherin resistance alleles (r1, r2 or r3) were required for resistance (Morin et al., 2003).

Laboratory colonies of *O. nubilalis* exhibit moderate resistance to Cry1Ab toxin in artificial diet (Bolin et al., 1999). Inheritance of resistance was autosomal and incompletely dominant in an *O. nubilalis* strain, when selected for survival on Dipel® Bt formulations (Huang et al., 1999). Laboratory selection with Cry1Ab produced >1000-fold resistance in two laboratory strains of *O. nubilalis*, which showed autosomal inheritance and a low initial frequency of major genes, which might be an important factor in delaying evolution of resistance to Bt corn in this pest (Alves et al., 2006).

It is important that selections in the laboratory are very different from selections that occur in the field. The laboratory colonies might have a considerably lower level of genetic diversity than the field populations (Schnepf et al., 1998). Resistance to Bt-toxins has been developed in the field in *P. xylostella*, however in this species no success in laboratory selections for Bt resistance have been achieved to date. The natural environment may contain factors affecting the viability or fecundity of resistant insects, factors excluded from the laboratory (Ferre and Van Rie, 2002). Moreover, resistance mechanisms can be associated with certain fitness costs that can be deleterious under natural conditions. In the field, natural enemies, such as predators and parasites, can influence the development of resistance to *B. thuringiensis* by preferring either the intoxicated, susceptible, or the healthy, resistant insects populations (Schnepf et al., 1998; Schuler et al., 2003).

Despite expectations that pests would rapidly evolve resistance to Bt Cry-toxins, increases in the frequency of resistance caused by exposure to Bt formulations or to Bt crops in the field are yet to be explored at a larger scale. Although many insects have shown resistance to Bt in laboratory selection, only *Plutella xylostella* has evolved resistance to Bt toxin in open field populations (Tabashnik et al., 2000; Ferre et al., 1991; Baxter et al., 2005), showing a 30-fold resistance to a Bt formulation (Dipel) and rapidly increased to >1,000-fold. This resistant strain showed a reduced binding of the Cry1Ac protein to gut BBMV (Tabashnik et al., 1994), the presence of at least some receptor molecules on the midgut (Escriche et al., 1995; Masson et al., 1995) and the inheritance of resistance is conferred largely by a single autosomal recessive locus (Tabashnik et al., 1997a). This was corroborated when a major autosomal locus inherited as a recessive trait was identified (Heckel et al., 1999). Moreover, the 'Hawaii' resistance allele simultaneously confers cross-resistance to Cry1Aa, Cry1Ab, Cry1Ac, Cry1Fa, and Cry1Ja but not to Cry1Ba, Cry1Bb, Cry1Ca, Cry1Da, Cry1Ia, or Cry2Aa (Granero et al., 1996). The toxins in this cross-resistance group have significant amino acid sequence similarity in Domain II, a region believed to be important for receptor binding in many species.

It is reported that binding to BBMV was reduced for Cry1Ab but almost unaffected for Cry1Aa, Cry1Ac, or Cry1C in a field population of diamondback moths from the Philippines, which showed partial resistance to Cry1Aa, Cry1Ab, and Cry1Ac, but sensitivity to Cry1C, Cry1F, and Cry1J, where inheritance of resistances to Cry1Aa and Cry1Ac was expressed in an autosomal dominant and semidominant fashion, respectively (Tabashnik et al., 1997b). A Cry1Ab resistance allele, associated with reduced binding to BBMV receptors, was partially responsible for a high level of resistance in a field collected Malaysian strain selected with *B. thuringiensis* subsp. *kurstaki* and *B. thuringiensis* subsp. *aizawai* (Wright et al., 1997).

A field-collected colony of *P. xylostella* showing a 31-fold resistance to Cry1C, survived and neonates completed their entire life cycle on transgenic broccoli expressing a Cry1C protein after 24 generations of laboratory selection with Cry1C (Zhao et al., 2000). Reduced binding is not the major mechanism of high level of resistance to Cry1C in this strain. Whatever the mechanism, it is inherited as an autosomal and incompletely recessive factor.

The first-instar larvae of *H. armigera* from a population selected for 25 generations with Cry1Ac (63-fold resistance) in laboratory conditions were able to complete their larval development on transgenic cotton expressing Cry1Ac and produced fertile adults (Akhurst et al., 2003). Resistance in this Australian strain is incompletely recessive and associated with lack of high affinity binding site. In a laboratory selection *H. armigera* developed resistance to Cry1Ac in India at comparable levels with the Australian strains (Kranthi et al., 2000). Additive gene action was predicted to mediate Bt resistance in *H. zea* (Burd et al., 2003)

It has been suggested that an increased rate of replacement or repair of damaged cells by Bt-resistant larvae compared to their susceptible counterparts is the basis of another mechanism of resistance in *H. virescens* larvae (Martinez-Ramirez et al., 1999). Sublethal doses of Bt toxins AA 1-9 and HD-73 cause dose-dependent destruction of cultured midgut cells from *H. virescens* larvae but regenerate at higher rates in resistant larvae than the susceptible ones (Loeb et al., 2001). This increasing rate of cell proliferation may involve in Bt resistance.

Lower glycosphingolipids content in BBMV is known to associate with very high-level resistance in a line of *P. xylostella* having a 130000-fold resistant to Cry1Ac (Maruyama et al., 1999). Receptor binding was not the core of resistance in this strain as Cry1Ac bound equally to a 120-kDa receptor protein of BBMV from both susceptible (PXS) and resistant (PXR) strains. Indeed, receptor binding must be accompanied by events such as membrane insertion and pore formation, and the evolution of resistance may be associated with appropriate changes in these post-binding events. Loss of glycolipid carbohydrates is known to involve in Bt resistance in nematodes *C. elegans* (Griffitts et al., 2001), where the Cry5B toxin which target nematodes directly and specifically binds glycolipids, and the binding is carbohydrate –dependent and relevant for toxin action *in vivo* (Griffitts et al., 2005).

Bt resistance mechanisms are not unique. For example, the case history of *P. xylostella* (DBM) and *H. virescens* resistance to the Cry toxins presents a cautionary tale for the wide range of resistance mechanisms within the species. Starting in the mid-1980s, with an intensive use of *B. thuringiensis* subsp. *kurstaki* in production agriculture, resistant DBM insects have evolved in numerous geographically isolated regions of the world, and *B. thuringiensis* subsp. *aizawai* resistance is beginning to appear even more rapidly. While various alleles showing cross-resistance, dominant inheritance, or stability in the absence of selection have been detected in case of very high levels of resistance in different insects, little is known about the mechanisms of low to medium level of resistance.

Inducible Tolerance to Bt-toxin in Insect Pests

Receptor inactivation by gene mutations (mode 1) provides immediate resistance to relatively high levels of the toxins. The case history of Bt-resistance demonstrates that the emergence of resistance through gene mutations is a rare event, given that only one pest insect species have developed genetic resistance in the field. Moreover, inactivation in one gene is not necessarily expressed as a resistant phenotype because some receptor genes comprise gene families with overlapping gene function. While mode 1 resistance is rare, the potential exists for a number of pest insects to become tolerant by other mechanisms.

Inducible Tolerance Mechanisms

Recent reports on inducible tolerance to low-to-medium levels of Bt-formulation in *E. kuehniella* (Rahman et al., 2004a) have highlighted possible alternative mode of resistance in a large number of pest insect species that have been selected in the laboratory for resistance against Bt-toxins. The observed tolerance in this strain is dependent on the induction of immune proteins, which occurs with the ingestion of sub-lethal doses of Bt-formulations and that both the immune induction and the Bt tolerance are transmitted to offspring by a maternal effect.

When larvae from a susceptible laboratory strain of *E. kuehniella* was selected for survival on progressively increasing levels of the Bt-formulation (Dipel) consisting of Cry1Aa, Cry1Ab, Cry1Ac, Cry2Aa, proteins, and spores for five generations, the resulting Bt-tolerant strain displayed a constitutively elevated immune response (Rahman et al., 2004a). Treating larvae of the susceptible strain with sublethal doses of a Bt-formulation led to an elevated melanization reaction in the hemolymph, which in turn, was correlated with an increase in tolerance against the toxin at a later stage of larval development. As susceptible larvae were immune-induced and exposed to Bt-toxin in the same generation, this experiment excludes the selection of a pre-existing resistance allele as a cause of increased tolerance to the toxin in pre-treated larvae.

Reciprocal crosses of tolerant and susceptible insects revealed a transmission of both the immune induction and tolerance to the toxin from one generation to the next by a maternal effect (Rahman et al., 2004a). Offspring of tolerant females and susceptible males (TxS) were found significantly more tolerant than the offspring from the susceptible females and tolerant males (SxT). A mechanism for this effect is the incorporation of an immune-elicitor into the oocyte by an immune-induced female. The elicitor could interact with embryonic tissues to induce the immune system of the neonate, so that by the time the neonate starts feeding; the insect is already induced thus increasing the chances of surviving the toxin.

The results from *E. kuehniella* were corroborated from the reports of a field-collected strain of *H. armigera* that has developed genetic resistance in the laboratory (Akhurst et al., 2003), comprising a complex mixture of putative receptor inactivation and elevated immune status (Ma et al., 2005) and *P. xylostella* (M. Sarjan, unpublished data), suggesting that the initial development of low-level Bt-tolerance through immune-related processes may be a common phenomenon. In *H. armigera*, the immune-induction was associated with an increased level of the larval serum protein hexamerin, which interacts with the toxin by forming an insoluble aggregate (Ma et al., 2005).

The inducible tolerance mechanisms are quite different from other Bt-resistance mechanisms (Gonzalez-Cabrera et al., 2001), where the observed reductions in Bt-toxicity are exclusively explained in terms of alterations to receptor properties in the gut epithelium (Darboux et al., 2002; Gahan et al., 2001). While inducible tolerance mechanisms may be effective only against low to medium levels of toxin and are no immediate threat to most transgenic crops expressing the Bt Cry-proteins, it may pose long-term threat through genotype selection, given that the toxin levels in transgenic crops may not be sufficient to kill tolerant insect populations in conjunction with semi-dominant receptor mutations.

The molecular basis of inducible tolerance mechanisms in *E. kuehniella* is not known; however, the evidence that the magnitude of resistance is determined by more than one gene suggests that it may involve multiple metabolic and regulatory pathways. For example, mutational changes in a gene product involved in post-translational modifications, such as glycosyltransferases, may potentially interfere with pro-coagulant, cell surface receptors and elicitor functions. If pro-coagulant molecules are transferred into the gut lumen as part of an immune response, it is possible that soluble immune components in the gut lumen interact with the mature toxin causing its inactivation by a coagulation or melanization reaction. Alternatively, lectin-like immune molecules that mimic the toxin may engage the receptor in internalization reactions without causing Osmo fragility. Likewise, the nature and molecular mechanism of transmission of the elicitor remains to be elucidated. It is known that the determination of dorsal-ventral polarity in the *Drosophila* embryo is mediated by a group of genes that are functionally involved in immune defense reactions of the adult insect (Lemaitre et al., 1997).

Summary

The success of *Bacillus thuringiensis* (Bt) toxins in controlling insect pests is overshadowed by documented cases of resistance in both laboratory-selected populations and field populations of various insect species. The emergence of resistance in diamondback moths and cabbage loopers emphasizes the inherent danger of relying on these environmentally compatible insecticides. Detecting resistance in the field relies on molecular and genetic knowledge, but surprisingly, the precise details of how

Bt toxins bind to insect gut cells are poorly understood. Although the general process involves proteolytic activation by midgut proteases and binding to membrane gut receptors, the specific mechanisms of resistance, beyond receptor loss or modification, remain unclear. The chapter emphasizes the urgent need to explore potential tolerance and resistance mechanisms in insect pests and vector mosquitoes for the long-term management of Bt Cry-toxins. Identifying major resistance genes and understanding variations in the genetic basis of resistance are crucial for developing effective strategies within integrated pest management and risk management approaches.

References

Abdullah, M.A.F., Alzate, O., Mohammad, M., McNall, R.J., Adang, M.J. and Dean, D.H. (2003). Introduction of *Culex* toxicity into *Bacillus thuringiensis* Cry4Ba by protein engineering. *Applied Environmental Microbiology*, 69: 5343–53.

Agaisse, H. and Lereclus, D. (1995). How does *Bacillus thuringiensis* produce so much insecticidal crystal protein? *Journal of Bacteriology*, *177*: 6027–32.

Akao, T., Mizuki, E., Yamashita, S., Saitoh, H. and Ohba, M. (1999). Lectin activity of *Bacillus thuringiensis* parasporal inclusion proteins. *FEMS Microbiology Letters*, *179*: 415–21.

Akhurst, R.J., James, W., Bird, L.J. and Beard, C. (2003). Resistance to the Cry1Ac delta-endotoxin of *Bacillus thuringiensis* in the cotton bollworm, *Helicoverpa armigera* (Lepidoptera: Noctuidae). *Journal of Economic Entomology*, *96*: 1290–99.

Alves, A.P., Spencer, T.A., Tabashnik, B.E. and Siegfried, B.D. (2006). Inheritance of resistance to the Cry1Ab *Bacillus thuringiensis* toxin in *Ostrinia nubilalis* (Lepidoptera: Crambidae). *Journal of Economic Entomology*, *99*: 494–501.

Aronson, A.I. and Shai, Y. (2001). Why *Bacillus thuringiensis* insecticidal toxins are so effective: Unique features of their mode of action. *FEMS Microbiology Letters*, *195*: 1–8.

Babu, R.M., Sajeena, A., Seetharaman, K. and Reddy, M.S. (2003). Advances in genetically engineered (transgenic) plants in pest management: An overview. *Crop Protection*, *22*: 1071–86.

Ballester, V., Granero, F., de Maagd, R.A., Bosch, D., Mensua, J.L. and Ferre, J. (1999a). Role of *Bacillus thuringiensis* toxin domains in toxicity and receptor binding in the diamondback moth. *Applied Environmental Microbiology*, *65*: 1900–1903.

Ballester, V., Granero, F., Tabashnik, B.E., Malvar, T. and Ferre, J. (1999b). Integrative model for binding of *Bacillus thuringiensis* toxins in susceptible and resistant larvae of the diamondback moth (*Plutella xylostella*). *Applied Environmental Microbio.*, *65*: 1413–19.

Bates, S.L., Zhao, J.Z., Roush, R.T. and Shelton, A.M. (2005). Insect resistance management in GM crops: Past, present, and future. *Nature Biotechnology*, *23*: 57–62.

Baxter, S.W., Zhao, J.Z., Gahan, L.J., Shelton, A.M., Tabashnik, B.E. and Heckel, D.G. (2005). Novel genetic basis of field-evolved resistance to Bt toxins in *Plutella xylostella*. *Insect Molecular Biology*, *14*: 327–34.

Berliner, E. (1911). Uber die schlaffsucht der mehlmottrnraupe (*Ephestia kuehniella* Zell), und ihren erreger *Bacillus thuringinsis* n . sp. *zeitschrift* fur angewandtes. *Entomol*, *2*: 29–56.

Binnington, K.C., Lehane, M.J. and Beaton, C.D. (1998). The peritrophic membrane. *In*: F.W. Harrison, M. Locke (Eds,), *Microscopic Anatomy of Invertebrates*, 747–58. Vol. 11B: Insecta. Wiley, New York.

Bolin, P.C., Hutchison, W.D. and Andow, D.A. (1999). Long-term selection for resistance to *Bacillus thuringiensis* Cry1Ac endotoxin in a Minnesota population of European corn borer (Lepidoptera: Crambidae). *Journal of Economic Entomology*, *92*: 1021–30

Bora, R.S., Murty, M.G., Shenbagarathai, R. and Sekar, V. (1994). Introduction of a lepidopteran-specific insecticidal crystal protein gene of *Bacillus thuringiensis* subsp *kurstaki* by conjugal transfer into a *Bacillus megaterium* strain that persists in the cotton phyllosphere. *Applied Environmental Microbiology, 60*: 214–22.

Bosch, D., Schipper, B., van der Kleij, H., de Maagd, R.A. and Stiekema, W.J. (1994). Recombinant *Bacillus thuringiensis* crystal proteins with new properties: Possibilities for resistance management. *Biotechnology (NY)*, *12*: 915–18.

Bravo, A., Sanchez, J., Kouskoura, T. and Crickmore, N. (2002). N-terminal activation is an essential early step in the mechanism of action of the *Bacillus thuringiensis* Cry1Ac insecticidal toxin. *Journal of Biological Chemistry*, *277*: 23985–23987.

Bravo, A. and Soberón, M. (2008). How to cope with insect resistance to Bt toxins? *Trends in Biotechnology,* 26(10): 573–579

Burd, A.D., Gould, F., Bradley, J.R., Van Duyn, J.W. and Moar, W.J. (2003). Estimated frequency of nonrecessive Bt resistance genes in bollworm, *Helicoverpa zea* (Boddie) (Lepidoptera: Noctuidae) in eastern North Carolina. *J. Economic Entomol.*, *96*: 137–42.

Burton, S.L., Ellar, D.J., Li, J. and Derbyshire, D.J. (1999. N-acetylgalactosamine on the putative insect receptor aminopeptidase N is recognized by a site on the domain III lectin-like fold of a *Bacillus thuringiensis* insecticidal toxin. *J. Molecular Biol.*, *287*: 1011–22.

Carroll, J., Convents, D., Van Damme, J., Boets, A., Van Rie, J. and Ellar, D.J. (1997). Intramolecular proteolytic cleavage of *Bacillus thuringiensis* Cry3A [delta]-endotoxin may facilitate its coleopteran toxicity. *Journal of Invertebrate Pathology*, *70*: 41–49.

Carroll, J. and Ellar, D.J. (1993). An analysis of *Bacillus thuringiensis* delta-endotoxin action on insect-midgut-membrane permeability using a light-scattering assay. *European Journal of Biochemistry*, *214*: 771–78.

Chandra, A., Ghosh, P., Mandaokar, A.D., Bera, A.K., Sharma, R.P., Das, S. and Kumar, P.A. (1999). Amino acid substitution in [alpha]-helix 7 of Cry1Ac [delta]-endotoxin of *Bacillus thuringiensis* leads to enhanced toxicity to *Helicoverpa armigera* Hubner. *FEBS Letters*, *458*: 175–79.

Crickmore, N., Zeigler, D.R., Bravo, A., Schnepf, E., Lereclus, D., Baum, J., Van Rie, J. and Dean, D.H. (2006). *Bacillus thuringiensis* Toxin Nomenclature. http://www.lifesci.sussex.ac.uk/home/Neil_Crickmore/Bt/. Access date May 20, 2006.

Crickmore, N., Zeigler, D.R., Feitelson, J., Schnepf, E., Van Rie, J., Lereclus, D., Baum, J. and Dean, D.H. (1998). Revision of the nomenclature for the *Bacillus thuringiensis* pesticidal crystal proteins. *Microbiology and Molecular Biology Reviews*, *62*: 807–13.

Darboux, I., Pauchet, Y., Castella, C., Silva-Filha, M.H., Nielsen-LeRoux, C., Charles, J.F. and Pauron, D. (2002). Loss of the membrane anchor of the target receptor is a mechanism of bioinsecticide resistance. *Proceedings of the National Academy of Sciences of the United States of America*, *99*: 5830–35.

de Maagd, R., Kwa, M., van der Klei, H., Yamamoto, T., Schipper, B., Vlak, J., Stiekema, W. and Bosch, D. (1996). Domain III substitution in *Bacillus thuringiensis* delta-endotoxin CryIA(b) results in superior toxicity for *Spodoptera exigua* and altered membrane protein recognition. *Applied Environmental Microbiology*, *62*: 1537–43.

de Maagd, R.A., Bravo, A. and Crickmore, N. (2001). How *Bacillus thuringiensis* has evolved specific toxins to colonize the insect world. *Trends in Genetics*, *17*: 193–99.

Deacon, J. (2003). *The Microbial World: Bacillus thuringiensis*. Institute of Cell and Molecular Biology, The University of Edinburgh.

Decker, H. and Terwilliger, N. (2000). Cops and robbers: Putative evolution of copper oxygen-binding proteins. *Journal of Experimental Biology*, *203*: 1777–82.

Dorsch, J.A., Candas, M., Griko, N.B., Maaty, W.S.A., Midboe, E.G., Vadlamudi, R.K. and Bulla, J.L.A. (2002). Cry1A toxins of *Bacillus thuringiensis* bind specifically to a region adjacent to the membrane-proximal extracellular domain of BT-R1 in *Manduca sexta*:: Involvement of a cadherin in the entomopathogenicity of *Bacillus thuringiensis*. *Insect Biochemistry and Molecular Biology*, *32*: 1025–36.

Escriche, B., Tabashnik, B.E., Finson, N. and Ferre, J. (1995): Immunohistochemical detection of binding of Cry1a crystal proteins of *Bacillus thuringiensis* in highly resistant strains of *Plutella xylostella* (L.) from Hawaii. *Biochemical and Biophysical Research Communications*, *212*: 388–95.

Estada, U. and Ferre, J. (1994). Binding of insecticidal crystal proteins of *Bacillus thuringiensis* to the midgut brush border of the cabbage looper, *Trichoplusia ni* (Hubner) (Lepidoptera: Noctuidae), and selection for resistance to one of the crystal proteins. *Applied and Environmental Microbiology*, *60*: 3840–46.

Ferre, J., Real, M.D., Van Rie, J., Jansens, S. and Peferoen, M. (1991). Resistance to the *Bacillus thuringiensis* bioinsecticide in a field population of *Plutella xylostella* is due to a change in a midgut membrane receptor. *Proceedings of the National Academy of Sciences of the United States of America*, *88*: 5119–23.

Ferre, J. and Van Rie, J. (2002). Biochemistry and genetics of insect resistance to *Bacillus thuringiensis*. *Annual Review of Entomology*, *47*: 501–33.

Forcada, C., Alcácer, E., Garcerá, M.D. and Martínez, R. (1996). Differences in the midgut proteolytic activity of two *Heliothis virescens* strains, one susceptible and one resistant to *Bacillus thuringiensis* toxins. *Archives of Insect Biochemistry Physiology*, *31*: 257–72.

Forcada, C., Alcácer, E., Garcerá, M.D., Tato, A. and Martínez, R. (1999). Resistance to *Bacillus thuringiensis* Cry1Ac toxin in three strains of *Heliothis virescens*: Proteolytic and SEM study of the larval midgut. *Archives of Insect Biochem. Physiology*, *42*: 51–63.

Francis, B.R., Bulla, J. and Lee A. (1997). Further characterization of BT-R1, the cadherin-like receptor for Cry1Ab toxin in tobacco hornworm (Manduca sexta) Midguts. *Insect Biochemistry and Molecular Biology*, *27*: 541–50.

Gahan, L.J., Gould, F. and Heckel, D.G. (2001). Identification of a gene associated with Bt resistance in *Heliothis virescens*. *Science*, *293*: 857–60.

Garner, K.J., Hiremath, S., Lehtoma, K. and Valaitis, A.P. (1999). Cloning and complete sequence characterization of two gypsy moth aminopeptidase-N cDNAs, including the receptor for *Bacillus thuringiensis* Cry1Ac toxin. *Insect Biochem. Mol. Biol.*, *29*: 527–35.

Gazit, E., Rocca, P.L., Sansom, M.S.P. and Shai, Y. (1998). The structure and organization within the membrane of the helices composing the pore-forming domain of *Bacillus thuringiensis* delta -endotoxin are consistent with an "umbrella-like" structure of the pore. *Proc. National Academy of Sciences of the United States of America*, *95*: 12289–12294.

Georghiou, G. and Wirth, M. (1997). Influence of exposure to single versus multiple toxins of *Bacillus thuringiensis* subsp. *israelensis* on development of resistance in the mosquito *Culex quinquefasciatus* (Diptera: Culicidae). *Applied Environ. Microb.*, *63*: 1095–1101.

Ghassemi, M. (1981): *Environmental Fates and Impact of Major Forest Use Pesticides*. Washington, D.C., U.S. Environmental Protection Agency.

Gill, S.S., Cowles, E.A. and Francis, V. (1995). Identification, isolation, and cloning of a *Bacillus thuringiensis* Cry1Ac toxin-binding protein from the midgut of the lepidopteran insect *Heliothis virescens*. *Journal of Biological Chemistry*, *270*,:27277–27282.

Gill, S.S., Cowles, E.A. and Pietrantonio, P.V. (1992). The mode of action of *Bacillus thuringiensis* endotoxins. *Annual Review of Entomology*, *37*: 615–34.

Gómez, I., Dean, D.H., Bravo, A. and Soberón, M. (2003). Molecular basis for *Bacillus thuringiensis* Cry1Ab toxin specificity: Two structural determinants in the *Manduca sexta* Bt-R1 receptor interact with loops -8 and 2 in domain II of Cy1Ab toxin. *Biochemistry*, *42*: 10482–10489.

Gomez, I., Sanchez, J., Miranda, R., Bravo, A. and Soberon, M. (2002): Cadherin-like receptor binding facilitates proteolytic cleavage of helix [alpha]-1 in domain I and oligomer pre-pore formation of *Bacillus thuringiensis* Cry1Ab toxin. *FEBS Letters*, *513*: 242–46.

Gonzalez-Cabrera, J., Herrero, S. and Ferre, J. (2001). High genetic variability for resistance to *Bacillus thuringiensis* toxins in a single population of diamondback moth. *Applied and Environmental Microbiology*, *67*: 5043–48.

Gould, F. (1998): Sustainability of transgenic insecticidal cultivars: Integrating pest genetics and ecology. *Annual Review of Entomology*, *43*: 701–26.

Gould, F., Anderson, A., Reynolds, A., Bumgarner, L. and Moar, W. (1995). Selection and genetic analysis of a *Heliothis virescens* (Lepidoptera: Noctuidae) strain with high levels of resistance to *Bacillus thuringiensis* toxins. *J. Economic Entomology*, *88*: 1545–59.

Gould, F., Martinez Ramirez, A., Anderson, A., Ferre, J., Silva, F.J. and Moar, W.J. (1992). Broad-spectrum resistance to *Bacillus thuringiensis* toxins in *Heliothis virescens*. *Proc. National Academy of Sciences of the United States of America*, *89*: 7986–90.

Granados, R.R., Fu, Y., Corsaro, B. and Hughes, P.R. (2001). Enhancement of *Bacillus thuringiensis* toxicity to lepidopterous species with the enhancin from *Trichoplusia ni* granulovirus. *Biological Control*, *20*: 153–59.

Granero, F., Ballester, V. and Ferre, J. (1996). *Bacillus thuringiensis* crystal proteins Cry1Ab and Cry1Fa share a high affinity binding site in *Plutella xylostella* (L.). *Biochemical and Biophysical Research Communications*, *224*: 779–83.

Griffitts, J.S. and Aroian, R.V. (2005). Many roads to resistance: How invertebrates adapt to Bt toxins. *BioEssays*, *27*: 614–24.

Griffitts, J.S., Haslam, S.M., Yang, T., Garczynski, S.F., Mulloy, B., Morris, H., Cremer, P.S., Dell, A., Adang, M.J. and Aroian, R.V. (2005). Glycolipids as receptors for *Bacillus thuringiensis* crystal toxin. *Science*, *307*: 922–25.

Griffitts, J.S., Whitacre, J.L., Stevens, D.E. and Aroian, R.V. (2001). Bt toxin resistance from loss of a putative carbohydrate-modifying enzyme. *Science*, *293*: 860–64.

Grochulski, P., Masson, L., Borisova, S., Pusztai-Carey, M., Schwartz, J.L., Brousseau, R. and Cygler, M. (1995): *Bacillus thuringiensis* CrylA(a) Insecticidal Toxin: Crystal Structure and Channel Formation. *Journal of Molecular Biology*, *254*: 447–64.

Gunning, R.V., Dang, H.T., Kemp, F.C., Nicholson, I.C. and Moores, G.D. (2005). New Resistance mechanism in *Helicoverpa armigera* threatens transgenic crops expressing *Bacillus thuringiensis* Cry1Ac toxin. *Applied Environ. Microbiology, 71:* 2558–63.

Haider, M., Knowles, B. and Ellar, D. (1986). Specificity of *Bacillus thuringiensis* var. colmeri insecticidal delta-endotoxin is determined by differential proteolytic processing of the protoxin by larval gut proteases. *European Journal of Biochemistry*, *156*: 531–40.

Heckel, D.G., Gahan, L.C., Gould, F. and Anderson, A. (1997). Identification of a linkage group with a major effect on resistance to *Bacillus thuringiensis* Cry1Ac endotoxin in the tobacco budworm (Lepidoptera: Noctuidae). *Journal of Economic Entomology*, *90*: 75–86.

Heckel, D.G., Gahan, L.J., Liu, Y.B. and Tabashnik, B.E. (1999). Genetic mapping of resistance to *Bacillus thuringiensis* toxins in diamondback moth using biphasic linkage analysis. *Proc. National Academy of Sci. of the United States of America*, *96*: 8373–77.

Herrero, S., Oppert, B. and Ferre, J. (2001). Different mechanisms of resistance to *Bacillus thuringiensis* toxins in the indianmeal moth. *Applied Environ. Microb*., *67*: 1085–89.

Hofmann, C., Vanderbruggen, H., Hofte, H., Rie, J. V., Jansens, S. and Mellaert, H.V. (1988). Specificity of *Bacillus thuringiensis* delta-endotoxins is correlated with the presence of high-affinity binding sites in the brush border membrane of target insect midguts. *Proc. National Academy of Sciences of the United States of America*, *85*: 7844–48.

Hofte, H. and Whiteley, H.R. (1989). Insecticidal crystal proteins of *Bacillus thuringiensis*. *Microbiological Reviews*, *53*: 242–55.

Hua, G., Masson, L., Jurat-Fuentes, J.L., Schwab, G. and Adang, M.J. (2001). Binding analyses of *Bacillus thuringiensis* Cry {delta}-endotoxins using brush border membrane vesicles of *Ostrinia nubilalis*. *Applied Environmental Microbiology*, *67*: 872–79.

Huang, F., Buschman, L.L., Higgins, R.A. and McGaughey, W.H. (1999). Inheritance of resistance to *Bacillus thuringiensis* toxin (Dipel ES) in the European corn borer. *Science*, *284*: 965–67.

Ishwata, S. (1901). One kind of severe flasherie (sotto disease). *Danihan Sanbshi Kaiho*, *9*: 1–5.

Jarman, M. (2006). Genetically altered cotton crop on rise: *State farmers increasingly make the switch. The Arizona Republic*.

Jurat-Fuentes, J.L. and Adang, M.J. (2004). Characterization of a Cry1Ac-receptor alkaline phosphatase in susceptible and resistant *Heliothis virescens* larvae. *European Journal of Biochemistry*, *271*: 3127–35.

Jurat-Fuentes, J.-L., Gould, F.-L. and Adang, M.-J. (2003). Dual resistance to *Bacillus thuringiensis* Cry1Ac and Cry2Aa toxins in *Heliothis virescens* suggests multiple mechanisms of resistance. *Applied Environmental Microbiology, 69*(10): 5898–5906.

Keeton, T. P., Bulla, L. A., Jr. (1997). Ligand specificity and affinity of BT-R1, the *Bacillus thuringiensis* Cry1A toxin receptor from *Manduca sexta*, expressed in mammalian and insect cell cultures. *Applied Environmental Microbiology*, *63*: 3419–25.

Keeton, T.P., Francis, B.R., Maaty, W.S.A. and Bulla, L.A. (1998). Effects of midgut-protein-preparative and ligand binding procedures on the toxin binding characteristics of BT-R-1, a common high-affinity receptor in *Manduca sexta* for Cry1A *Bacillus thuringiensis* toxins. *Applied Environmental Microbiology*, *64*: 2158–65.

Knight, P.J.K., Crickmore, N., and Ellar, D.J. (1994). The receptor for *Bacillus thuringiensis* CryIA(c) delta-endotoxin in the brush border membrane of the lepidopteran *Manduca sexta* is aminopeptidase N. *Molecular Microbiology*, *11*: 429–36.

Knight, P.J.K., Knowles, B.H., and Ellar, D. J. (1995): Molecular cloning of an insect aminopeptidase N that serves as a receptor for *Bacillus thuringiensis* CryIA(c) toxin. *Journal of Biological Chemistry*, *270*: 17765–17770.

Knowles, B.H. (1994). Mechanism of action of *Bacillus thuringiensis* insecticidal delta-endotoxins. *Advance Insect Physiology*, *24*: 275–308.

Knowles, B.H. and Dow, J.A.T. (1993). The crystal endotoxins of *Bacillus thuringiensis*: Models for their mechanism of action on the insect gut. *BioEssays*, *15*: 469–76.

Knowles, B.H. and Ellar, D.J. (1987). Colloid-osmotic lysis is a general feature of the mechanism of action of *Bacillus thuringiensis* [delta]-endotoxins with different insect specificity. *Biochimica et Biophysica Acta (BBA) - General Subjects*, *924*: 509–18.

Kranthi, K.R., Kranthi, S., Ali, S. and Banerjee, S.K. (2000). Resistance to Cry1Ac d-endotoxin of *Bacillus thuringiensis* in a laboratory selected strain of *Helicoverpa armigera* (Hubner). *Current Science*, *78*: 1001–1004.

Kumaraswami, N.S., Maruyama, T., Kurabe, S., Kishimoto, T., Mitsui, T. and Hori, H. (2001). Lipids of brush border membrane vesicles (BBMV) from *Plutella xylostella* resistant and susceptible to Cry1Ac [delta]-endotoxin of *Bacillus thuringiensis*. *Comparative Biochemistry and Physiology Part B: Biochemistry and Molecular Biology*, *129*: 173–83.

Lee, M., Rajamohan, F., Gould, F. and Dean, D. (1995). Resistance to *Bacillus thuringiensis* Cry1A delta-endotoxins in a laboratory-selected *Heliothis virescens* strain is related to receptor alteration. *Applied Environmental Microbiology*, *61*: 3836–42.

Lee, M.K., Rajamohan, F., Jenkins, J.L., Curtiss, A.S. and Dean, D.H. (2000b). Role of two arginine residues in domain II, loop 2 of Cry1Ab and Cry1Ac *Bacillus thuringiensis:* Endotoxin in toxicity and binding to *Manduca sexta* and *Lymantria dispar* aminopeptidase N. *Molecular Microbiology*, *38*: 289–98.

Lehane, M.J. (1997). Peritrophic matrix structure and function. *Ann. Rev. Entomol.*, *42*: 525–50.

Lemaitre, B., Reichhart, J.-M. and Hoffmann, J.A. (1997). *Drosophila* host defense: Differential induction of antimicrobial peptide genes after infection by various classes of microorganisms. *Proc. Nat. Acad. Sci. United States of Am.*, *94*: 14614–14619.

Li, H., Oppert, B., Higgins, R.A., Huang, F., Zhu, K.Y. and Buschman, L.L. (2004). Comparative analysis of proteinase activities of *Bacillus thuringiensis*-resistant and -susceptible *Ostrinia nubilalis* (Lepidoptera: Crambidae). *Insect Biochem. Molecular Biology*, *34*: 753–62.

Li, J., Carroll, J. and Ellar, D.J. (1991). Crystal structure of insecticidal-endotoxin from *Bacillus thuringiensis* at 2.5 Å resolution. *Nature*, *353*: 815–21.

Li, J., Derbyshire, D.J., Promdonkoy, B. and Ellar, D.J. (2001). Structural implications for the transformation of the *Bacillus thuringiensis* delta-endotoxins from water-soluble to membrane-inserted forms. *Biochemical Society Transactions*, *29*: 571–77.

Li, J., Koni, P.A. and Ellar, D.J. (1996). Structure of the mosquitocidal delta-endotoxin CytB from *Bacillus thuringiensis* sp. *kyushuensis* and implications for membrane pore formation. *Journal of Molecular Biology*, *257*: 129–52.

Loeb, M.J., Martin, P.A.W., Hakim, R.S., Goto, S. and Takeda, M. (2001). Regeneration of cultured midgut cells after exposure to sublethal doses of toxin from two strains of *Bacillus thuringiensis*. *Journal of Insect Physiology*, *47*: 599–606.

Loseva, O., Ibrahim, M., Candas, M., Koller, C.N., Bauer, L.S. and Bulla Jr., L.A. (2002). Changes in protease activity and Cry3Aa toxin binding in the Colorado potato beetle: implications for insect resistance to *Bacillus thuringiensis* toxins. *Insect Biochemistry and Molecular Biology*, *32*: 567–77.

Luo, K., Sangadala, S., Masson, L., Mazza, A., Brousseau, R., and Adang, M.J. (1997): The *Heliothis virescens* 170 kDa aminopeptidase functions as "receptor A" by mediating specific *Bacillus thuringiensis* Cry1A delta-endotoxin binding and pore formation. *Insect Biochemistry and Molecular Biology*, *27*: 735–43.

Ma, G., Roberts, H., Sarjan, M., Featherstone, N., Lahnstein, J., Akhurst, R. and Schmidt, O. (2005). Is the mature endotoxin Cry1Ac from *Bacillus thuringiensis* inactivated by a coagulation reaction in the gut lumen of resistant *Helicoverpa armigera* larvae? *Insect Biochemistry and Molecular Biology*, *35*: 729–39.

MacIntosh, S., Stone, T., Jokerst, R. and Fuchs, R. (1991). Binding of *Bacillus thuringiensis* proteins to a laboratory-selected line of *Heliothis virescens*. *Proceedings of the National Academy of Sciences of the United States of America*, *88*: 8930–33.

Martinez-Ramirez, A.C., Gould, F. and Ferre, J. (1999). Histopathological effects and growth reduction in a susceptible and a resistant strain of *Heliothis virescens* (Lepidoptera: Noctuidae) caused by sublethal doses of pure Cry1A crystal proteins from *Bacillus thuringiensis*. *Biocontrol Science and Technology*, *9*: 239–46.

Maruyama, T., Hama, H. and Asano, S. (1999). Establishment and maintenance of high resistance to *Bacillus thuringiensis* formulation in diamondback moth, *Plutella xylostella* (Lepidoptera: Yponomeutidae). *Japanese Journal Applied Entomology and Zoology*, *43*.

Masson, L., Lu, Y.-j., Mazza, A., Brousseau, R. and Adang, M.J. (1995). The Cry1A(c) Receptor purified from *Manduca sexta* displays multiple specificities. *Journal of Biological Chemistry*, *270*: 20309–20315.

Masson, L., Tabashnik, B.E., Liu, Y.-B., Brousseau, R. and Schwartz, J.-L. (1999). Helix 4 of the *Bacillus thuringiensis* Cry1Aa toxin lines the lumen of the ion channel. *Journal of Biological Chemistry*, *274*: 31996–32000.

McGaughey, W.H. (1985). Insect resistance to the biological insecticide *Bacillus thuringiensis*. *Science*, *229*: 193–95.

McGaughey, W.H. and Johnson, D.E. (1994). Influence of crystal protein composition of *Bacillus thuringiensis* strains on cross-resistance in indianmeal moths (Lepidoptera: Pyralidae). *Journal of Economic Entomology*, *87*: 535–40.

Meadows, M., Ellis, D., Butt, J., Jarrett, P. and Burges, H. (1992): Distribution, frequency, and diversity of *Bacillus thuringiensis* in an animal feed mill. *Applied Environmental Microbiology*, *58*: 1344–50.

Moar, W., Pusztai-Carey, M., Van Faassen, H., Bosch, D., Frutos, R., Rang, C., Luo, K. and Adang, M. (1995). Development of *Bacillus thuringiensis* Cry1C resistance by *Spodoptera exigua* (Hubner) (Lepidoptera: Noctuidae). *Applied Environ. Microb.*, *61*: 2086–92.

Mohammed, S., Johnson, D. and Aronson, A. (1996). Altered binding of the Cry1Ac toxin to larval membranes but not to the toxin-binding protein in *Plodia interpunctella* selected for resistance to different *Bacillus thuringiensis* isolates. *Applied Environ. Microb.*, *62*: 4168–73.

Morin, S., Biggs, R.W., Sisterson, M.S., Shriver, L., Ellers-Kirk, C., Higginson, D., Holley, D., Gahan, L.J., Heckel, D.G., Carriere, Y., Dennehy, T.J., Brown, J.K. and Tabashnik, B.E. (2003). Three cadherin alleles associated with resistance to *Bacillus thuringiensis* in pink bollworm. *Proc. Nat. Acad. Sciences of the United States of America*, *100*: 5004–5009.

Morse, R.J., Yamamoto, T. and Stroud, R.M. (2001). Structure of Cry2Aa suggests an unexpected receptor binding epitope. *Structure*, *9*: 409–17.

MullerCohn, J., Chaufaux, J., Buisson, C., Gilois, N., Sanchis, V. and Lereclus, D. (1996). *Spodoptera littoralis* (Lepidoptera: Noctuidae) resistance to Cry1C and cross-resistance to other *Bacillus thuringiensis* crystal toxins. *J. Economic Entomology*, *89*: 791–97.

Nagamatsu, Y., Toda, S., Koike, T., Miyoshi, Y., Shigematsu, S. and Kogure, M. (1998). Cloning, sequencing, and expression of the *Bombyx mori* receptor for *Bacillus thuringiensis* insecticidal Cry1A(a) toxin. *Bioscience Biotechn. Biochemistry*, *62*: 727–34.

Oppert, B., Kramer, K.J., Beeman, R.W., Johnson, D. and McGaughey, W.H. (1997). Proteinase-mediated insect resistance to *Bacillus thuringiensis* toxins. *Journal of Biological Chemistry*, *272*: 23473–23476.

Oppert, B., Kramer, K.J., Johnson, D.E., MacIntosh, S.C. and McGaughey, W.H. (1994). Altered protoxin activation by midgut enzymes from a *Bacillus thuringiensis* resistant strain of *Plodia interpunctella. Biochemical Biophysical Res. Commun.*, *198*: 940–47.

Rahman, M.M., Roberts, H.L.S., Sarjan, M., Asgari, S. and Schmidt, O. (2004a). Induction and transmission of *Bacillus thuringiensis* tolerance in the flour moth *Ephestia kuehniella. Proc. Nat. Acad. Sciences of the United States of America*, *101*: 2696–99.

Rajamohan, F., Alzate, O., Cotrill, J.A., Curtiss, A. and Dean, D.H. (1996). Protein engineering of *Bacillus thuringiensis* delta-endotoxin: Mutations at domain II of CryIAb enhance receptor affinity and toxicity toward gypsy moth larvae. *Proceedings of the National Academy of Sciences of the United States of America*, *93*: 14338–14343.

Rausell, C., Garcia-Robles, I., Sanchez, J., Munoz-Garay, C., Martinez-Ramirez, A.C., Real, M.D. and Bravo, A. (2004a): Role of toxin activation on binding and pore formation activity of the *Bacillus thuringiensis* Cry3 toxins in membranes of *Leptinotarsa decemlineata* (Say). *Biochimica et Biophysica Acta (BBA) - Biomembranes, 1660*: 99–105.

Rausell, C., Pardo-Lopez, L., Sanchez, J., Munoz-Garay, C., Morera, C., Soberon, M. and Bravo, A. (2004b). Unfolding events in the water-soluble monomeric Cry1Ab toxin during transition to oligomeric pre-pore and membrane-inserted pore channel. *Journal of Biological Chemistry*, *279*: 55168–55175.

Schmidt, O. and Theopold, U. (2004). An extracellular driving force of endocytosis and cell-shape changes (Hypothesis). *BioEssays*, *26*: 1344–50.

Schnepf, E., Crickmore, N., Van Rie, J., Lereclus, D., Baum, J., Feitelson, J., Zeigler, D.R. and Dean, D.H. (1998). *Bacillus thuringiensis* and its pesticidal crystal proteins. *Microbiology and Molecular Biology Reviews*, *62*: 775–806.

Schnepf, H.E. and Whiteley, H.R. (1981). Cloning and expression of the *Bacillus thuringiensis* crystal protein gene in *Escherichia coli. Proceedings of the National Academy of Sciences of the United States of America*, *78*: 2893–97.

Schuler, T.H., Potting, R.P.J., Denholm, I., Clark, S.J., Clark, A.J., Stewart, C.N. and Poppy, G.M. (2003): Tritrophic choice experiments with Bt plants, the diamondback moth (*Plutella xylostella*) and the parasitoid *Cotesia plutellae*. *Transgenic Res.*, *12*: 351–61.

Schwartz, J.-L., Lu, Y.-J., Sohnlein, P., Brousseau, R., Laprade, R., Masson, L. and Adang, M.J. (1997). Ion channels formed in planar lipid bilayers by *Bacillus thuringiensis* toxins in the presence of *Manduca sexta* midgut receptors. *FEBS Letters*, *412*: 270–76.

Shelton, A.M., Zhao, J.-Z., and Roush, R.T. (2002). Economic, ecological, food safety, and social consequences of the development of Bt transgenic plants. *Annual Review of Entomology*, *47*: 845–81.

Shelton, M.A., Hossain, M.J., Paranjape, V., Azad, A.K., Rahman, M.L., Khan, A.S.M.M.R., Prodhan, M.Z.H., Rashid, M.A., Majumder, R., Hossain, M.A., Hussain, S.S., Huesing, J.E., and McCandless, L. (2018). Bt Eggplant Project in Bangladesh: History, Present Status, and Future Direction. *Frontiers in Bioengineering and Biotechnology*, 6(106): 1–6

Tabashnik, B., Finson, N., Groeters, F., Moar, W., Johnson, M., Luo, K. and Adang, M. (1994). Reversal of resistance to *Bacillus thuringiensis* in *Plutella xylostella. Proceedings of the National Academy of Sciences of the United States of America*, *91*: 4120–24.

Tabashnik, B.E. (1994). Evolution of resistance to *Bacillus thuringiensis. Annual Review of Entomology*, *39*: 47–79.

Tabashnik, B.E., Carrière, Y., Dennehy, T.J., Morin, S., Sisterson, M.S., Roush, R.T., Shelton, A.M. and Zhao, J.-Z. (2003). Insect resistance to transgenic Bt crops: Lessons from the laboratory and field. *Journal of Economic Entomology*, *96*: 1031–38.

Tabashnik, B.E., Finson, N. and Johnson, M.W. (1991). Managing resistance to *Bacillus-thuringiensis*: Lessons from the diamondback moth (Lepidoptera, Plutellidae). *Journal of Economic Entomology*, *84*: 49–55.

Tabashnik, B.E., Liu, Y.-B., Finson, N., Masson, L. and Heckel, D.G. (1997a). One gene in diamondback moth confers resistance to four *Bacillus thuringiensis* toxins. *Proceedings of the National Academy of Sciences of the United States of America*, *94*: 1640–44.

Tabashnik, B.E., Liu, Y.-B., Malvar, T., Heckel, D.G., Masson, L., Ballester, V., Granero, F., Mensua, J.L. and Ferre, J. (1997b). Global variation in the genetic and biochemical basis of diamondback moth resistance to *Bacillus thuringiensis*. *Proceedings of the National Academy of Sciences of the United States of America*, *94*: 12780–12785.

Tabashnik, B.E., Liu, Y.-B., Unnithan, D.C., Carrière, Y., Dennehy, T.J. and Morin, S. (2004). Shared genetic basis of resistance to Bt toxin Cry1Ac in independent strains of pink bollworm. *Journal of Economic Entomology*, *97*: 721–26.

Tabashnik, B.E., Roush, R.T., Earle, E.D., Shelton; A.M., Huang, F., Buschman, L., Higgins, R. and McGaughey, W. (2000). Resistance to Bt toxins. *Science*, *287*: 287(5450): 41.

Tang, J., Shelton, A., Van Rie, J., De Roeck, S., Moar, W., Roush, R. and Peferoen, M. (1996). Toxicity of *Bacillus thuringiensis* spore and crystal protein to resistant diamondback moth (*Plutella xylostella*). *Applied Environmental Microbiology*, *62*: 564–569.

Terra, W.R. (2001): The origin and functions of the insect peritrophic membrane and peritrophic gel. *Archives of Insect Biochemistry and Physiology*, *47*: 47–61.

Tigue, N.J., Jacoby, J. and Ellar, D.J. (2001). The alpha-helix 4 residue, Asn135, is involved in the oligomerization of Cry1Ac1 and Cry1Ab5 *Bacillus thuringiensis* toxins. *Applied Environmental Microbiology*, *67*: 5715–20.

Vadlamudi, R.K., Weber, E., Ji, I., Ji, T.H., Bulla Jr., L.A. (1995). Cloning and expression of a receptor for an insecticidal toxin of *Bacillus thuringiensis*. *Journal of Biological Chemistry*, *270*: 5490–94.

Vaeck, M., Reynaerts, A., Höfte, H., Jansens, S., Beuckeleer, M.D., Dean, C., Zabeau, M., Montagu, M.V. and Leemans, J. (1987), Transgenic plants protected from insect attack. *Nature*, *328*: 33–37.

Van Rie, J., Jansens, S., Hofte, H., Degheele, D. and Van Mellaert, H. (1989). Specificity of *Bacillus thuringiensis* delta-endotoxins. Importance of specific receptors on the brush border membrane of the mid-gut of target insects. *Eur. J. Biochem.*, *186*: 239–47.

Van Rie, J., McGaughey, W.H., Johnson, D.E., Barnett, B.D. and Van Mellaert, H. (1990). Mechanism of insect resistance to the microbial insecticide *Bacillus thuringiensis*. *Science*, *247*: 72–74.

Wang, P. and Granados, R.R. (1997). An intestinal mucin is the target substrate for a baculovirus enhancin. *Proc. Nat. Acad. Sciences of the United States of Am.*, *94*: 6977–82.

Wang, P. and Granados, R.R. (2000). Calcofluor disrupts the midgut defense system in insects. *Insect Biochemistry and Molecular Biology*, *30*: 135–43.

Whalon, M.E., Miller, D.L., Hollingworth, R.M., Grafius, E.J. and Miller, J.R. (1993). Selection of a Colorado potato beetle (Coleoptera:Chrysomelidae) strain resistant to *Bacillus thuringiensis*. *Journal of Economic Entomology*, *86*: 226–33.

Worthington, C.R. (1991). *The Pesticide Manual*. Surrey, British Crop Protection Council.

Wright, D., Iqbal, M., Granero, F. and Ferre, J. (1997). A change in a single midgut receptor in the diamondback moth (*Plutella xylostella*) is only in part responsible for field resistance to *Bacillus thuringiensis* subsp. *kurstaki* and *B. thuringiensis* subsp. *aizawai*. *Applied Environmental Microbiology*, *63*: 1814–19.

Zhao, J.Z., Collins, H.L., Tang, J.D., Cao, J., Earle, E.D., Roush, R.T., Herrero, S., Escriche, B., Ferre, J. and Shelton, A.M. (2000). Development and characterization of diamondback moth resistance to transgenic broccoli expressing high levels of Cry1C. *Applied Environmental Microbiology*, *66*: 3784–89.

Zheng, S.J., Henken, B., de Maagd, R.A., Purwito, A., Krens, F.A. and Kik, C. (2005): Two different *Bacillus thuringiensis* toxin genes confer resistance to beet armyworm (*Spodoptera exigua* Hubner) in transgenic Bt-shallots (*Allium cepa* L.). *Transgenic Research*, *14*: 261–72.

Chapter 21

Climate Change Challenges and Strategic Research on Bangladesh Agriculture

Md. Abdul Karim[1]*

Introduction

Climate Change Concern

Anthropogenic activities of the past century have led to a concerning increase in greenhouse gas emissions into the atmosphere, with current levels reaching approximately 388 ppm for CO_2 and 1.8 ppm for CH_4. This surge in emissions has resulted in a heightened re-radiation of heat from the atmosphere, contributing to an unprecedented warming of the Earth. The correlation between the concentration of CO_2 and the rise in global temperatures is depicted in Fig. 1, illustrating a linear relationship between CO_2 concentration and the upward trend in global temperatures.

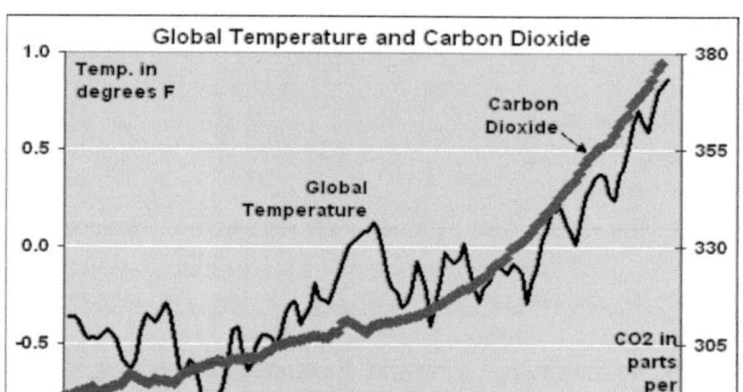

Fig. 1 Correlation between CO_2 and temperature rise. (Image: courtesy of zFacts.com)

[1] Bangabandhu Sheikh Mujibur Rahman Agricultural University, Bangladesh.
* Corresponding author: akarim1506@gmail.com

The IPCC's 6th assessment report for policymakers (AR6) in 2022 highlights the potential rise in global mean temperature, projecting a range of 3.7°C–4.8°C by the end of the century if greenhouse gas emissions are not curtailed. This forecast signals a doubling of the temperature increase witnessed over the past century. The year-to-year and intra-seasonal changes in climatic factors, particularly rainfall onset and intensity, and temperature fluctuations, are expected to become more pronounced.

According to the IPCC (2022), climate variability associated with global warming manifests in various ways, including:

i. Escalation of extreme heat events and heatwaves in summer, coupled with extreme cold events in winter.
ii. Augmentation of average summer precipitation, contrasted by a reduction in the lean period.
iii. Amplification of tropical storm wind strength.
iv. A 20 cm increase in sea levels since 1870.
v. A 0.1 pH unit rise in ocean acidification since the preindustrial period.

These environmental shifts are occurring at an alarming rate, accelerated by the surge in anthropogenic activities linked to uncontrolled industrial expansion, increased automobile usage, and unplanned infrastructural development.

Concerns about climatic variability's impact on Bangladesh's agriculture are underscored by the AR6, which identifies 14 climatic hazards associated with climate change. These hazards include heightened risks of monsoon floods, flash floods, salinity, storm surges, prolonged heat periods, extended drought periods, reduced rainy days but increased daily rainfall, river erosion, landslides, lightning, cold spells, sea-level rise, urban flooding, and ocean acidification.

Bangladesh's unique geophysical location renders it highly vulnerable to the effects of global warming and climatic variability, as evident in the year-to-year temperature and rainfall variations observed in the Kurigram district. The data from Kurigram, one of the country's most vulnerable districts, illustrates significant fluctuations in maximum temperatures, ranging from 29°C–33°C over the years 2012, 2013, and 2014, despite consistent minimum temperatures of 21°C during the same period (Fig. 2a, b).

The year to year inter- and intra-seasonal variations in temperature were specifically prominent (Fig. 2a, b). The minimum temperatures in April (when most of the kharif 1 crops are at active vegetative stage) of 2012, 2013, 2014, and 2015 in Kurigram were 22.3°C, 21.47°C, 21.27°C, and 20°C, while the maximum temperatures were 30.87°C, 35.07°C, 37.30°C, and 30°C, respectively. Therefore, inter-annual variation in maximum temperature is more evident than that of minimum temperature.

The total annual rainfall during 2010–2014 ranged from 1771 mm in 2013 to 2263 in 2012 (Fig. 3a, b). The rainfall in April of 2012–2015 were 209 mm, 72 mm, 10.74 mm, and 212 mm, respectively. Consequently, crops in kharif 1 season of 2014 were mostly damaged due to drought in the charlands (newly developed land due to siltation) of Kurigram (Karim, 2014a). The findings indicated that despite higher total rainfall in 2014 (1913 mm) than in 2011 and 2013, kharif 1 crops were damaged due to low rainfall in April 2014.

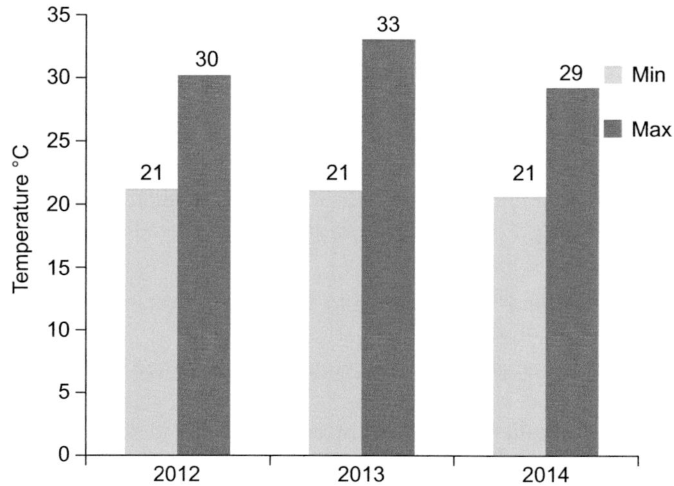

Fig. 2a Average max. and min. temperatures during 2012–2015 (May) in Kurigram.

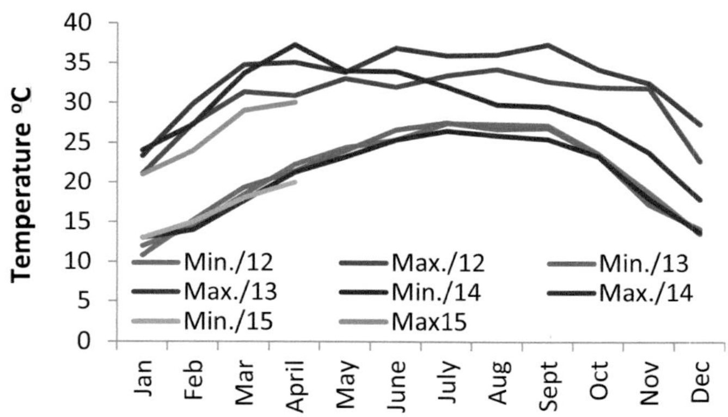

Fig. 2b Average max. and min. temperatures during 2012–2015 (May) in Kurigram.

Kharif-1 agricultural season (March–June) is the most vulnerable one to climatic variability in Bangladesh. The crop success is dependent on onset and intensity of rainfall at that time. Farmers grow mostly Aus rice, sesame, mungbean, and jute in kharif 1. As mentioned that in 2014 there was a serious drought in April and the reason for crop damage was the delay in onset of rainfall. The crop success was also dependent on prevailing soil surface temperature. The surface temperature largely depends on soil properties. As shown in Fig. 4 that when the air, soil surface, and root zone (15 cm depth) temperatures measured on 24 April 2014 at Bangabandhu Agricultural University (BSMRAU), Gazipur were 41, 49, and 33°C, respectively, those were 39, 59, and 33°C, respectively on 25 April 2014 at Char Rasulpur, Kurigram (Karim, 2014a; Karim, et al., 2018). That indicated that despite the 2°C prevailing high air temperature at BSMRAU, the soil surface temperature was 10°C lower than that at Char Rasulpur. The root zone temperature was the same at both the locations despite an extremely high difference in soil properties. Such a big difference in soil

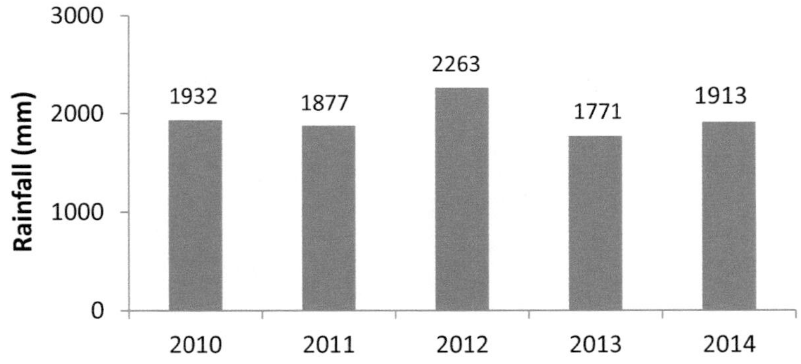

Fig. 3a Average rainfall in Kurigram during 2012–2015 (May).

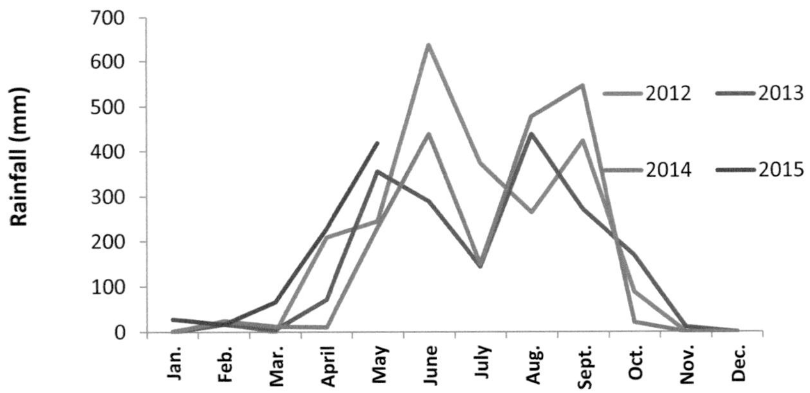

Fig. 3b Average rainfall in Kurigram during 2012–2015 (May)

surface temperature was mostly due to loamy sand soil of Char Rasulpur, while that at BSMRAU was sandy loam.

Climatic variability is also causing unpredicted floods in the northern and middle part of Bangladesh. For example, heavy late shower during August–September in the same year, 2014, caused two consecutive long-term, about four weeks, floods. Therefore, both serious drought and long-term late flood damaged the crops in kharif 1 and kharif 2 seasons, especially in the charlands of northern and middle regions of the country. The flood happened to occur in June in 2012 due to heavy rainfall at that time, while it occurred in August–September in 2014. Thus, the climatic variability posed a big threat for making effective agricultural planning by the farmers living in the unfavorable ecosystems.

Since the recent past, the agricultural sector of Bangladesh has been facing a big dilemma. While the country is bearing the world's most thickly and ever-increasing population with a density of 1100 persons/km^2, it is losing almost 0.8 million hectares of agricultural land per year due to its unplanned industrialization, urbanization, and infrastructural development activities. The declining agricultural land in one hand, and the rising climatic variability on the other is putting serious pressure on agricultural

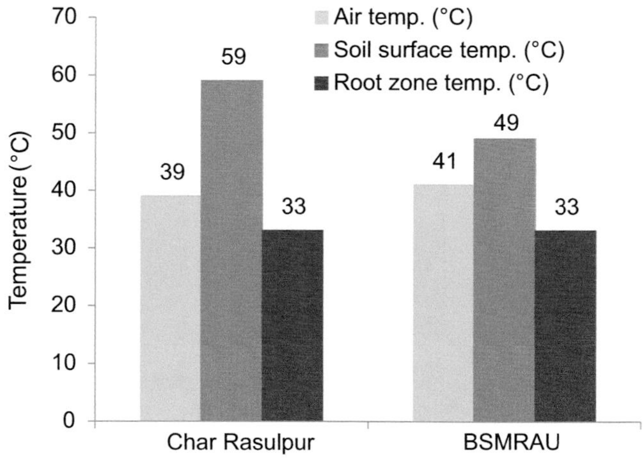

Fig. 4 Temperatures at different levels at BSMRAU and Char Rasulpur at 12.30 pm on 24, 25 April 2014, respectively.

productivity as a whole for a particular ecosystem. The following are the major challenges for crop production in the context of climatic variability in Bangladesh.

Temperature rise: As predicted by 6th IPCC (2022), the mean temperature may increase by 1–2°C as against 1850–1900; and number of days with maximum temperatures above 35°C would likely be 90–120 days in a year in case of 1.5–2°C global warming, while that may be more than 180 days in the case of 4°C global warming. It is predicted that sea- level rise may continue throughout the 21st century. Category 4 and 5 cyclones will increase in the coastal region of Bangladesh. It is predicted that even a 1°C rise in global mean temperature may inundate 18% land in the south of Bangladesh. The rise in global mean temperature has already increased the climatic variability in the country, which makes unpredictable high temperature extremes during March–May and is threatening establishment of kharif 1 crop (Karim, et. al., 2014a; Karim et al., 2018).

Salinity: The land of the southern coastal area of Bangladesh is inherently affected by natural salinity. However, due to long-time siltation on the riverbed in summer, the smooth flow of river water is being hindered to a great extent. Since 1973, the salinity has increased to 27% and the total area by now exceeds one million ha at various levels. More areas are under threat for salinization due to the combined effects of sea-level rise, increased tidal effect, introduction of brackish water shrimp cultivation, continuous reduction of river flow particularly during the dry period, capillary upward movement of soluble salts due to the presence of high saline ground water table at shallower depths and faulty management of sluice gates in the south and southwestern part of Bangladesh (Aziz, 2003; Mannan, 2009; Khan, 2013; Karim, et al., 2017).

Drought: The increase in climatic variability resulted in an exceedingly unpredictable nature of rainfall, especially during March–May. Due to such an erratic rainfall, drought affects almost 3.7 m ha, mostly during the kharif 1 season, in the northwestern part and charlands of different districts of the country. Crops grown in the rabi season suffer from water shortage at the reproductive stage, while crops are difficult to establish in the kharif 1 season due to the erratic nature of rainfall (Islam, 2006; Chowdhury, 2009; Siddique, 2014).

Flooding/Waterlogging: Due to climatic variability, the erratic nature of rainfall in Bangladesh has become more unpredictable. According to IPCC (2013, 2022) the intensity of rainfall in summer may increase noticeably, while it will decrease in the dry period, and the changed pattern of rainfall will cause flash flooding to a great extent. About 2.6 m ha of agricultural land is affected by floods, which is almost a regular phenomenon of the country. About 0.4 m ha crop land is waterlogged due to faulty floodgates/dikes in south and southwestern Bangladesh. The flood may occur due to either sudden heavy rainfall during June–September which is called flashflood and may last for 3–14 days, even more, or due to the tropical monsoon summer that may last for a month or more (Islam, 2003; Amin, 2014). The frequency of flood and cyclones have been increased recently. In 2021, the flood lasted for about 40 days which was the 2nd longest and worst flood after 1998. Moreover, the highest flood water level in history was in 2018.

Charland agriculture: About one million ha of land is developed newly on the riverbed and banks of the rivers of all over the country due to siltation, which is called charland, except in the hilly eastern part (Karim, et. al., 2017). The soil of charland is very heterogeneous as the physical properties vary even within a meter, texture varies from sandy to loamy sand or silty loam; is deficient of all kinds of plant nutrients, has very low organic matters and minimum moisture holding capacity. It is very difficult to organize the crop during kharif 1 season without repeated irrigations in the sand-dominated chars of northern Bangladesh, which increases cost of production substantially (Karim, et al., 2014; Karim, et al., 2018).

Organic matter depletion: The organic matters deplete fast in tropical conditions. However, the matters are seemingly depleted faster, especially in the sandy/silty loam soils of the northern and middle regions of the country, probably due to increase in maximum temperature event. The intensive cropping (3–4 crops/year) is putting a question mark on the depleting plant nutrients, especially in the medium high and highland under irrigated agriculture (Haque, 2012).

Some Technological Achievements in Bangladesh

The agricultural technologies that developed under a favorable environment do not work properly in an unfavorable ecosystem. Though not commendable, but some progress has been made on the development of technologies that are suitable for a specific problem. However, only some of the technologies the farmers are practicing depict the weakness of technology transfer mechanism in the agricultural sector.

***Elevated CO_2 and crop response*:** Studies on the effects of rise in the atmospheric CO_2 concentration on rice and mungbean were carried out at the Bangabandhu Sheikh Mujibur Rahman Agricultural University using Open Top Chamber Technology. Along with a clear varietal difference, the growth and yield of both the crops responded positively to the elevated concentration of CO_2 (ambient and elevated CO_2

concentrations were 360± 10 ppm and 560± 10 ppm, respectively). Root growth was much higher under elevated CO_2 conditions than normal growth conditions (Fig. 5).

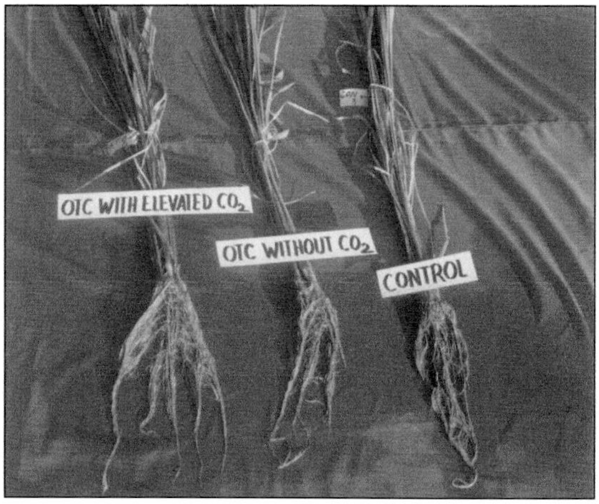

Fig. 5 Differences in root growth at different CO_2 levels (Hamid, et al., 2003)

However, flowers' drop in mungbean and the number of unfilled grains in rice increased enormously under elevated CO_2 due to concomitant rise in temperature (Hamid et al., 2003; Haque, et al., 2005). Therefore, the positive impact of elevated CO_2 on C_3 crops might be nullified by the associated negative effect of rise in temperature. Intensive research should get due priority for identifying varieties that show high tolerance to concomitant elevated CO_2 and high temperature conditions.

***Salinity tolerance*:** Significant progress has been made in developing two salt tolerant mungbean genotypes, namely MB1 and BM8 (Aziz, 2003), three salt tolerant soybean genotypes, namely AGS313 (Mannan, 2009), BD2330 (Khan, 2013), and G00028 (Tareq, 2022), which are in demonstration in the farmers' field (Fig. 6). Wheat genotype BAW 1147 was found to be a fairly tolerant one to salinity when 50 genotypes were screened against salinity stress (Table 1; personal communication). Rice varieties such as BRRI dhan47, BRRI dhan97, BRRI dhan99 (8–14 dS/m), and BINA dhan-23, soybean varieties, BU Soybean 2, and BARI Soybean 4, BARI Mung 6, BINA Gom, BARI Barley7, Binachinabadam-6, BARI Alu-72, BARI Barley7 and 8 are salt tolerant (Mondal et al., 2012; BRRI, 2020).

Sorjan methods of crop cultivation in the coastal area, seed priming, using mulches, etc. are favorable technologies to control salinity. Despite selection of the salt tolerant genotypes of different crops at the laboratory and control-field level, very few could be made available to the farmers. Therefore, research attention should be more pragmatic considering "laboratory to field" concept of technology development.

***Submergence tolerance*:** Two rice varieties, BRRI dhan51 and BRRI dhan52, have been released as submergence tolerant (BRRI, 2013) and six mungbean genotypes GK-48, IPSA 13, VC-6173A, BU mug2, BARI mug5, and IPSA-12 have been identified as submergence tolerant materials (Islam, 2003; Amin, 2014). The results

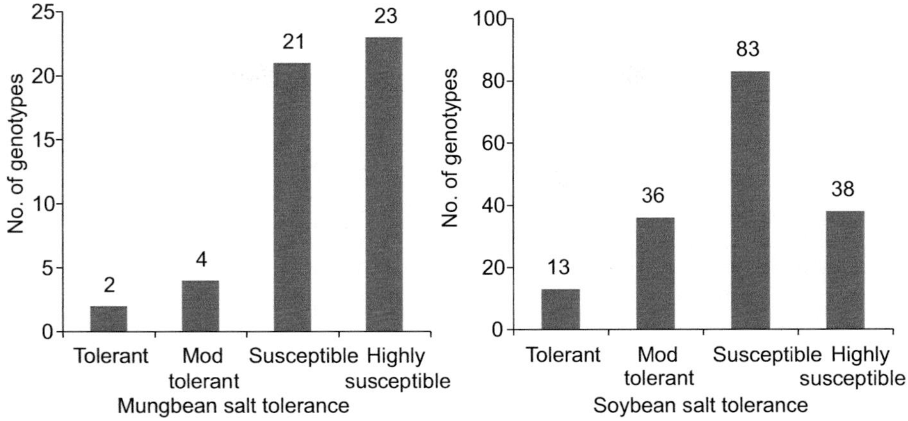

Fig. 6 Screening of mungbean and soybean genotypes against salinity stress

Table 1 Salt tolerance of 50 wheat genotypes.

Rank Order	*Salt Tolerance % Control*	*Genotypes*
Tolerant	More than 80.00	BAW 1147
Moderately Tolerant	60.01–80.00	BAW 1140, BARI Gom 25, BAW 1135, BAW 1142, BAW 1143, BAW 1150, BARI Gom 28, BAW 1157, BAW 1159
Moderately Susceptible	50.01–60.00	Shatabdi, BARI Gom 27
Susceptible	Less than 70.01	Rest 38 genotypes

in Table 2 indicated a wide range of yield reduction, from 25–50%, of mungbean genotypes due to 4-day soil flooding in kharif 1 season. Therefore, the yield level of mungbean can be increased substantially under waterlogging conditions by making available the right mungbean variety. There is high genetic variability in submergence tolerance of rice (Abedin et al. 2019; Islam et al., 2019), and submergence at 10 days after transplanting (DAT) is more detrimental than that at 30 DAT as shown in Table 3 (Abedin et al., 2019). Sac gardening and floating agriculture are found very effective for bringing waterlogged/flooded soil into production.

Table 2 Effect of 4-day soil flooding on seed yield of mungbean genotypes (Amin, 2014).

Genotype	*Seed Yield (g plant^{-1})*		*% Relative to Control*
	Control	*4-day Soil Flooding*	
IPSA-13	4.81 ± 0.11	2.40 ± 0.29	50.0
VC-6173A	4.04 ± 0.72	2. 99 ± 0.40	75.0
BU mug 2	3.79 ± 0.57	2.46 ± 0.34	65.0
BARI mung 5	3.95 ± 0.34	1.96 ± 0.40	50.0
IPSA-12	3.79 ± 0.48	2.16 ± 0.52	57.0

Table 3 Effect of submergence at 10 and 30 days after transplanting (DAT) for 14 days on yield of 21 Aman rice cultivars (Abedin, 2018).

Genotypes	*Grain Yield (g hill^{-1})*		
	Non-flooded	*Flooding on 10 DAT*	*Flooding on 30 DAT*
BRRI dhan33	23.51 bcd	12.14 abcd (52)	19.82 a-f (84)
BRRI dhan51	25.80 bcd	6.12 ab (62)	21.64 a- f (84)
BRRI dhan56	23.38 bcd	10.75 bcd (46)	17.34 b-f (74)
BRRI dhan57	18.75 d	6.03 d (32)	13.57 f (72)
BINA dhan11	19.22 d	9.11 bcd (47)	14.65 cdef (76)
BU dhan1	20.04 d	9.11 bcd (45)	14.65 cdef (73)
Guti shorna	19.93 d	7.41 cd (37)	14.12 def (71)
Malshira	19.05 d	10.12 bcd (53)	15.55 b-f (82)
Joldepa	36.51 ab	12.18 abcd (33)	19.38 a-f (53)
Sadapajam	23.53 bcd	12.90 abcd (55)	18.70 a-f (79)
Lalpayka	40.63 a	14.13 abcd (35)	25.75 abc (63)
Mutha	22.64 cd	14.68 abc (65)	18.91 a-f (84)
Dudmona	28.10 abcd	13.37 abcd (48)	22.63 a-f (81)
Jalkucha	29.80 abcd	11.62 abcd (39)	25.07 a-e (84)
Lalchikon	36.03 aba	11.33 bcd (31)	26.61 ab (74)
Achin	31.27 abcd	14.52 abcd (46)	30.14 a (96)
Nakochimota	28.79 abcd	11.99 abcd (42)	25.56 abc (89)
Razashail	18.12 d	7.88 bcd (43)	13.64 ef (75)
Kutiagoni	23.31 bcd	12.13 abcd (52)	17.47 b-f (75)
Sadamota	34.57 abc	20.31 a (59)	26.97 ab (78)
Vog	20.05 d	9.95 bcd (50)	17.97 b-f (90)
CV (%)	19.6		

Drought tolerance: Drought tolerant genotypes of wheat such as BARI Gom 26, Sourav, BAW 1169, and BAW 1158 (Table 4, Bazzaz, 2014), French bean namely BB24 and BB43 (Chowdhury, 2009), mungbean such as BARI mug2 and BU mug2 (Islam, 2006) and soybean namely BARI soybean5, BARI soybean6, Shohag and BD-2331 (Table 5; Chowdhury, 2014) have been identified at BSMRAU. BU Soybean2, Binadhan-20, Binadhan-2, BINA Moog11, BARI cheenabadam-10, BARI hybrid Butta-13, BARI Barley-11 are also reported to be drought tolerant.

Table 4 Ranking of 35 wheat genotypes based on their drought tolerance (Bazzaz, 2014).

Rank Order	*Yield Reduction (%)*	*Genotypes*
Tolerant	Less than 30.00	BARI Gom 26, Sourav, BAW 1169, and BAW 1158
Moderately tolerant	30.01–40.00	BAW 1151, BAW 1157, BAW 1159, BAW 1161, BAW 1165, and BAW 1170
Moderately susceptible	40.01–50.00	Prodip, Shatabdi, Gourav, Sufi, Kanchan, Barkat, Balaka, Aghrani, Akbar, Protiva, Ananda, Bijoy, BAW 1160, BAW 1162, BAW 1163, BAW 1164, BAW 1168, and BAW 1172
Susceptible	Above 50.01	Seri, Pavon, BARI Gom 25, BAW 1166, BAW 1167, BAW 1171, and BAW 1173

Table 5 Ranking of 50 soybean genotypes based on their drought tolerance (Chowdhury, 2014).

Ranking	*Yield Reduction (%)*	*Genotypes*
Tolerant	Less than 50	BARI Soybean-6, BARI Soybean-5, Shogah, and BD2331
Moderately Tolerant	51–60	BD2336 and BD2329
Moderately susceptible	61–70	G00015 and BGM2093
Susceptible	Above 70%	Rest 42 genotypes

Therefore, a high variability exists in drought tolerance of different crops and the variability can be exploited for developing high-yielding variety through genetic manipulation. Mini pond as found in the Barind tract and other places is an immensely useful practice for saving the crops from terminal drought. Seed priming helps in faster germination and roots' development that increase crop drought tolerance. Mulching is also advocated for saving the crops from drought stress.

Other technologies: Strategic research to develop technologies to face the challenges of climatic variability effect on agriculture has been conducted at various levels in Bangladesh. IPSA 12 has been found as a low N and P required mungbean genotype (Razzaque, 2014). A model for cultivation of three crops per year, lentil/wheat – mungbean – short duration Aman rice in charlands of Kurigram (Karim, 2014a, b, c) and some other technologies for unfavorable conditions have been identified (Mondal, 2012; Siddique, 2014). Growing four crops (e.g., mustard/potato – mungbean – Aus rice (short duration, cv. Parija) – Aman rice (short duration) are getting popularity in the high and medium high land (Haque, 2011; Mondal, 2012). Diffusion of those technologies to the farmers should get due priority to combat the climate change impact on agriculture.

Suggested Strategic Agricultural Research

Adaptation and mitigation strategies are crucial for sustaining crop production in the face of climate change in Bangladesh. While some mitigation programs involve significant costs and regular maintenance, adopting practical and sustainable adaptation strategies through breeding programs for developing environmentally stress-tolerant crop varieties and appropriate management has been emphasized (Karim, 2006; Karim et al., 2014).

Several strategic research areas and approaches can be considered for developing technologies related to crop production in the context of climate change in Bangladesh:

i. **Exploiting the agricultural potential of unfavorable ecosystems:** Research should focus on generating cost-effective technologies to enhance crop productivity under adverse conditions such as drought, flooding/waterlogging, salinity, and high temperatures. Dryland, coastal, charland, haor, and hill agriculture deserve special attention.
ii. **Fine-tuning technologies to ecological conditions:** Technologies developed in research institutes and universities should undergo testing in specific ecological conditions before being made available to end-users. For instance, technologies targeted for drought-prone areas should be thoroughly tested in those areas for at least 2–3 years before release.
iii. **Flood-tolerant short-duration rice varieties:** Research should explore flood-tolerant short-duration rice varieties to ensure successful cultivation in flood-prone areas, such as charlands, with a focus on achieving multiple crops per year.
iv. **Improving water-use efficiency:** Systematic studies on improving water-use efficiency for intensifying upland crop cultivation, emphasizing low-water requiring crops like pulses, wheat, and oilseeds during the dry season, are essential. This can be achieved through modern irrigation and drainage channels, appropriate irrigation timing, and water-conserving practices.
v. **Developing winter crops with climate resilience:** With the projected shorter but colder winter durations due to global warming, research should focus on developing winter crop varieties tolerant to higher temperatures at the reproductive stage while being cold-tolerant at the seedling stage.
vi. **Enhancing crop establishment during kharif 1 season:** Research on improving organic matter content to support better crop establishment, especially in silty loam or sandy loam soils, under drought conditions during the kharif 1 season is recommended.
vii. **Addressing grain fertility under elevated CO_2:** Systematic research on improving grain fertility of C_3 crops under increased CO_2 concentration and rising temperatures, along with selecting high-responsive crop varieties to elevated CO_2, is necessary.
viii. **Harvesting and conserving monsoon rainwater:** Given the projected increase in summer rains and decrease during the lean period, efforts should be directed towards harvesting and conserving monsoon rainwater through large-scale water reservoir construction. However, a thorough analysis of the ecological and environmental impact of such projects is essential before implementation.

Conclusions

According to the 6th IPCC Report (AR6, 2022) the following 14 climatic hazards due to climate change are identified for Bangladesh: increase in monsoon flood, flash flood, salinity, storm surge, excess heat period, longer drought period, shorter rainy days but excess rain/day, river erosion, land slide, lightning, cold spell, sea-level rise, urban flood, and ocean acidification. All those hazards are related to agricultural production. Nonetheless, good agricultural lands are decreasing due to developmental activities at the rate of almost 0.7%. Despite the challenges of climate change, much has been achieved in all the sectors of agricultural production. Tolerant varieties, especially tolerant to drought, salinity, flooding, and low nutrition have been identified for several crops. However, the increasing population, decreasing agricultural lands, and the challenge of climate change appears as threat for food security. Though technologies are developing at the research level, few are adopted by the farmers. Agricultural research strategy should therefore be oriented considering climate change issues and involving end users in a participatory approach.

References

Abedin, M.H. (2018). *Effect of Flooding at Vegetative Stages on the Performance of Rice Cultivars*. MS Thesis submitted at the Department of Agronomy, BSMRAU.

Abedin, M.H., Mamun, M.A.A., Mia, M.A.B. and Karim, M.A. (2019). Evaluation of submergence tolerance in landrace rice cultivars by various growth and yield parameters. *Journal of Crop Science and Biotechnology*, *22*: 335–44.

Amin, M.R. (2014). *Minimizing Flooding Damage through N and K Management in Mungbean.* A PhD dissertation. Department of Agronomy, Bangabandhu Sheikh Mujibur Rahman Agricultural University, Gazipur.

Aziz M.A. (2003). *Growth, Yield, and Some Physiological Mechanisms of Salinity Tolerance in Mungbean.* A PhD dissertation. Department of Agronomy, Bangabandhu Sheikh Mujibur Rahman Agricultural University, Gazipur.

Bazzaz, M.M. (2014). *Morpho-physiology and Yield Response of Wheat to Variable Water Regimes.* A PhD dissertation. Department of Agronomy, Bangabandhu Sheikh Mujibur Rahman Agricultural University, Gazipur.

BRRI. (2013). *Adhunik Dhaner Chash.* Bangladesh Rice Research Institute, Gazipur-1701.

Chowdhury, A.K. (2009). *Studies on Water Stress Tolerance of French Bean (Phaseolus vulgaris* L.). A PhD dissertation. Department of Agronomy, Bangabandhu Sheikh Mujibur Rahman Agricultural University, Gazipur.

Chowdhury, J.A. (2014). *Yield Performance and Some Physiological Mechanisms of Drought Tolerance in Soybean.* A PhD dissertation. Department of Agronomy, Bangabandhu Sheikh Mujibur Rahman Agricultural University, Gazipur.

Hamid, A., Haque, M.M., Khanam, M., Hossain, M.A., Karim, M.A., Khaliq, Q.A., Biswas, D.K., Gomosta, A.R., Chowdhury, A.M. and Uprety, D.C. (2003). Photosynthesis, growth, and productivity of rice under elevated CO_2. *Ind. J. Plant Physiol.*, *8*: 253–58.

Haque, M.M. (2012). *Crop Intensification in Northern Region of Bangladesh through Up-scaling the Production of Short Duration Rice and Mungbean.* A report of KGF-World Bank funded project.

Haque, M.S., Karim, M.A., Haque, M.M. Hamid, M.A. and Nawata, E. (2005). Effect of elevated CO_2 concentration on growth, chlorophyll content, and yield of mungbean (*Vignaradiata* L. Wilczek) genotypes. *Japan J. Trop. Agric.*, *49*: 189–96.

Intergovernmental Panel on Climate Change (IPCC). (2013). *IPCC 5th Assessment Report, AR5.*

Intergovernmental Panel on Climate Change (IPCC). (2022). *IPCC 6th Assessment Report, AR6.*

Islam, M.R. (2003). *Ecophysiology of Soil Flooding Tolerance of Mungbean.* A PhD dissertation. Department of Agronomy, Bangabandhu Sheikh Mujibur Rahman Agricultural University, Gazipur.

Islam, M.S. (2006). *Water Stress in Mungbean (Vigna radiata L. Wilczeck) and Its Response to Application of Plant Growth Hormones.* A PhD dissertation. Department of Agronomy, Bangabandhu Sheikh Mujibur Rahman Agricultural University, Gazipur.

Islam, M.Z., Islam, M.R., Mamun, M.A.A.A., Haque, M.M., Ahmed, J.U., Akter, N. and Karim, M.A. (2019). Genetic variability of submergence tolerance in rice related to yield and yield contributing traits. *Journal of Plant Stress Physiology*, *5*: 22–27. doi:10.25081/jpsp.2019.v5.5493

Karim, M.A. (2006). Crop production in Bangladesh: Status, challenges ahead, and strategic research that needs to be strengthened. *South Pacific Newsletter*, *17*: 1–5.

Karim, M.A. (2014a). Report of the project *Up-scaling Mungbean and Rice Cultivation in the Charlands of Kurigram*. KGF-World Bank supported project.

Karim, M.A. (2014b). *Dharla O Dhudkumar Nadir Chare Dhaner Chash* (in Bengali). A publication of the project on Up-scaling Mungbean and Rice Cultivation in the Charlands of Kurigram. KGF-World Bank supported project.

Karim, M.A. (2014c). *Dharla O Dhudkumar Nadir Chare Mung Daler Chash* (in Bengali). A publication of the project on Up-scaling Mungbean and Rice Cultivation in the Charlands of Kurigram. KGF-World Bank supported project.

Karim, M.A., Haque, M.M. and Hamid, A. (2014). Concern, Opportunity, and Strategic Research in Crop Sector under Climate Change Context in Bangladesh. *In: Rising Atmospheric CO_2 and Crop Response*, 1–64. A compendium of published works at BSMRAU, Bangladesh.

Karim, M.A., Quayuum, M.A., Samsuzzaman, S., Higuchi, H. and Nawata, E. (2017). Challenges and opportunities in crop production in different types of char lands of Bangladesh: Diversity in crops and cropping. *Tropical Agric. Develop.*, *61*: 77–93.

Karim, M.A., Higuchi, H. and Nawata, E. (2017). Technological options for sustaining crop production under climate change in Bangladesh. *Research Tropical Agric.*, *10*: 33–34.

Karim, M.A., Higuchi, H. and Nawata, E. (2018). A Participatory Research Experience on the Introduction of Mungbean and Short Duration Aman Rice as Summer Crops in the Char Lands of Northern Bangladesh. *Tropical Agriculture and Development*, *62*: 14–23.

Khan, M.S.A. (2013). *Evaluation of Soybean Genotypes in Relation to Yield Performance, Salinity and Drought Tolerance*. A PhD dissertation. Department of Agronomy, Bangabandhu Sheikh Mujibur Rahman Agricultural University, Gazipur.

Mannan, M.A. (2009). *Response of Soybean Genotypes to Salinity Stress*. A PhD dissertation. Department of Agronomy, Bangabandhu Sheikh Mujibur Rahman Agricultural University, Gazipur.

Mondal, R.I. (2012). Advances in Agronomic Research under Changing Environment in Bangladesh. Keynote paper: *11th Conference of Bangladesh Society of Agronomy* held on 06 October 2012 at BARI, Joydebpur, Gazipur-1701.

Razzaque, M.A. (2014). *Genotypic Variability in Yield and Nutrient Acquisition of Mungbean under Low Nitrogen and Phosphorus Conditions*. A PhD dissertation. Department of Agronomy, Bangabandhu Sheikh Mujibur Rahman Agricultural University, Gazipur.

Siddique, S.A. (2014). *Mineral Nutrition in Mitigating Water Stress Effects on Morpho-physiology and Yield in Mungbean*. A PhD dissertation. Department of Agronomy, Bangabandhu Sheikh Mujibur Rahman Agricultural University, Gazipur.

Tareq, M.S. (2022). *Growth, Yield, and Physiological Responses of Soybean Genotypes to Salinity*. MS thesis submitted to Department of Agronomy, Banagabandhu Sheikh Mujibur Rahman Agricultural University, Gazipur.

Chapter 22

Climate Change and Bangladesh Initiatives

Dr. Md. Billal Hossain[1]*

Introduction

Climate change is an unprecedented global challenge that has far-reaching implications for the socio-economic and environmental fabric of nations around the world. Bangladesh is one of those countries that has been severely hit by climate change. Bangladesh is regarded as one of the world's most climate-vulnerable countries in light of different research findings and national and international reports (Hossain et al., 2020; Islam et al., 2016; Ahmed and Haque, 2002; McCarthy et al., 2001), even though it contributes very little to global warming (0.98 tons CO2-eq per person, or less than 0.47% of global emissions). The biggest threat to the nation's overall development has emerged as climate change, which is also getting worse. Its widespread and unprecedented impact has disproportionately burdened the country's poorest and most marginalized populations (MoEFCC, 2021).

Bangladesh, a country in South Asia, is particularly vulnerable because to its location, topography, and high population. Bangladesh is the seventh most vulnerable country to climate change in the world having seen the greatest impact from climate change between 2000–2019 according to German Watch Global Climate Risk Index 2021 (Eckstein et al., 2021). The country's susceptibility to the impacts of climate change, such as, high temperatures, erratic and unpredictable rainfall, widespread floods and droughts, powerful tropical cyclones, sea level rise, river erosion, and significant siltation are just a few of the natural calamities that have made Bangladesh's people more vulnerable and has led to a complex web of challenges that threaten the well-being and livelihoods of its population (Fig. 1).

Out of all the external factors, Sea Level Rise (SLR) is Bangladesh's biggest concern due to climate change as about 10% of the country is hardly 1 meter above the mean sea level (MSL), and one-third is under tidal excursions. According to a study

[1] Bangladesh Hi-Tech Park Authority, Information and Communication Technology Division, ICT Tower, Dhaka-1207, Bangladesh.

* Corresponding author: mdbhossain05@gmail.com

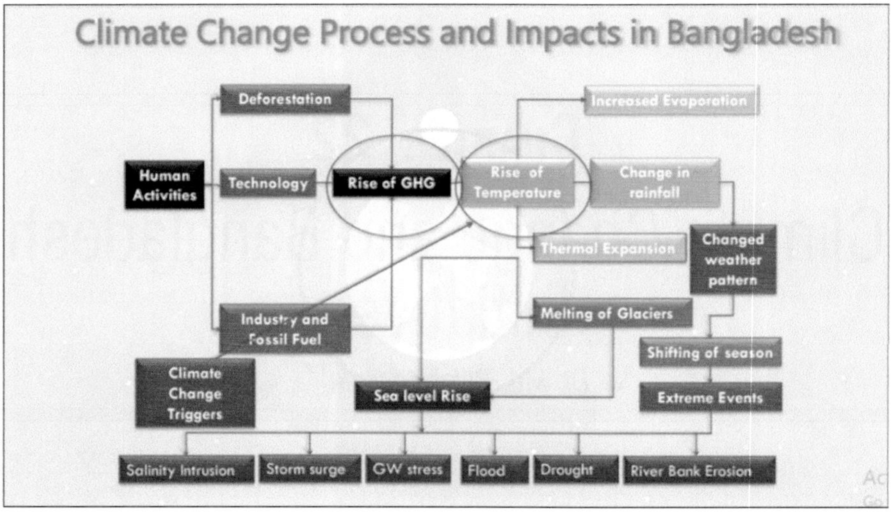

Fig. 1 Climate Change Process and impact in Bangladesh (*Source:* CEGIS, 2011)

by the Department of Environment (DOE), Sea level rise in Bangladesh's coastal regions has averaged between 3.8–5.8 millimeters each year over the past 30 years. The study shows that about 12.34–17.95% of the coastal area will be submerged due to sea-level rise by this century. Sea level rise leads to salinity and saltwater intrusion. Increased soil salinity is expected to affect coastal areas of Barisal, Chattogram, and Khulna area of Bangladesh. Increases in soil salinity are expected to vary from 26–55% in most affected areas, by 2050 (World Bank, 2015). Due to increased saline intrusion into freshwater sources during the low flow circumstances, there will be less freshwater available. This is caused by saline water intrusion into estuaries and groundwater in coastal locations due to sea level rise. The effects are made worse by increased freshwater evaporation and evapotranspiration when temperatures rise, as well as increased demand for freshwater during periods of water stress.

The northwestern districts of Bangladesh are more susceptible to drought, which will result in more crop loss as climate change intensifies. The Fourth Assessment Report (AR4) of the IPCC predicts that by 2050, Bangladesh's production of rice and wheat may decline by 8% and 32%, respectively (Faisal and Parveen, 2004). Due to increasingly erratic rainfall, the northeastern and southeastern hilly regions of Bangladesh have also been affected by landslides. By the end of this century, Bangladesh is predicted to experience an annual loss of up to 9% of its economy. If global behavior does not change, the nation's annual economic spending will reach 2% of its GDP by 2050 and might reach 9.4% by 2100 (ADB, 2014).

Bangladesh already faces a number of interconnected difficulties, such as the depletion of natural resources, environmental pollution, and the harmful effects of climate change. The Ganges, Brahmaputra, and Meghna (GBM) basins include Bangladesh as one of their lowermost riparian nations (Fig. 2). The main factor causing riverine floods in Bangladesh is excessive rainfall in GBM basins. There are 57 rivers that have their beginnings outside Bangladesh's borders. Annually, around 1.18 trillion cubic meters of water flow to the sea, of which 1.07 trillion cubic meters,

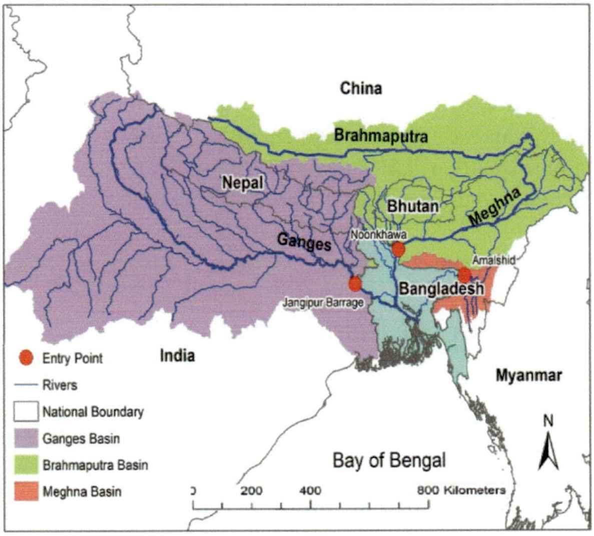

Fig. 2 The South Asian River Basins of Ganges, Brahmaputra, and Meghna. The location of entry points into Bangladesh indicated by the red circle (*Source:* Akbor et al., 2014).

or 91%, enter Bangladesh from upstream catchments and the remaining portion is provided by total internal rainfall. The three basins' combined area is 1.72 million km^2. Just 7% of it is in Bangladesh. As a result, the country is susceptible to floods from runoff from upper co-riparian's as well as from local precipitation (MoEFCC, 2012).

The flood gets much worse when all three river systems rise at once. Even when two rise at the same time, the ensuing flood is extremely disastrous. Adding fuel to the fire is the phenomenon of climate change. Recent decades have seen an increase in the frequency of large floods; as a result of climate change, future floods in the nation may be more violent, last longer, and occur more frequently. One report published in *Daily Star* outlined that, coastal flooding will affect 42 million people in Bangladesh by 2050 (Roy, P., 2019). Another report published in *Dhaka Tribune* quoting World the Economic Forum (WEF), 1 in 5 Bangladeshi will be landless due to climate change (*Dhaka Tribune*, 2019). Another report in the *Daily Star* cited a study by the Grantham Research Institute on Climate Change and the Environment and the Center for Climate Change Economics as showing that peak river flows in Bangladesh could increase by 36% between 2070–2099 if global emissions continue. Even if emissions remain low, floods will increase by 16% during this period compared to 1971–2000 (*Daily Star*, 2023).

The Bangladeshi parliament passed a resolution in November 2019 describing climate change as a "planetary emergency" and urging the international community, particularly the UN, to take immediate action. The parliament unanimously approved a historic motion that proclaimed the climate situation a global emergency and urged the international community to take immediate and decisive action to achieve net zero (carbon) emissions long before 2050 (Prothom Alo English Desk, 2019).

Materials and Methods

The information for this study was gathered from a variety of online publications, national and international papers, books, and documents about climate change in Bangladesh. In preparing this article, presentation slides from several trainings, workshops, and seminars on Bangladesh-related climate change were also analyzed. Additionally, a number of articles and stories about Bangladesh and climate change that were published in daily newspapers are also examined.

Results and Discussion

Government of Bangladesh's Efforts to Combat Climate Change

In response to these pressing concerns, Bangladesh has embarked on a series of ambitious climate change initiatives that exemplify its commitment to adaptation, mitigation, and resilience-building. These initiatives reflect a comprehensive and multifaceted approach, ranging from community-driven projects to national policy frameworks, all aimed at addressing the multifaceted challenges posed by a changing climate.

In the face of escalating climate challenges, Bangladesh has emerged as a global frontrunner in devising and implementing innovative climate change initiatives. Situated at the confluence of multiple environmental vulnerabilities, including sea-level rise, extreme weather events, and shifting precipitation patterns. Bangladesh has not only recognized the urgency of addressing the impacts of climate change but has also embarked on a remarkable journey towards resilience and sustainability. With a determined focus on adaptation, mitigation, and fostering community engagement, Bangladesh's climate change initiatives stand as a beacon of inspiration and a testament to the nation's unwavering commitment to safeguarding its people and the environment.

The country has launched a series of innovative and forward-thinking climate change initiatives aimed at mitigating environmental impacts and adapting to changing conditions. These initiatives, rooted in a comprehensive understanding of the local context and driven by a commitment to safeguarding its population and ecosystems, exemplify Bangladesh's determination to be a proactive player in the global effort to combat climate change.

Bangladesh is well known throughout the world for its achievements in disaster risk reduction and climate change adaptation. In instance, Cyclone Amphan a few years ago had a considerable impact on people's livelihoods, especially those who depend on fisheries and farms, even though it killed only a few people (less than 30). In the past, severe cyclones had claimed the lives of more people. For instance, Bangladesh has made significant investments to ensure that people are safe during storms after losing at least 300,000 people to the Great Bhola Cyclone in 1970 (Reliefweb, 2020).

Bangladesh Climate Change Strategy and Action Plan

In support of the Bali Action Plan of COP13, a plan to coordinate initiatives addressing climate change Bangladesh created its National Climate Change Strategy and Action Plan (BCCSAP) in 2009 to describe its strategy for addressing climate change impacts. One of the first nations to develop a climate change strategy and action plan was Bangladesh, which was updated in 2023 and has since acted as a model for other nations that are vulnerable to climate change. This comprehensive plan focuses on a number of industries, including agriculture, water resources, health, disaster management, food security, social protection, infrastructure like coastal and river embankments, research and knowledge management, mitigation, and capacity building and institutional strengthening, management of natural resources, gender, and the urban aspect of climate change. In the BCCSAP, it is made very clear that the fundamental stance of national policy is one of an integrated approach to climate change and development issues. This has received praise from all across the world as a significant, original step taken by Bangladesh. It demonstrates the government of Bangladesh's commitment to doing everything in its power to address climate change, both in terms of adaptation and mitigation (Ahmad, 2017).

Bangladesh Climate Change Trust Fund (BCCTF)

The government established the Bangladesh Climate Change Trust Fund (BCCTF) as part of the BCCSAP in order to manage the country's own budgetary contribution for combating climate change. Under Bangladesh Climate Change Trust Fund, using its own funds in FY2009–2010, and has so far allocated $490 million to support 856 projects to carry out the BCCSAP's strategic activities. In addition, the climate-relevant budgetary allocation in the Annual Development Plan has doubled over the past seven years, going from around US$ 1.44 billion in FY2015–2016 to roughly US$ 2.96 billion in FY2021–2022 for developing resilience to the effects of climate change (MoEFCC, 2021).

The government developed the Bangladesh Delta Plan 2100, a comprehensive 100-year strategic plan, with the aim of fostering the country's progressive, sustainable growth through delta management practices. Bangladesh Delta Plan 2100 is envisioned as a long-term holistic plan that takes the water resource management, climate change and environmental challenges for supporting development of the country. Through strong, adaptive, and integrated strategies, as well as fair water governance, ensure long-term food and water security, economic growth, and environmental sustainability while successfully addressing natural catastrophes, climate change, and other delta concerns. The plan's specific objectives include ensuring safety from floods and disasters associated to climate change, improving water security and efficiency of water use, ensuring sustainable and integrated river and estuary management, conserving and preserving wetlands and ecosystems, and encouraging their sensible use. To accomplish the best and integrated use of land and water resources, create efficient institutions and equitable governance for managing domestic and international water resources (GED, 2018).

Nationally Determined Contribution (NDC)

In August 2021, Bangladesh revised its NDC with measurable mitigation targets increasing both unconditional and conditional contributions from 15% below Business as Usual (BAU) to 21.85% by 2030. Bangladesh proposed quantified emission reduction commitments of 6.73%, or reductions of 27.56 M tons of CO_2 equivalent, in the unconditional scenario and an additional 15.12%, or reductions of 61.91 M tons of CO_2 equivalent, in the conditional scenario by 2030. The first NDC for Bangladesh's energy, industrial, and transportation sectors was submitted to the UNFCCC in 2015. By 2030, the NDC suggested a 12 Mt (5%) unconditional reduction in greenhouse gas emissions compared to "business as usual" and a 24 Mt (10%) conditional reduction with international help. In case of mitigation initiatives, the country has registered a total of 24 Clean Development Mechanism (CDM) projects to the CDM Executive Board. So far, these projects have generated 18.56 Mt CO_2e and have the potential to reduce 118 Mt CO_2e by 2030.

National Adaptation Plan (NAP)

On November 2, 2022, the government revealed Bangladesh's NAP (2023–2050) to the public with the goal of reducing climate risks and vulnerabilities and establishing a realistic route to climate-resilient development (NAP, 2022). The NAP will be the main vehicle for addressing country adaptation to the UNFCCC process. NAP will be Bangladesh's key planning and investment document to adapt to climate change in the coming years. With an underlying objective of reducing risks and vulnerabilities to climate change impacts, the NAP envisions building a climate-resilient nation through effective adaptation strategies that foster a robust society and ecosystems and stimulate sustainable economic growth. Alongside conserving water bodies and protecting embankments, it suggests building more shelters and creating livelihood opportunities, and recommends improvements to irrigation and crop varieties to increase agricultural production. In some regions, it also proposes adopting a "living with floods" strategy, such as preventing the development of homes in high-risk areas and supplying building supplies that are resistant to water.

As part of the Glasgow Leaders' Declaration on Forests and Land Use, Bangladesh has promised to cooperate in order to prevent and reverse land degradation and forest loss by 2030. To lessen the effects of climate change and boost adaptive capacity, the nation has set a target of 50,000 hectares for mangrove afforestation in coastal areas under the 8th Five Year Plan (EFYP, 2020). A statement made during the UN Climate Summit COP26 stated that signatory nations committed to working together to stop and reverse land degradation and forest loss by 2030 through fostering sustainable development and inclusive rural transformation. In the statement, the nations also pledged to collaborate on six topics, such as preserving forests, promoting sustainable development policies, improving rural livelihoods, and putting agricultural policies into practice (Illius, 2021).

In order to reduce GHG emissions and advance green energy, the country has developed certain renewable energy initiatives. In the past several years, around 65 lakhs (6.5 M) solar-home systems (SHS) have been installed throughout the nation to support renewable energy. As a result, over 18 people or about 12% of Bangladesh's population are currently receiving the benefits of solar power. The largest national program in the world for off-grid electrification is the **Solar Home Systems (SHS) Program** in Bangladesh. It is the world's longest continuously running off-grid electrification initiative. SHS was mostly utilized in rural households for illumination, charging cell phones, and running TVs and radios. As per Cabraal et al. (2021), there were employment in over 200,000 rural businesses and places of worship.

To assist lower pollutant emissions from burning biomass, more than 8.5 M Improve Cook Stoves (ICSSs) have been supplied in rural regions in recent years. As a result, indoor air pollution levels will be reduced. Up until December 2022, IDCOL installed a total of 3.7 M stoves, and the program is on track to reach the target of 4.0 M by 2023 (IDCOL, 2022). Department of Environment (DoE) under the Ministry of Environment of Forests has been working with GIZ in the dissemination of ICS. Till March 2023, 5.0 M ICS have been installed (BONDHU, 2023).

Bangladesh has been working to diversify its energy mix by expanding its renewable energy capacity. In order to boost the share of renewable energy in its total energy consumption, the government has set ambitious goals for solar power generation. This move not only helps reduce greenhouse gas emissions but also improves energy access, particularly in rural areas. Bangladesh has set an ambitious goal of generating more than 4,100 MW of electricity from renewable energy sources by 2030 as the county looks to cut greenhouse gas emissions significantly. Solar power will account for half of the energy, at 2,277 MW, followed by hydropower 1,000 MW and wind 597 MW (*Daily Star*, 2021).

A silent transformation is on the way in transport sector in Bangladesh through the introduction of Mass Rapid Transit (MRT) and Elevated Expressway in Capital Dhaka. In December 2022 and in September 2023, Bangladeshi Prime Minister Sheikh Hasina publicly opened the nation's first metro rail network and MRT system, respectively. Both systems would contribute to the improvement of the traffic congestion in capital Dhaka and reduce carbon emissions and curb sound and air pollution, driving positive environmental impacts.

The National Plan for Disaster Management (NPDM), which combines disaster risk reduction and management, is an example of the government of Bangladesh's strategic goal to create a resilient society. In each stage of the disaster cycle, including Disaster Risk Reduction (DRR), Humanitarian Response, and Emergency Recovery Management, the fundamental objectives of NPDM 2021–2025 are to preserve lives and lessen economic losses. The NPDM follows a phase-wise strategy with 34 main aims to be implemented in collaboration with pertinent stakeholders and based on Government of Bangladesh's prior disaster risk management lessons learned and worldwide DRR frameworks (NPDM, 2020). NPDM 2021–2025 acknowledged the need for tackling developing concerns in national policies and programs, such as climate change, drought, desertification, and human-induced hazards. The plan emphasizes the connection between disaster risk reduction and fast urbanization and climate change as well as the need for DRR for sustainable development (NPDM, 2020).

Bangladesh is globally recognized for its success in disaster management. The success has not come overnight, throughout the coast, the government has built 4291 cyclone shelters. In addition to these, 523 other flood shelters have been built throughout the country in flood-prone locations. There have also been built 550 "Mujib Killa" (specially planned raised land) to protect animals in coastal areas from storm and tidal surge. Under the Cyclone Preparedness Program (CPP), some 56,000 volunteers work to reduce catastrophe risk by awaking and protecting the public against cyclonic disasters. Another vital aspect of Bangladesh's climate change initiatives is its emphasis on community-based adaptation. Recognizing the importance of involving local communities, the country has implemented projects that empower people at the grassroots level to tackle climate-related challenges effectively. These projects include efforts to improve agricultural practices, build climate-resilient infrastructure, and enhance disaster preparedness. Bangladesh actively engages in global climate cooperation and negotiations in order to obtain financial and technical support for its programs. The country leverages partnerships with international organizations, donor countries, and climate funds to implement projects that address climate vulnerabilities.

Summary and Conclusions

Despite various efforts, Bangladesh still faces numerous challenges in effectively combating climate change. Limited financial resources, inadequate infrastructure, and the need for technological advancements remain significant barriers. International support and cooperation are crucial for Bangladesh to scale up its climate change initiatives and build resilience against the impacts of global warming. The country's commitment to adaptation, renewable energy, and international cooperation demonstrates its recognition of the urgent need to address climate change and protect its citizens. Continued efforts, both domestically and internationally, are essential to support Bangladesh in its fight against climate change and ensure a sustainable future for its people.

The following questions related to climate change initiatives in Bangladesh can be framed as follows:

- **Assessment of Climate Change Impacts:** Understanding the specific ways in which climate change is affecting Bangladesh.
- **Vulnerability and Adaptation:** Determining the susceptibility of various groups and geographical areas in Bangladesh to these effects of climate change.
- **Mitigation Strategies:** Evaluating strategies to reduce greenhouse gas emissions within the country promoting different strategies and action plan.
- **Socio-economic Implications:** Studying the socioeconomic implications of climate change, including its impact on livelihoods, agriculture, water resources, and health, to develop targeted responses.
- **Policy and Governance:** Understanding how policy frameworks and governance structures impact climate change initiatives in Bangladesh.

Bangladesh is one of the nations most exposed to the consequences of climate change, which is a significant worldwide issue. A number of Bangladeshi initiatives are cutting-edge, significant, and timely in tackling the problems caused by climate change:

Policy and Other Legal Frameworks

- Bangladesh Climate Change Trust Fund, using its own funds
- Adaptation Strategies

- Community-Based Approaches
- Renewable Energy
- Mangrove Reforestation
- Climate Finance
- Early Warning Systems
- Climate-Resilient Infrastructure
- Disaster Risk Reduction and Disaster Management
- River Basin Management
- Implementing Bangladesh Delta Plan, 2100
- Research and Innovation
- National Adaptation Plan (NAP)
- Introduction of Mass Rapid Transit (MRT) and Elevated Expressway in Capital Dhaka
- International Collaboration.

Here are some key findings and initiatives related to climate change in Bangladesh:

- **Vulnerability to Sea-Level Rise:** Bangladesh is highly susceptible to rising sea levels, which can lead to coastal erosion, saltwater intrusion into freshwater sources, and displacement of coastal communities.
- **Increased Flooding:** Climate change has led to more frequent and severe flooding in Bangladesh, affecting agriculture, infrastructure, and livelihoods.
- **Extreme Weather Events:** The country experiences an increase in extreme weather events like cyclones and hurricanes, which can cause extensive damage and loss of life.
- **Agricultural Impacts:** Climate change affects agriculture, which is a major source of livelihood in Bangladesh. Erratic rainfall, temperature extremes, and increased salinity can reduce crop yields.

The main conclusions regarding climate change and Bangladesh's initiatives are as follows:

- **Adaptation Efforts:** Bangladesh has taken the initiative to put adaptation plans into place in order to deal with climate change. This includes initiatives such as building cyclone shelters, embankments, and early warning systems to protect its coastal communities from the increasing intensity of cyclones and flooding events.
- **Climate Finance:** Bangladesh has been actively seeking international climate finance to support its adaptation and mitigation efforts.
- **Community-based Approaches:** Many initiatives in Bangladesh are community-driven and aim to empower local populations in adapting to climate change.
- **Renewable Energy:** To lessen its dependency on fossil fuels and cut greenhouse gas emissions, Bangladesh has also made progress in encouraging renewable energy sources like solar power.
- **Challenges Remain:** Despite these initiatives, Bangladesh still has a lot of work to do to address climate change.

Limitations of the Study

As the study is based on the initiatives taken by Bangladesh government to reduce the impact of climate change in the country. Hence, emphasis has been given on relevant government information and data. Moreover, there are very few journals that publish data based on this topic. As a result, research focuses on official documents and reports.

References

ADB. (2014). *Bangladesh Could See Climate Change Losses Reach over 9% of GDP*. Asian Development Bank, ADB News Release 19 August. (https://www.adb.org/news/b angladesh-could-see-climate-change-losses-reach-over-9-gdp-report).

Ahmad, Q.K. (2017). *Environment, Climate Change, and Water Resources*, 137 pp. Pathak Shamabesh. 17 Aziz Market, Shahbag, Dhaka.

Ahmed, A.U. and N. Haque. (2002). Climate Change and Sustainable Development. Paper presented at the Dhaka Meet on Sustainable Development, March 14–18. Incorporated in Q.K. Ahmad, A.U. Ahmed (Eds.), *Citizens' Perspectives on Sustainable Development*. Bangladesh Unnayan Parishad (BUP).

Akbor, Siddique-E., A.H.M., Hossain, F., Sikder, S., Shum, C.K., Tseng, S., Yi. Yuchan, Turk, F.J. and Limaye, A. (2014). Satellite precipitation data–driven hydrological modeling for water resources management in the Ganges, Brahmaputra, and Meghna basins. *Earth Interactions*, *18*: 17.

BONDHU. (2023). Bangladesh Bondhu Foundation 2023. (https://www.bondhufoundation.org/ongoing-project/bondhu-chula/).

Cabraal. A., William A., Ward, V., Susan Bogach and Amit Jain. (2021). *Living in the Light: The Bangladesh Solar Home Systems Story*. International Bank for Reconstruction and Development/The World Bank Group e-Library, 24 March.

CEGIS. (2011). *Technical Report on Climate Change Process*. Center for Environmental and Geographic Information Services (CEGIS), Bangladesh.

Daily Star. (2021). Renewable Energy, Target to generate 4,100MW by 2030. *The Daily Star*, October 31.

Daily Star. (2023). Study on Climate Change: Flooding may rise by 36pc. *The Daily Star*, August 24.

Dhaka Tribune. (2019). WEF (World Economic Forum): 1 in 5 Bangladeshis to be landless due to climate change. *Dhaka Tribune*, 12 Nov.

Eckstein, D., Kunzel, V. and Schafer, L. (2021). *Global Climate Risk Index 2021*. German watch e.V., Office Bonn, Dr. Warner-Schuster-Haus, Kaiserstr. 201, Bonn.

EFYP. (2020). *EFYP* (*Eighth Five Year Plan*) (*July 2020-June 2025*), 507 pp. General Economics Division, Bangladesh Planning Commission, Government of the People's Republic of Bangladesh, 29 December.

Faisal, I.M. and Parveen, S. (2004). Food security in the face of climate change, population growth and resource constraints: Implications for Bangladesh. *Environmental Management*, *34*: 487–98.

GED. (2018). *Bangladesh Delta Plan 2100*. General Economics Division (GED), Bangladesh Planning Commission, October.

Hossain, B., Sohel, M.S. and Ryakitimbo, C.M. (2020). Climate change induced extreme flood disaster in Bangladesh: Implications on people's livelihoods in the Char Village and their coping mechanisms. *Progress in Disaster Science*, *6*: 100079.

IDCOL. (2022). Infrastructure Development Company Limited. *Annual Report 2022*. (https://idcol.org/annual_report/IDCOL_Annual_Report_2022.pdf).

Illius, S. (2021). Bangladesh joins pledge to end global deforestation. *Earth Journalism Network-Glasgow Leader*, 07 November.

Islam, M., Kotani, K. and Managi, S. (2016). Climate perception and flood mitigation cooperation: A Bangladesh case study. *Economic Analysis and Policy*, *49*: 117–33.

McCarthy, J.J., Canziani, O.F., Leary, N.A., Dokken, D.J., K.S. (Eds.). (2001). *Climate Change 2001: Impacts, Adaptation. and Vulnerability*. Inter-Governmental Panel on Climate Change, Work Group-II Input to the Third Assessment Report, Cambridge University Press, Cambridge.

MoEFCC. (2012). *Bangladesh National Report on Sustainable Development*. Ministry of Environment and Forest and Climate Change. Peoples' Republic of Bangladesh.

MoEFCC. (2021). *Climate Change Initiatives of Bangladesh Achieving Climate Resilience*. Ministry of Environment, Forest, and Climate Change, Dhaka, Government of Bangladesh.

NAP. (2022). *National Adaptation Plan of Bangladesh* (*2023–2050*). Ministry of Environment, Forest, and Climate Change, Government of the People's Republic of Bangladesh.

NPDM. (2020). *Action for Disaster Risk Management Towards Resilient Nation*. Government of the People's Republic of Bangladesh, Dhaka, March.

Prothom Alo English Desk. (2019). Bangladesh parliament adopts motion declaring 'planetary emergency'. *Prothom Alo English Desk*, 14 Nov.

Reliefweb. (2020). Cyclone Amphan. *In*: *Bangladesh, Preparedness Paid Off*. Reliefweb news and press release, 29 May. https://reliefweb.int/report/bangladesh/cyclone-amphan-bangladesh-preparedness-paid.

Roy, P. (2019). Climate Change Fallout in Bangladesh, 42m people to be hit by 2050. *The Daily Star*, Dhaka, Oct 31.

World Bank. (2015). *Salinity Intrusion in a Changing Climate Scenario Will Hit Coastal Bangladesh Hard*. February 17.

Index

For Product Safety Concerns and Information please contact our EU representative GPSR@taylorandfrancis.com Taylor & Francis Verlag GmbH, Kaufingerstraße 24, 80331 München, Germany